AF799611

Atmospheric Perspectives

Vlado Spiridonov • Mladjen Ćurić
Nenad Novkovski

Atmospheric Perspectives

Unveiling Earth's Environmental Challenges

Springer

Vlado Spiridonov
Institute of Physics, Faculty of Natural Sciences and Mathematics
Ss. Cyril and Methodius University in Skopje
Skopje, North Macedonia

Mladjen Ćurić
Department of Meteorology, Faculty of Physics
University of Belgrade
Belgrade, Serbia

Nenad Novkovski
Institute of Physics, Faculty of Natural Science and Mathematics
Ss. Cyril and Methodius University in Skopje
Skopje, North Macedonia

ISBN 978-3-031-86756-9 ISBN 978-3-031-86757-6 (eBook)
https://doi.org/10.1007/978-3-031-86757-6

© The Editor(s) (if applicable) and The Author(s), under exclusive license to Springer Nature Switzerland AG 2025

This work is subject to copyright. All rights are solely and exclusively licensed by the Publisher, whether the whole or part of the material is concerned, specifically the rights of translation, reprinting, reuse of illustrations, recitation, broadcasting, reproduction on microfilms or in any other physical way, and transmission or information storage and retrieval, electronic adaptation, computer software, or by similar or dissimilar methodology now known or hereafter developed.
The use of general descriptive names, registered names, trademarks, service marks, etc. in this publication does not imply, even in the absence of a specific statement, that such names are exempt from the relevant protective laws and regulations and therefore free for general use.
The publisher, the authors and the editors are safe to assume that the advice and information in this book are believed to be true and accurate at the date of publication. Neither the publisher nor the authors or the editors give a warranty, expressed or implied, with respect to the material contained herein or for any errors or omissions that may have been made. The publisher remains neutral with regard to jurisdictional claims in published maps and institutional affiliations.

This Springer imprint is published by the registered company Springer Nature Switzerland AG
The registered company address is: Gewerbestrasse 11, 6330 Cham, Switzerland

If disposing of this product, please recycle the paper.

Preface

Welcome to *Atmospheric Perspectives: Unveiling Earth's Environmental Challenges*, a comprehensive exploration of the dynamic and interconnected systems shaping our extraordinary planet. This book embarks on a multidisciplinary journey through Earth science, offering insights into the complex workings of our globe. It is dedicated to Earth—our greatest challenge and inspiration—and reflects humanity's enduring fascination with the universe. This book aims to bridge modern science, technology, and public understanding by showcasing how Earth science advances our knowledge of the planet. From the origins of Earth and the solar system to the present era marked by accelerating climate change, this work examines the essential components of the Earth system and their intricate interactions. It covers physical processes operating from local to global scales and over timeframes ranging from seconds to centuries. The journey begins in Chap. 1, where readers are introduced to the foundational aspects of Earth's complex systems. This introductory chapter sets the stage for understanding the interdependence of the geosphere, hydrosphere, biosphere, cryosphere, and atmosphere. In Chap. 2, the focus shifts to the geophysical principles underpinning Earth's composition, structure, and processes, providing a robust framework for appreciating the planet's dynamic nature. Subsequent chapters delve into specialized domains of Earth science. Chap. 3 explores the geosphere's role in shaping Earth's surface and influencing its processes. Chap. 4 focuses on the biosphere, highlighting the interactions within ecosystems and their significance for biodiversity and ecological balance. Water in its various states takes center stage in Chap. 5, which examines the hydrosphere and cryosphere, emphasizing their roles in sustaining life and driving climate systems. Chap. 6 then turns to the atmosphere, the vital layer enveloping our planet, explaining its structure, composition, and key functions. The energy that powers atmospheric dynamics is explored in Chap. 7, which covers solar radiation and its role in driving weather, climate, and planetary change. This leads naturally into Chap. 8, which investigates global temperature trends and the impacts of planetary warming. The interplay between the atmosphere and oceans is the subject of Chap. 9, which examines integrated mechanisms of global circulation and their effects on climate and weather systems. Atmospheric systems and air masses are explored further in

Chap. 10, highlighting their complex interactions and influence on weather patterns. Natural hazards and their underlying geophysical forces are addressed in Chap. 11, providing insights into phenomena such as earthquakes, floods, and beyond. Moving beyond physical processes, Chap. 12 emphasizes the need for sustainable interactions with Earth's natural environment, presenting pathways to balance ecological resilience and human needs. In Chap. 13, this book takes on the urgent challenges of the twenty-first century, including climate change, environmental degradation, and the pressures of human activity. The closing chapter, Chap. 14, highlights the transformative role of advanced technologies in monitoring Earth's systems and gaining insights into weather, water, and climate.

Atmospheric Perspectives: Unveiling Earth's Environmental Challenges is not just a book; it is an invitation to engage with the wonders of our home planet, comprehend the forces shaping it, and appreciate the delicate equilibrium sustaining life. Whether you are a student, scientist, or curious learner, this book aspires to inspire a profound connection with the dynamic and resilient world we call home.

Skopje, North Macedonia — Vlado Spiridonov
Belgrade, Serbia — Mladjen Ćurić
Skopje, North Macedonia — Nenad Novkovski

Acknowledgments

The authors express their deepest gratitude to Dr. Aaron Schiller, editor at Springer, for his continuous support and guidance, as well as to the anonymous reviewers whose valuable suggestions greatly influenced the conception of this book. We also extend our sincere thanks to the main editorial board at Springer in New York for recognizing the true value of this work and for their enthusiasm in approving and initiating its preparation.

The authors are especially grateful to Ms. Tina Silva and Ms. Cineha, the Production Editor and the Project manager of this book, for their exceptional dedication, professionalism, and tireless efforts in bringing this project to fruition. Their invaluable assistance, meticulous attention to detail, and steadfast commitment to excellence have been instrumental in transforming this book from a manuscript into the extraordinary work it has become.

We also deeply appreciate the special commitment and professionalism of all other contributors from Springer.

We are sincerely grateful to the wide network of friends, esteemed colleagues, and global experts who have enriched our journey with insightful contributions and constructive dialogues during invited talks, visits, seminars, and various joint ventures over the years. Your generosity in sharing original materials, photographs, and vivid graphic illustrations has significantly added to the depth and richness of this work.

Special thanks are due to organizations such as NASA, NOAA, NCAR, EUMETSAT, AMS, UCAR Comet, WMO, ZAMG, IMGW, and Harvard University, whose permissions have allowed us to reproduce figures that greatly enhance the visual narrative of this book.

Finally, we owe an immense debt of gratitude to our families for their unwavering support, boundless patience, and deep understanding throughout the complex and demanding process of producing this book. Your steadfast encouragement has been the cornerstone of our perseverance.

With sincere thanks,
The Authors

Competing Interests The authors have no competing interests to declare that are relevant to the content of this manuscript.

Contents

About the Authors

Vlado Spiridonov born in Skopje, completed his primary education at "Grigor Prlichev" and high school at "R.J."-UKIM. He earned his specialist, master's, and doctoral degrees in meteorology from the Institute of Meteorology, Faculty of Physics, University of Belgrade. In 2004, he was awarded a postdoctoral scholarship by the Government of Canada but was appointed as Director of the Administration for Hydrometeorological Affairs and Macedonia's permanent representative to the World Meteorological Organization (WMO) the same year.

Dr. Spiridonov was elected Appointed Assistant Professor of Meteorology in 2007 and promoted to Appointed Associate Professor in 2012 and 2017. He represents the Faculty of Natural Sciences and Mathematics (PMF) on the National Council for Climate Change and is a member of the WMO Commission for Atmospheric Sciences. From 2017 to 2019, he served as a Visiting Professor at the University of Vienna, Austria, and was appointed as an Honorary Visiting Professor at FINKI from 2019 to 2020. In December 2020, he was elected as a Full Professor of Meteorology and Atmospheric Physics at the Faculty of Natural Sciences and Mathematics, Institute of Physics, "St. Cyril and Methodius" University in Skopje. In 2024, the President of North Macedonia appointed him to the National Security Council as an expert in climate change, extreme weather events, and environmental air pollution.

Dr. Spiridonov has made significant contributions to atmospheric science through his participation in international conferences, the organization of global events and symposiums, and plenary presentations. He has published extensively in international journals and authored books in meteorology, including with Springer Nature. His work has earned numerous awards, including "Patent of the Year 2000" in Macedonia, a gold medal at EUREKA 2000 in Brussels, the Hungarian Genius Award, and the GRAND PRIX at the 4th Budapest International Exhibition of Inventions for his fog-scattering method.

Mladjen Ćurić was born in Žabljak, Montenegro. He graduated with top marks from the Hydrometeorological School in Belgrade. He earned his bachelor's degree in meteorology from the University of Belgrade, where he also completed his postgraduate studies and doctoral thesis. Between 1977 and 1980, he participated in study visits to several institutions, including Colorado State University in Fort Collins, USA; Imperial College in London, England; the University of Manchester in England; and the Meteorological Office in Bracknell. From 1972 to 1978, Ćurić served as an assistant at the Department of Physics and Meteorology at the University of Belgrade. Since 1990, he has been a full professor at the Institute of Meteorology at the University of Belgrade. He served as the director of the Institute of Meteorology from 1979 to 1982 and as vice-dean of the Faculty of Physics from 1982 to 1984. He was also a member of the Executive Committee of the International Commission on Clouds and Precipitation (ICCP) from 1988 to 1996. Since 1990, Ćurić has been a member of the National Association for Environmental Protection's Executive Committee. He served as vice-dean of the Faculty of Physics from 1996 to 1998 and again from 2004 to 2007.

His professional and scientific interests include fundamental topics in meteorology, such as atmospheric dynamics, cloud physics, applied meteorology and hydrology, weather modification, and environmental protection. Mladjen Ćurić is the author of several books and has published over 200 papers, primarily in international peer-reviewed journals, including the *Journal of Atmospheric Sciences*, the *Quarterly Journal of the Royal Meteorological*

Society, *Tellus*, *Journal of Applied Meteorology*, *Atmosphere-Ocean*, *Theoretical and Applied Climatology*, *Atmospheric Research*, *Meteorology and Atmospheric Physics*, and *Journal of Geophysical Research*.

Nenad Novkovski was born on May 14, 1958, in Kriva Palanka. He completed his primary and secondary education in his hometown, graduating in 1976. He then enrolled in the Faculty of Physics in Skopje, graduating in 1980. In 1983, he furthered his studies at the Faculty of Electrical Engineering at the University of Belgrade. He earned his master's degree in 1985 with a thesis titled "A New Approach to the Theory of BSF Solar Cells."

In 1987, Novkovski was awarded a scholarship by the Swiss government for scientific training in Lausanne. He was appointed as an Assistant Professor at the Institute of Physics, Faculty of Natural Sciences (PMF), in 1991. By 1996, he was promoted to Associate Professor of Condensed Matter Physics and Fundamentals of Solid-State Physics, and in 2001, he achieved the rank of Full Professor, a title he was reappointed to in 2006.

Throughout his career, Novkovski has held various positions, including Head of Postgraduate Studies at the Institute of Physics, Head of the Institute for Condensed Matter Physics, Deputy Chief, and Editor-in-Chief of the political magazine Glas. From 1998 to 2002, he served as the Minister of Education and Science in the Government of North Macedonia.

Novkovski is a member of several professional societies, including the Society of Physicists of the Republic of Macedonia (SFRM), the European Society of Physicists (IEEE), the American Physical Society, and the World Academic Association "Plato." He has authored numerous scientific publications, books, brochures, and articles and has participated in various congresses, scientific conferences, seminars, and workshops. He has received a lot of recognition and certificates for his contributions to the field.

Chapter 1
Introduction: Understanding Earth's Complex Systems

Atmospheric Perspectives: Unveiling Earth's Environmental Challenges is a heartfelt dedication to the remarkable marvel that is our planet Earth—a dynamic and resilient sphere teeming with life, yet increasingly vulnerable to the pressures of human activity and natural forces. This book celebrates Earth's intrinsic beauty and complexity, presenting it as both a source of immense challenges and boundless inspiration.

The narrative unfolds as a natural extension of humanity's enduring fascination with the vastness of the universe, where Earth occupies a unique position as a cradle

© The Author(s), under exclusive license to Springer Nature Switzerland AG 2025
V. Spiridonov et al., *Atmospheric Perspectives*,
https://doi.org/10.1007/978-3-031-86757-6_1

of life amidst a cosmos of seeming desolation. Within its pages, we endeavor to illuminate the milestones and achievements of the new millennium, bridging the divide between modern scientific advancements and public understanding. By connecting scientific inquiry with pressing global issues, this book underscores the transformative potential of knowledge and innovation in shaping a sustainable future.

At its core, this effort captures the essence of the emerging field of atmospheric perspectives, which has become indispensable in unraveling the complexities of our home. From the origins of Earth and the solar system to the intricate interplay of atmospheric, oceanic, and terrestrial systems, this book provides a holistic framework for understanding the planet's past, present, and future.

Through pivotal chapters, it examines the geophysical and environmental processes that shape Earth, explores the evolution of life and the biosphere, and delves into the urgent challenges of contemporary climate change. By weaving together scientific insights, technological advancements, and actionable solutions, this book serves as both a guide and a call to action—empowering readers to engage with the critical environmental issues of our time.

1.1 Understanding Earth and the Solar System

This book explores the origin and evolution of the Earth, including its atmosphere, oceans, and life on our planet, which is under increasing pressure from both natural processes and human activities. Understanding the Earth and the solar system is fundamental to comprehending the processes that formed the Earth and made life possible (Kasting, 1993; Taylor & McLennan, 2009).

The Earth is a dynamic system where interactions between the lithosphere, hydrosphere, atmosphere, and biosphere have shaped its history over billions of years. These interconnections continue to influence the planet's present and future. From the accumulation of primordial elements in the protoplanetary disk to the formation of a habitable environment, the story of the Earth and its place in the solar system provides critical insights into planetary evolution and sustainability.

Our ability to predict the impact of human activities, including deforestation, urbanization, and greenhouse gas emissions, relies on a comprehensive understanding of the Earth system. This knowledge extends to assessing the resilience of ecosystems, monitoring biodiversity loss, and forecasting global climate changes. Additionally, the capacity to manage natural hazards depends on our understanding of the frequency and intensity of earthquakes, volcanic eruptions, floods, and storms, as well as the atmospheric processes that drive weather and climate patterns (Houghton, 2002).

These insights not only enhance our ability to mitigate risks but also deepen our appreciation of the intricate balance that sustains life on Earth. By examining Earth's past and present within the broader context of the solar system, we aim to inspire a renewed commitment to safeguarding the planet for future generations.

1.2 Earth System Science and Atmospheric Dynamics

The atmosphere is an integral part of the Earth's climate and (bio)geochemical cycles, and studying its composition and dynamics is crucial for Earth system science (Jacobson, 2014). Earth system science (ESS) is a relatively new interdisciplinary field that examines the complex interactions and feedback among the Earth's spheres, the atmosphere, hydrosphere, lithosphere, biosphere, and cryosphere. This holistic approach provides a framework for understanding the processes that sustain life and drive planetary changes.

Although the energy density of processes occurring in the atmosphere is much lower than that of the solid parts of the Earth system, atmospheric dynamics can profoundly influence living conditions, ecosystems, and economic activities. Weather events such as hurricanes, droughts, and extreme temperature fluctuations directly impact agriculture, water resources, and infrastructure, underscoring the atmosphere's critical role in human and environmental systems (Sellers, 1997). Moreover, the atmosphere acts as a medium for energy transfer and the exchange of gases, aerosols, and moisture, making it central to understanding global climate patterns and cycles. Processes such as atmospheric circulation, cloud formation, and the transport of heat and pollutants highlight the intricate linkages between regional and global scales. These dynamics are pivotal for addressing key challenges, including climate change, air quality, and the mitigation of natural disasters.

As ESS evolves, it provides valuable tools and models for predicting atmospheric behavior and its cascading effects on other Earth systems. By integrating advanced observational techniques, satellite data, and computational simulations, scientists can better understand and anticipate atmospheric changes (Ahrens & Henson, 2016), offering critical insights for policymaking, sustainable development, and disaster preparedness. The atmosphere's role in regulating the Earth's energy balance, supporting ecosystems, and influencing human activities highlights its indispensable place in Earth system science.

1.3 Physical Geography and Geographic Grid Systems

Before delving into the description of Earth's integral components, it is essential for readers to briefly familiarize themselves with physical geography and its significance in understanding the natural world. Physical geography examines the physical processes and features that shape the planet, including landforms, climates, ecosystems, and hydrological systems. This exploration begins with an introduction to the planet Earth, its place within the solar system, the characteristics of terrestrial planets, and the contrasting features of the giant outer planets (Lutgens & Tarbuck, 2015).

A fundamental aspect of this understanding is the shape and dimensions of the Earth. While it is often modeled as a perfect sphere, the Earth is an oblate spheroid, slightly flattened at the poles and bulging at the equator due to its rotation. This shape affects phenomena such as gravity, ocean circulation, and atmospheric dynamics. Equally crucial is grasping the geographic grid system, which provides a framework for locating points on the Earth's surface. This system involves understanding great and small circles, which are essential for navigation, mapping, and geospatial analysis. Latitude and longitude form the backbone of this grid, enabling the precise determination of positions globally. Latitude lines run parallel to the equator, while longitude lines converge at the poles, creating a coordinate system that aids in describing geographic locations (Peel & McGregor, 2002).

The International Date Line, another key element of the geographic grid, marks the transition between calendar days and serves as a reference point for global timekeeping. By following a path roughly along the 180th meridian, it underscores the interplay between Earth's rotation, time zones, and human activity. Understanding these concepts is not merely academic; they underpin many aspects of modern science, navigation, and environmental studies.

This foundational knowledge of physical geography and the geographic grid system sets the stage for a deeper exploration of Earth's systems and their interconnections, providing a basis for comprehending the dynamic processes that shape our planet and influence life on its surface.

1.4 Earth's Systems and Processes

The introduction to Earth begins with its relationship with the Sun, exploring how solar radiation drives the planet's energy balance and influences phenomena such as the seasons, weather patterns, and climatic variability. The physical effects of Earth's revolutions, including its axial tilt and orbital dynamics, are fundamental to understanding the distribution of solar energy and the resulting environmental conditions across different latitudes (Lamb, 1991).

Subsequent chapters delve deeper into Earth's intricate systems, providing a comprehensive understanding of the geophysical, atmospheric, and environmental processes that shape its past, present, and future. These interconnected systems—geosphere, hydrosphere, atmosphere, and biosphere—form the foundation of Earth's functionality and its capacity to sustain life.

Starting with Earth's physical characteristics, this book examines the geosphere, highlighting its dynamic role in shaping the planet. Topics such as plate tectonics, volcanic activity, mountain building, and mineral cycles are explored, emphasizing their profound impact on both the natural environment and human societies (McPhee, 1998). From the formation of Earth's crust to the processes driving earthquakes and resource distribution, the geosphere is a key driver of planetary change.

The hydrosphere is explored in depth, with a focus on the global water cycle and its influence on climate regulation and ecosystem dynamics. Ocean currents, driven by temperature and salinity gradients, are integral to redistributing heat and sustaining marine and terrestrial ecosystems. The critical role of water in supporting life, maintaining biodiversity, and shaping landscapes is underscored, along with discussions of freshwater availability and its implications for human development (Stommel & Stommel, 1983).

The biosphere, a vibrant network of life on Earth, is presented as a dynamic system that encompasses all living organisms and their interactions with the physical environment. The rich biodiversity within the biosphere supports ecosystems that provide essential services such as carbon sequestration, nutrient cycling, and habitat formation. By examining the delicate balance between biological and abiotic processes, this book highlights the vulnerabilities and resilience of life on Earth (Chapin et al., 2011). Together, these systems and processes offer a holistic perspective on Earth's complexity, illustrating how they interact to create the conditions necessary for life while also posing challenges such as climate change, resource depletion, and biodiversity loss. This exploration sets the stage for a deeper understanding of the planet's dynamic nature and the steps needed to preserve its equilibrium for future generations.

1.5 Atmospheric Science and Climate Systems

The atmosphere, a dynamic and life-sustaining layer enveloping the Earth, is meticulously analyzed in this book, with a particular focus on its composition, structure, and behavior. Its vital role in governing weather, climate, and air quality is explored, emphasizing how these elements affect ecosystems, human health, and economic activities (Atkinson & Davies, 2018). Central to this discussion is solar radiation, the primary energy source driving Earth's climate system. Its influence extends beyond surface warming, affecting atmospheric and oceanic circulation, weather patterns, and climate variability. Key topics include the absorption, reflection, and scattering of solar radiation and its role in maintaining the energy balance critical to sustaining life on Earth (Oke, 1987).

Global temperature trends, accelerated by human activities, are examined in depth. This book scrutinizes the causes, consequences, and scientific evidence underpinning climate change, drawing on the latest findings from the Intergovernmental Panel on Climate Change (IPCC, 2021). Particular attention is paid to the impact of greenhouse gas emissions, deforestation, and industrialization, which collectively drive global warming and disrupt natural systems. A comprehensive analysis of global ocean-atmosphere circulation reveals the interconnected mechanisms that regulate weather patterns and climate. Topics such as the El Niño-Southern Oscillation (ENSO), the Atlantic Meridional Overturning Circulation (AMOC), and the monsoon systems are discussed, highlighting their profound influence on regional climates and their susceptibility to anthropogenic perturbations (Hurrell et al., 2003).

The dynamics of air masses and atmospheric systems, including weather fronts, cyclones, anticyclones, and jet streams, are explored in detail. Their role in shaping global and regional weather patterns is examined, alongside their implications for storm formation, precipitation, and temperature variability. These systems are central to understanding both everyday weather and extreme events (Stull, 2017).

Natural hazards and disasters, ranging from earthquakes and floods to volcanic eruptions, are also discussed in the context of their interactions with atmospheric processes (Andrews, 2010). This multidisciplinary approach underscores the interconnectedness of Earth's systems and the importance of mitigating risks through scientific understanding and planning (Schwartz, 2006).

In an era marked by rapid climate change and environmental stress, this book tackles the increasing frequency and intensity of weather and climate extremes. Destructive floods, prolonged droughts, intense heat waves, and powerful storms are examined as manifestations of a warming globe, with a focus on their socioeconomic impacts and the need for adaptive strategies (Trenberth et al., 2014).

A special chapter is devoted to global environmental impacts, detailing anthropogenic influences such as land-use changes, air and water pollution, soil degradation, and the depletion of natural resources. The depletion of the ozone layer, the disruption of the carbon cycle, and the role of uncontrolled greenhouse gas emissions in accelerating climate change are critically analyzed. This section underscores the urgent need for sustainable practices and global cooperation to mitigate these challenges (Crutzen & Stoermer, 2000).

This comprehensive examination of atmospheric science and climate systems provides readers with a robust understanding of Earth's interconnected processes, equipping them with the knowledge necessary to address current and future environmental challenges.

1.6 The Anthropocene and Future Outlook

This chapter introduces the concept of the Anthropocene, a proposed geological epoch defined by significant and lasting human influence on the Earth's systems and environment (Will et al., 2011). From the Industrial Revolution to the digital age, human activities have left an indelible mark on the geosphere, biosphere, hydrosphere, and atmosphere, signaling an era of accelerated environmental change and ecological transformation (Steffen et al., 2011).

Key topics include the ongoing challenges of climate change, with a focus on the vulnerabilities of natural and human systems in a world warming by more than 1 °C above pre-industrial levels. The chapter examines the interrelated issues of rising sea levels, extreme weather events, biodiversity loss, and shifts in ecosystems. Adaptive and mitigation strategies, ranging from renewable energy adoption to reforestation and carbon capture technologies, are explored. The discussion is enriched with future climate projections based on varying emission pathways and environmental protection scenarios, illustrating the profound choices humanity faces (Ritchie & Roser, 2020).

Advancing the conversation, this book highlights the role of modern environmental monitoring techniques and technologies, such as satellite-based observations, remote sensing, and advanced modeling systems. These tools are indispensable for tracking changes in weather, water resources, and climate, providing critical data for policymakers, scientists, and communities to foster resilience and sustainability (NASA, 2023).

This book integrates essential chapters on the climate system, eco-climatology, biodiversity, biogeochemical cycles, ecosystems, and global environmental factors. These sections weave a narrative that unpacks the intricate web of processes sustaining life on Earth. Topics include the mysteries of solar and terrestrial radiation, the global heat and energy budgets, and the dynamic interactions between atmospheric and oceanic circulation patterns. A detailed examination of atmospheric systems, natural disasters, and weather and climate extremes offers insights into their causes, impacts, and mitigation strategies.

By framing the Anthropocene within the context of these interconnected systems, this book fosters a holistic understanding of humanity's role and responsibility in shaping the planet's future. *Atmospheric Perspectives: Unveiling Earth's Environmental Challenges* equips readers with the scientific knowledge and practical tools to navigate the environmental challenges of the twenty-first century. It inspires a deeper appreciation of Earth's complexity and the urgency of adopting sustainable practices to safeguard its equilibrium for future generations.

References

Ahrens, C. D., & Henson, R. (2016). *Meteorology today: An introduction to weather, climate, and the environment*. Cengage Learning.

Andrews, D. G. (2010). *An introduction to atmospheric physics*. Cambridge University Press.

Atkinson, B. J., & Davies, H. C. (2018). *The atmosphere: An introduction to meteorology*. Wiley.

Chapin, F. S., Matson, P. A., & Mooney, H. A. (2011). *Principles of terrestrial ecosystem ecology*. Springer.

Crutzen, P. J., & Stoermer, E. F. (2000). The "Anthropocene". *Global Change Newsletter, 41*, 17–18.

Houghton, J. D. & Neck, C. P. (2002) The Revised Self-Leadership Questionnaire: Testing Hierarchical Factor Structure for Self-Leadership. *Journal of Managerial Psychology*, 17, 672–691. https://doi.org/10.1108/02683940210450484.

Hurrell, J. W., et al. (2003). The North Atlantic Oscillation: Climate significance and environmental impact. In *Climate variability and ecosystem response at long-term ecological research sites*. Oxford University Press.

IPCC. (2021). *Climate change 2021: The physical science basis*. Cambridge University Press.

Jacobson, M. Z. (2014). *Fundamentals of atmospheric modeling*. Cambridge University Press.

Kasting, J. F. (1993). Earth's early atmosphere. *Science, 259*(5097), 920–926.

Lamb, H. H. (1991). *Climate, history and the modern world*. Routledge.

Lutgens, F. K., & Tarbuck, E. J. (2015). *The atmosphere: An introduction to meteorology*. Pearson.

McPhee, J. (1998). *The control of nature*. Farrar, Straus and Giroux.

NASA. (2023). *Fire information for resource management system (FIRMS)*. https://firms.modaps.eosdis.nasa.gov/

Oke, T. R. (1987). *Boundary layer climates*. Routledge.

Peel, M. C., & McGregor, J. (2002). *Geographical information systems for environmental management*. Routledge.

Ritchie, H., & Roser, M. (2020). *Climate change*. Our World in Data. https://ourworldindata.org/climate-change

Schwartz, S. E. (2006). The atmospheric sciences: A brief overview. *Reviews of Geophysics, 44*(2), RG2001.

Sellers, W. D. (1997). *Physical climatology*. University of Chicago Press.

Steffen, W., Grinevald, J., Crutzen, P. & McNeil, J. (2011) The Anthropocene: Conceptual and Historical Perspectives. *Philosophical Transactions of Royal Society A*, 369, 842–867. https://doi.org/10.1098/rsta.2010.0327.

Stommel, H., & Stommel, E. (1983). *Atmospheric and oceanic circulation*. University of Washington Press.

Stull, R. B. (2017). *Meteorology for scientists and engineers*. Brooks/Cole.

Taylor, S. R., & McLennan, S. M. (2009). *Planetary geochemistry*. Blackwell Publishing.

Trenberth, K. E., et al. (2014). *Climate change 2014: Impacts, adaptation, and vulnerability*. Cambridge University Press.

Will, S., Jacques, G., Paul, C., & John, M. N. (2011). The Anthropocene: Conceptual and historical perspectives. *Philosophical Transactions: Mathematical, Physical and Engineering Sciences, 369*, 842–867. https://doi.org/10.1098/rsta.2010.0327

Chapter 2
Geophysical Basis for Understanding Planet Earth's Systems

Volcanic activity

In this book, we explore the solar system, the origin and evolution of the Earth, and its atmosphere, oceans, and life. Understanding the Earth and the solar system is fundamental to grasping the processes that have shaped our planet and made life possible. Our ability to predict the impacts of human activities depends on our

© The Author(s), under exclusive license to Springer Nature Switzerland AG 2025
V. Spiridonov et al., *Atmospheric Perspectives*,
https://doi.org/10.1007/978-3-031-86757-6_2

comprehension of the Earth's systems. Managing natural hazards requires knowledge of the frequency and intensity of earthquakes, eruptions, and floods, as well as an understanding of the atmosphere, weather, and climate (Lovelock, 2000). The atmosphere, integral to the Earth's climate and biogeochemical cycles, plays a crucial role in Earth system science. Studying its composition and dynamics is key to understanding the Earth's system (Jacob, 1999). Earth system science (ESS), a scientific discipline established in 1988, examines the interactions between the Earth's spheres (Steffen et al., 2004). This discipline is foundational to understanding the planet's various components, including the origin of the solar system, the formation of planets, and the geographic grid system, which includes large and small circles, latitude, longitude, and the concept of international date and time (Strahler, 2013).

2.1 Physical Geography

Physical geography, a key branch of natural science, delves into the study of the Earth's physical features and processes. As illustrated in Fig. 2.1, this field involves studying landscapes, landforms, climate systems, oceans, and ecosystems to reveal the intricate interactions between the Earth's surface and its atmosphere (Christopherson, 2012). This discipline bridges the natural sciences and human activities, emphasizing the impact of physical processes on both natural systems and societal development. Through its insights, physical geography enables us to better predict and mitigate the effects of natural hazards such as floods, earthquakes, and hurricanes. Moreover, it supports sustainable resource management and helps address the profound challenges posed by climate change and the expanding

Fig. 2.1 Illustration of physical geography

footprint of human activities (McKnight & Hess, 2014). By understanding the Earth's physical systems, we can design more effective policies and practices to safeguard ecosystems, ensure resource availability, and enhance human resilience.

2.1.1 Geographic Grid System: Great and Small Circles

A foundational concept in physical geography is the geographic grid system, which provides a framework for mapping and describing the Earth's surface. This system consists of an imaginary network of intersecting lines—latitude and longitude—spanning the globe. It serves as a reference system for pinpointing locations, defining boundaries, and analyzing spatial relationships (Snyder, 1987).

Great Circles: Great circles are the largest possible circles that can be drawn on the surface of a sphere, with their center coinciding with the sphere's center. On Earth, examples include the equator and all meridians. Great circles are significant in navigation and geodesy because they represent the shortest path between two points on the globe, making them critical for air and sea travel.

Small Circles: In contrast, small circles are circles on the Earth's surface whose centers do not coincide with the Earth's center. Parallels of latitude (except the equator) are examples of small circles. These lines are essential for defining zones of latitude and mapping the Earth's diverse climatic and ecological regions.

This grid system forms the backbone of geographic analysis, enabling the precise measurement of distance, direction, and area. Additionally, it facilitates the study of global phenomena, such as climate patterns, ocean currents, and migratory routes. By mastering the geographic grid, we gain a deeper understanding of Earth's spatial dimensions and the interplay of its physical and human systems.

2.2 Definition of Geography

Geography is the scientific discipline that explores the Earth's physical features, climate systems, weather patterns, and the distribution of life, including plants, animals, and human populations, along with the complex interactions among them. It investigates how these elements are interconnected, focusing on spatial patterns, processes, and dynamics that shape the planet's surface and atmosphere (Johnston et al., 2000).

Geography also delves into the relationships between people and their environment, examining how human activities influence and are influenced by the physical world. This includes understanding how urbanization, resource use, and climate change impact ecosystems, societies, and natural systems. Physical geographers specifically study the processes that form and alter the Earth's spheres, providing essential insights into how natural and human systems are intertwined and constantly evolving (Smithson et al., 2008). Through these studies, geography helps us better understand the interconnectedness of global systems and the challenges posed by environmental and societal changes.

2.3 Earth and Solar System

The universe encompasses an incomprehensible vastness, containing galaxies, stars, planets, and all forms of matter and energy (Fig. 2.2). It extends across all space and time, from the tiniest subatomic particles to the most massive celestial structures. Modern cosmology provides the framework for understanding the universe's origin, its complex structure, ongoing evolution, and eventual fate. This scientific field, through observations and theoretical models, explores phenomena such as the Big Bang, dark matter, dark energy, and the expansion of the universe (Moore et al., 2012).

At the heart of the universe's grand scale are galaxies, the primary building blocks, consisting of vast collections of stars, interstellar gas, dust, and dark matter bound together by gravity. An estimated 2 trillion galaxies are in the observable universe, with some containing over 100 billion stars.

An estimated 2 trillion galaxies are in the observable universe, with some containing over 100 billion stars. Galaxies exhibit a range of forms, including spiral, elliptical, and irregular shapes, each contributing to the dynamic nature of the cosmos. Our solar system resides in the Milky Way, a spiral galaxy approximately 100,000 light-years in diameter. Positioned in one of its spiral arms, called the Orion

Fig. 2.2 The universe as a vast expanse within the cosmos

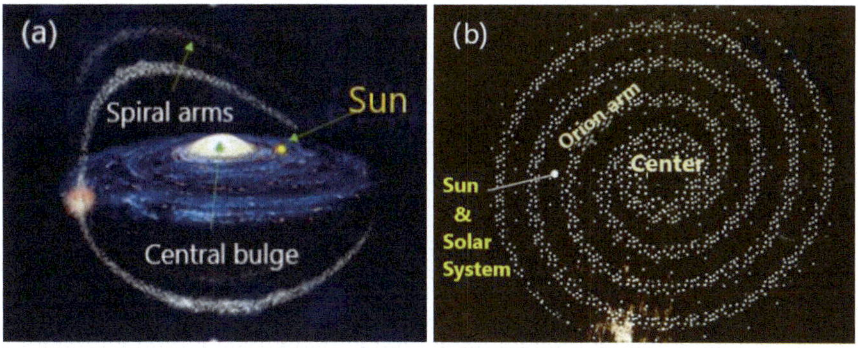

Fig. 2.3 The sun and the solar system. (**a**) Spiral arm of Orion; (**b**) disk and central bulge

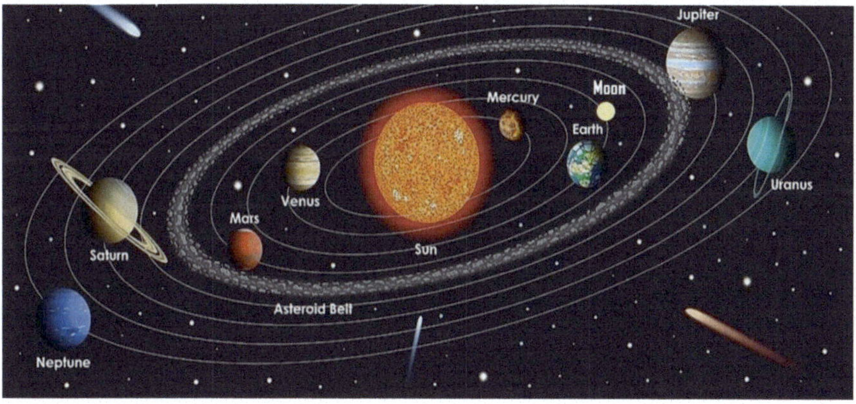

Fig. 2.4 An illustration of the solar system depicting terrestrial (inner) planets and outer (giant) planets

Arm (Fig. 2.3a), our Sun is just one of the galaxy's hundreds of billions of stars. The Milky Way itself is a barred spiral galaxy—characterized by a bright central bulge and a flat, rotating disk of stars, gas, and dust (Fig. 2.3b). This central bulge is believed to contain a supermassive black hole, which influences the galaxy's structure and rotation (Sparrow, 2006). Understanding our galaxy's place in the larger cosmic context deepens our appreciation of Earth's location in the universe. The solar system—with its intricate arrangement of planets, moons, and smaller celestial bodies—is a unique environment within the vast expanse of the Milky Way, providing the conditions necessary for life (Fig. 2.4). Exploring these cosmic relationships helps us trace Earth's formation and its ongoing evolution as part of a much larger, dynamic universe.

2.3.1 Asteroids

Asteroids, also known as minor planets, are rocky, metallic, or icy bodies that orbit the Sun, ranging in size from small rocks to bodies nearly 1000 km in diameter. Lacking atmospheres, they are remnants from the early solar system, offering insights into its formation (Fig. 2.5). Most asteroids are in the asteroid belt between Mars and Jupiter, with others, like Near-Earth Asteroids (NEAs), having orbits that bring them close to Earth. Some NEAs termed potentially hazardous asteroids (PHAs) pose a collision risk (Lewis, 1997).

Besides the asteroid belt, Trojan asteroids share orbits with bigger planets, like Jupiter, at stable Lagrange points. Recent space missions, such as NASA's OSIRIS-REx to Bennu and Japan's Hayabusa2 to Ryugu, have enhanced our understanding of asteroid composition and potential resource utilization. These small celestial bodies hold significance for planetary defense, space exploration, and the future of resource mining.

2.3.2 Meteorites

Meteorites are solid fragments of asteroids, comets, or other celestial bodies that survive their passage through Earth's atmosphere and land on the surface. They offer crucial insights into the early solar system and are valuable for scientific research. Meteorites are classified into three main types shown in Fig. 2.6:

Fig. 2.5 View of asteroid diversity

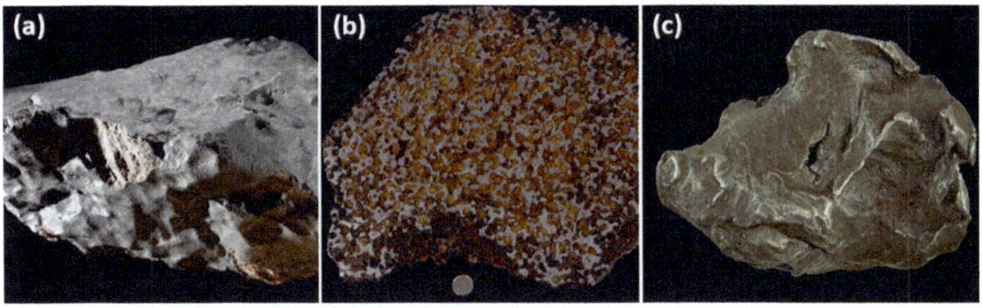

Fig. 2.6 Types of meteorites: (**a**) Stony meteorite, (**b**) stony-iron meteorite, (**c**) iron meteorite

Stony Meteorites: Primarily composed of silicate minerals. This category includes chondrites, which contain spherical chondrules, and achondrites, which do not.

Iron Meteorites: Made mainly of metallic iron and nickel, often displaying distinctive crystalline patterns.

Stony-Iron Meteorites: A mix of metal and silicate minerals, representing the transitional zones within asteroids.

Studying meteorites helps scientists learn about the early solar system, the formation of planets, and the potential for life beyond Earth. Notable meteorites like the Allende and Tissint provide valuable data and insights, and their study aids in planetary defense and resource exploration (Hutchison, 2004).

2.3.3 Comets

Comets are icy bodies that originate from the outer regions of the solar system. As they approach the Sun, the heat causes their ice to sublimate, creating a glowing coma (a cloud of gas and dust) and often forming a visible tail that points away from the Sun due to solar wind and radiation pressure. Comets are composed primarily of water ice, frozen gases, and organic compounds. They are often categorized into two main types based on their orbits: short-period comets, which take less than 200 years to orbit the Sun and typically originate from the Kuiper Belt, and long-period comets, which have orbits lasting thousands of years and come from the distant Oort Cloud (Lamy et al., 2004).

The most famous comet, Halley's Comet, is a short-period comet that becomes visible from Earth approximately every 76 years. Comets like Halley's have been observed and recorded by astronomers for centuries, providing valuable information about the early solar system's composition and dynamics. Recent space missions, such as ESA's Rosetta mission to Comet 67P/Churyumov–Gerasimenko, have significantly advanced our understanding of comets. These missions have analyzed cometary materials and discovered complex organic molecules, offering insights into the conditions and processes that might have contributed to the origin of life on Earth. Notably, Comet C/2017K2 (PANSTARRS) shown in Fig. 2.7 is one

Fig. 2.7 K2, the brightest comet in our solar system. (Credit: Pixabay/CC0 Public Domain)

of the largest comets observed, with an estimated nucleus diameter of about 18 km, making it a significant object for studying the structure and behavior of cometary bodies.

2.3.4 *Natural Satellites or Moons*

Many authors define a "satellite" or "natural satellite" as an object orbiting a planet or minor planet, using the term synonymously with "moon." By this definition, all natural satellites are considered moons, while Earth and other planets are not classified as satellites. Natural satellites or moons, numbering over 150, are integral components of the solar system, formed alongside the eight planets from the same gases and dust particles within nebulae (Fig. 2.8). These moons vary greatly in size, composition, and origin, showcasing a wide range of geological features and evolutionary histories. While some moons formed in orbit around their parent planets during the early solar system, others were likely to capture asteroids or remnants of early planetary collisions (Rothery, 1999). There is no clearly defined lower limit for what qualifies as a "moon." Any natural celestial object with a confirmed orbit around a planet in the solar system is generally regarded as a moon, even if it is only a kilometer in size. However, even smaller objects—some just a fraction of that size, about a tenth of a kilometer, have been identified within Saturn's rings and are referred to as moonlets, though many of these have not been directly observed.

The Moon was the first natural satellite known to humanity. However, it was long regarded as a "planet" until Copernicus redefined celestial understanding in 1543 (Koyré, 1957). For example, Earth's moon is thought to have formed because of a massive impact between a Mars-sized body and the early Earth (Canup & Asphaug, 2001). Moons play a crucial role in the dynamics of their planetary systems. They can influence planetary tides, stabilize axial tilts, and even contribute to the geological activity of their parent planets. For instance, Jupiter's moon Io is the most volcanically active body in the solar system, with its intense volcanic activity driven by tidal forces from Jupiter and its other large moons (Krupp & Dougherty, 2001). Saturn's moon Enceladus, however, harbors a subsurface ocean beneath its icy

Fig. 2.8 The Moon, Earth's closest celestial companion, has captivated civilizations since ancient times. As the brightest object in the night sky, it has inspired awe and wonder, influencing cultural myths, calendars, and navigation. This natural satellite, remarkable for its phases and tidal effects, plays a vital role in scientific exploration and our understanding of the universe. (Credit: https://techfruit.com/focus/the-moon-is-more-than-a-natural-satellite/)

crust, ejecting water vapor and organic materials into space, which suggests the possibility of life-sustaining environments (Porco et al., 2006). Many of the outer planets, particularly Jupiter and Saturn, host vast systems of moons. Some of these moons, like Ganymede, the largest moon in the solar system, are even larger than the planet Mercury. In contrast, the inner planets (Mercury and Venus) lack moons, which raises questions about the different formation conditions or gravitational dynamics in that region of the solar system.

In recent years, missions like NASA's Galileo, Cassini, and the upcoming Europa Clipper have provided and continue to offer new insights into the nature of these moons, revealing not only their unique characteristics but also their potential to host environments suitable for life (NASA, 2021). As we continue to explore and study these distant worlds, the moons of our solar system remain vital to understanding the broader processes that have shaped planetary bodies and the potential for habitability beyond Earth.

2.4 Origin of the Solar System

The origin of the solar system is a fundamental topic in astrophysics and planetary science, crucial to understanding the processes that shaped not only our cosmic neighborhood but also planetary systems across the universe. Approximately 4.6 billion years ago, our solar system began its formation, a process extensively explained by the nebular hypothesis. This prevailing theory posits that the solar system emerged from a vast, rotating cloud of gas and dust known as the solar nebula (Lynden-Bell & Pringle, 1974). As this nebula collapsed under the force of its gravity, it spun faster, flattening into a rotating disk. The immense concentration of mass at the center of this disk eventually ignited to form the Sun, while the remaining material coalesced to form the planets, moons, and other smaller bodies such as asteroids and comets.

The formation of these planets occurred through the process of accretion, where dust particles clumped together into larger planetesimals, which then merged to form protoplanets. Over time, these protoplanets cleared their orbits and developed into the planetary bodies we observe today. This theory also accounts for the compositional differences between the inner and outer planets. The inner planets—Mercury, Venus, Earth, and Mars—are rocky and smaller due to the higher temperatures near the Sun, which prevents lighter elements like hydrogen and helium from condensing. In contrast, the outer planets—Jupiter, Saturn, Uranus, and Neptune—are gas giants, composed mostly of these lighter elements, as lower temperatures in the outer regions of the nebula allowed for the accumulation of gas.

Understanding the solar system's origin offers profound insights into the formation of other planetary systems. Observations of protoplanetary disks around young stars in distant regions of the galaxy provide critical evidence that supports the nebular hypothesis, suggesting that our solar system's formation process is not unique but part of a broader, universal mechanism for star and planet formation (Hartmann et al., 1998). Moreover, the study of how planets and moons developed within this system sheds light on the conditions that made Earth uniquely hospitable for life. Examining the influences of planetary migration, asteroid impacts, and the distribution of water and organic compounds across the early solar system provides vital clues to the development of Earth's environment, offering key perspectives on planetary habitability in other star systems. As space missions, telescopes, and computer simulations continue to refine our understanding of the Solar System's origin, they open new possibilities for identifying planets orbiting distant stars, contributing to the broader search for life beyond Earth.

2.5 Nebula Theory

The nebula theory, also known as the nebular hypothesis, stands as one of the foremost explanations for the origin of the solar system. This theory was initially proposed by the French mathematician and philosopher Pierre-Simon Laplace in the late eighteenth century and has since undergone refinement and expansion by subsequent scientists. The nebular hypothesis can be summarized as follows: it offers a comprehensive and widely accepted explanation for the formation of the solar system, supported by astronomical observations, computer simulations, and the study of celestial bodies within our solar system (Lynden-Bell & Pringle, 1974). Remnants of the solar nebula are still observable today in the form of asteroids, comets, and objects within the Kuiper Belt, providing valuable insights into the early stages of the solar system's formation. According to the nebula hypothesis: The solar system originated from a rotating cloud of dust and gases known as a nebula, primarily composed of hydrogen and helium formed from the primordial nucleosynthesis following the Big Bang. The nebula began to coalesce around five billion years ago, flattening into a disk with a protosun at its center. This protosun represents the initial condensation within the giant molecular cloud, from which the Sun and the protoplanetary disk, the precursor to the planets, asteroids, and other small bodies of the solar system, formed.

Intense solar winds removed lighter gases such as hydrogen and helium from the inner planets, leaving behind primarily rocky materials. In contrast, the larger outer planets began to form from lighter gases with a higher percentage of frozen gases, including water, carbon dioxide, ammonia, and methane (Weidenschilling, 1977).

2.5.1 Stages of the Formation of the Solar System

The nebular theory remains the most widely accepted explanation for the formation of the solar system. This process unfolds over several stages, each marked by intricate physical and chemical interactions that gave rise to the Sun, planets, moons, and smaller celestial bodies. These stages are outlined below:

1. Nebular Cloud Formation:

   The formation of the solar system began within a vast, rotating cloud of gas and dust, primarily composed of hydrogen and helium, along with trace elements heavier than helium produced by nucleosynthesis following the Big Bang. This nebular cloud, enriched by heavier elements from previous generations of stars through supernova explosions, sets the stage for the solar system's birth. Some theories propose that shock waves from a nearby supernova may have initiated the gravitational collapse of this nebula, triggering the initial contraction of the cloud.
2. Gravitational Collapse and Protostar Formation:

   As gravity caused the nebular cloud to collapse inward, a dense central core, known as the "protosun" or "solar protostar," began to form. The collapse accelerated the rotation of the cloud due to the conservation of angular momentum, flattening it into a spinning disk. At this stage, the central region became increasingly hotter and denser, ultimately igniting nuclear fusion reactions that would give birth to the Sun. Meanwhile, in the outer regions of the cloud, temperatures remained lower, allowing solid materials to coalesce.
3. Accretion of Planetary Material:

   As the proto-Sun continued to form at the center of the collapsing cloud, surrounding material began to orbit it, forming the initial stages of a protoplanetary disk. This orbiting material began to collide and aggregate, forming small, solid particles known as planetesimals. Accretion, a process where these planetesimals grew by accumulating dust, gas, and other debris through gravitational attraction and collisions, began to dominate, setting the stage for the formation of larger planetary bodies.

4. Formation of the Protoplanetary Disk:

   The rotating disk of gas and dust surrounding the proto-Sun is referred to as the "protoplanetary disk." This disk is crucial as the birthplace of planets. Within this flattened disk, temperature gradients played a significant role: the inner regions were too hot for volatile compounds to condense, leading to the formation of rocky planets, while the cooler outer regions allowed gas and ice giants to develop. This differentiation in temperature across the disk sets the foundation for the structure of the solar system.

5. Planet Formation and Differentiation:

   Over time, planetesimals continued to collide and merge, gradually forming larger bodies known as "protoplanets." These protoplanets underwent further accretion, clearing their paths of debris as they grew. In the inner solar system, rocky planets like Earth and Mars formed, while the outer regions saw the birth of gas giants like Jupiter and Saturn.

   The gravitational influence of these larger bodies began shaping the overall architecture of the solar system, influencing the orbits of other bodies and even ejecting some debris into interstellar space.

6. Clearing of the Solar Nebula:

   As the newly formed planets continued to grow, they began to carve out gaps in the protoplanetary disk by sweeping up material along their orbital paths. This process led to the gradual clearing of the solar nebula. In addition, solar winds from the young, active Sun helped expel any remaining gas and dust from the system, particularly in the inner regions. This marked the final stage of the planet-forming process, as the planets settled into stable orbits.

7. Final Differentiation and Formation of Celestial Bodies:

   The protoplanets continued to evolve, undergoing differentiation—a process where heavier elements sank to the core, and lighter materials rose to the surface. Depending on their size, composition, and distance from the Sun, these celestial bodies developed into distinct types. In the inner solar system, differentiation led to the formation of solid, rocky terrestrial planets with metal-rich cores, while in the outer solar system, protoplanets accumulated vast amounts of gas, forming the gas giants. Smaller celestial bodies, such as moons, asteroids, and comets, also emerged during this stage.

8. Completion of the Formation Process:

   For millions of years, the solar system reached a state of relative equilibrium. The inner region became populated by terrestrial planets like Mercury, Venus, Earth, and Mars, while the outer solar system saw the formation of gas giants such as Jupiter and Saturn, as well as ice giants like Uranus and Neptune. Other remnants from the solar nebula—asteroids, comets, and dwarf planets—settled into regions such as the asteroid belt between Mars and Jupiter and the Kuiper Belt beyond Neptune. This distribution reflects the complex interplay of gravitational forces and accretion during the solar system's formation.

The formation and evolution of the solar system phase, as depicted in Fig. 2.9, begins with the collection of the nebula into a rotating disk, which heats up as gravitational energy is converted into thermal energy. Subsequently, it is assumed that the cooling nebula condenses and forms small rocks and solid metal particles. In the concluding stages of this process, as dust-sized particles collide, they amalgamate to form asteroids and eventually coalesce to form planets. The larger outer planets began to form from lighter gases with a high percentage of frost or frozen gases such as water, carbon dioxide, ammonia, and methane. The shared characteristics of the eight planets are outlined in Table 2.1.

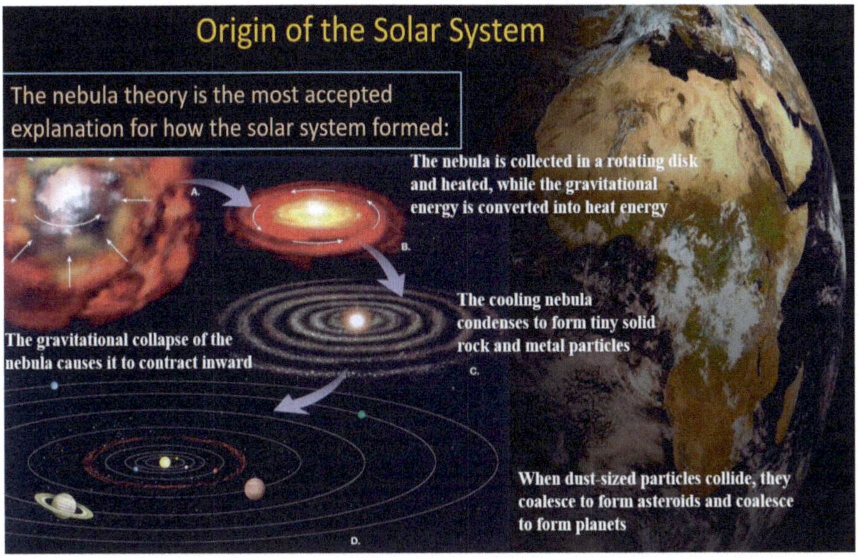

Fig. 2.9 The nebula theory is the most accepted explanation for the formation of the solar system

Table 2.1 Physical characteristics of the planets

Planet	Rotation time (days)	Equator diameter (km)	Average density ($g \bullet cm^{-3}$)
Terrestrial planets			
Mercury	58.7	4.880	5.43
Venus	243	12.104	5.24
Earth	1	12.760	5.52
Mars	1.03	6.787	3.98
Giant planets			
Jupiter	0.41	142,796	1.33
Saturn	0.43	120,660	0.69
Uranus	0.72	51,200	1.27
Neptune	0.67	49,500	1.76
Other			
Pluto	6.39	2300	2.03

The terrestrial planets (Mercury, Venus, Earth, and Mars), which belong to the inner solar system, are composed primarily of minerals and rocky materials. They exhibit relatively high-density $\rho > 3$ (g • cm^{-3}). These planets are approximately spherical (slightly flattened) and have a relatively slow rotation.

The giant Jovian planets (Jupiter, Saturn, Uranus, and Neptune), part of the outer solar system, are significantly larger than terrestrial planets. They are entirely composed of gases, characterized by low density, extreme flattening, and rapid rotation. These planets possess dense and turbulent atmospheres.

2.6 The Shape of the Earth

Ancient Greek scientists were among the first to propose that the Earth has a spherical shape. In 200 BC, Eratosthenes made the groundbreaking estimate of the Earth's circumference. Several pieces of evidence support the spherical shape of the Earth:

- Darkening of the Moon
- Circular horizon around the Earth
- Satellite photos of the Earth
- Variations in the force of gravity around the Earth

The Earth's spherical shape is primarily caused by the gravitational attraction of its materials. However, the centrifugal force resulting from the Earth's rotation distorts its shape from a perfect sphere. As a result, the Earth bulges at the equator and flattens at the poles, leading to a shape described as an oblate spheroid. Key dimensions of the Earth include:

- Polar diameter: 12,714 km
- Equatorial diameter: 12,756 km
- Polar circumference: 39,943 km
- Equatorial circumference: 40,007.6 km

The polar diameter is shorter than the equatorial diameter by 42 km (see Fig. 2.10). The value of gravity at the equator is relatively small compared to the larger value at the poles.

Geographical Width—Latitude: The imaginary lines that trace the true east–west direction on the Earth's surface are lines of latitude (Fig. 2.11a). These geographic latitudes are also referred to as parallels of latitude because they do not intersect. The circles of latitude on Earth are called parallels or lines of latitude. Latitude is the angular distance of a location north or south of the Earth's equator, which is defined as a latitude of 0°. The circle at latitude 0° (equator) lies halfway between the poles and forms the largest circle known as the "great circle." Other circles of latitude are termed "small circles."

Geographical Longitude—Longitude: Lines extending from pole to pole, from north to south, are longitudinal and intersect geographic latitudes at right angles

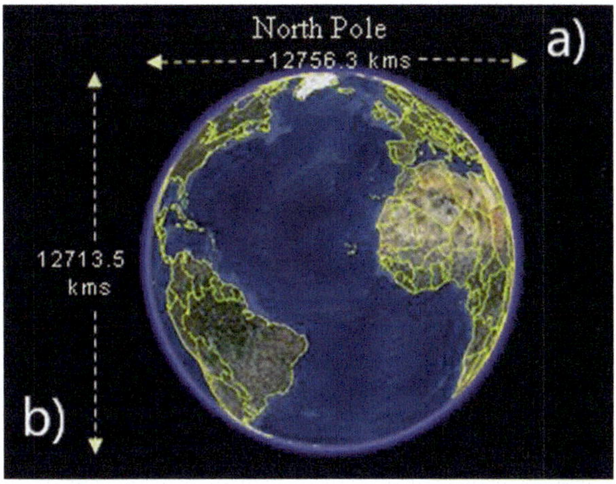

Fig. 2.10 Dimensions of the Earth. (**a**) Polar diameter; (**b**) equatorial diameter

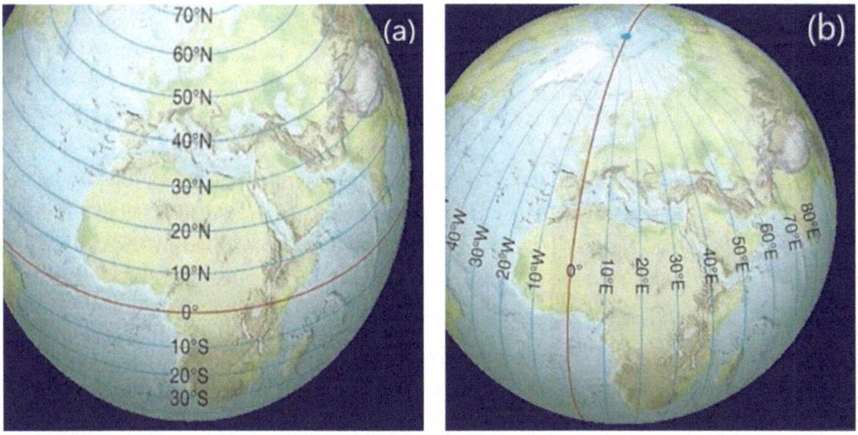

Fig. 2.11 Global view: (**a**) polar view at latitude and (**b**) equatorial view at longitude

of 90° (Fig. 2.11b). Longitude is measured in angular degrees east or west of the prime meridian, also known as the Greenwich meridian (i.e., longitude 0°). These lines are also called meridians or meridian lines because all places sharing the same geographic longitude have the same mean.

The local solar time on the prime meridian was selected as the standard for the entire system and became the reference for the center of the time zone, extending 7.5° of longitude to the west and east of it (Fig. 2.12). Additionally, 23 other standard meridians were established, each spanning more than 15° of longitude.

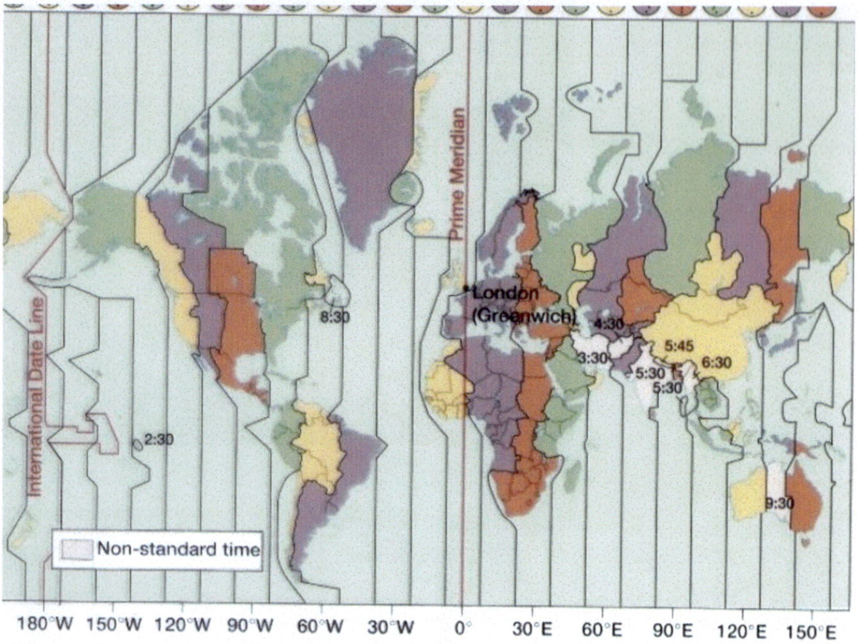

Fig. 2.12 Longitudes and standard time zones

2.7 Earth–Sun Relations and Seasons

The Earth's rotation around its axis is a fundamental movement that significantly influences our daily and seasonal experiences. The Earth's axis tilted at an angle of approximately 23.5° relative to its orbit around the Sun (not 66.5° from the ecliptic plane as sometimes misstated) is directed toward the North Star (Polaris). This tilt is crucial for the seasonal variations experienced on Earth.

The rotation occurs from west to east when viewed from the sides, giving the appearance of a counterclockwise movement as seen from the North Pole. During this rotation, all points on the Earth's surface (except those at the poles) travel in a circular path around the axis. This path, or circle of rotation, is defined by the latitude of each point. The Earth completes one full rotation around its axis in approximately 24 h, defining the length of a solar day. However, when measured relative to the fixed stars, the Earth takes about 23 h, 56 min, and 4.099 s to complete one rotation, known as a sidereal day. This discrepancy arises because the Earth is also orbiting the Sun, causing the rotational period to differ slightly when measured against distant stars.

At the equator, the Earth's rotational speed reaches a maximum of approximately 465 m/s (or 1670 km/h), while at the poles, the rotational speed effectively drops to 0 m/s. This variation in speed is because the Earth rotates in such a way that points on the equator travel a greater distance in the same amount of time compared to points closer to the poles.

Additionally, the Earth's axial tilt and orbit around the Sun lead to the changing seasons. As the Earth orbits the Sun throughout the year, the tilt causes different latitudes to receive varying amounts of solar energy, resulting in the distinct seasons of spring, summer, autumn, and winter. This variation in solar energy is further modulated by the Earth's elliptical orbit, which affects the intensity and duration of seasonal changes.

2.8 Rotation of the Earth Around Its Axis

Rotation is the movement of the Earth around its axis, with the Earth's axis directed toward the North Star (Polaris) and inclined at 66.5° from the ecliptic plane. The rotation is from west to east when viewed from the side, appearing counterclockwise when observed from the North Pole (Fig. 2.13). During rotations, all points on the surface of the Earth move in a circle around the axis (except the poles). The circle of rotation is defined by the latitude of the points. The Earth completes one full rotation around its axis in 24 h, constituting one solar day.

The Earth completes one full rotation of the stars in 861,640,905 s = 239,344,696 h (a sidereal day). The maximum rotational speed on Earth is 465 m/s or 1670 km/h at the equator. The rotational speed is 0 km/h at both poles.

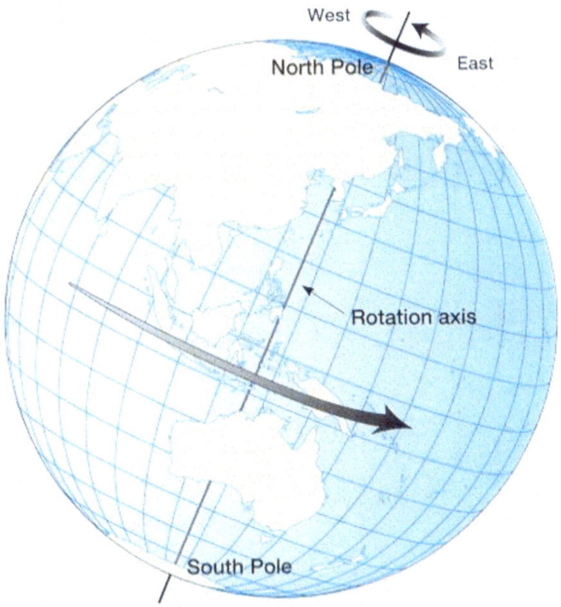

Fig. 2.13 Representation of the Earth's rotation around its axis

2.9 Physical Effects of Rotations on Earth

The Earth's rotation on its axis defines the North and South Poles and plays a crucial role in shaping various physical phenomena on our planet (Peixoto & Oort, 1992). The circle of rotation for any point on Earth's surface is determined by its latitude. This rotation introduces several key effects:

Day and Night Cycle: The rotation of the Earth causes the alternating periods of day and night. As the Earth spins on its axis, different regions are exposed to sunlight or darkness, leading to the cyclical nature of day and night. This cycle affects daily routines and the biological rhythms of living organisms.

Daily Variations: The rotation of the Earth leads to daily variations in temperature, humidity, and wind patterns. As different areas of the Earth's surface rotate into and out of sunlight, they experience fluctuations in temperature and humidity. These variations also influence wind movements as the differential heating of the Earth's surface generates pressure differences that drive atmospheric circulation.

Earth's Shape: The rotation contributes to the Earth's oblate spheroid shape—a slightly flattened shape at the poles and bulging at the equator. This is a result of the centrifugal force generated by the Earth's rotation, which causes the equatorial region to bulge outward.

Coriolis Effect: The rotation of the Earth induces the Coriolis effect, a crucial phenomenon that affects the movement of air and water masses. As the Earth rotates in an eastward direction, the Coriolis effect causes moving air and water to be deflected to the right in the Northern Hemisphere and the left in the Southern Hemisphere (Fig. 2.14). This deflection influences weather patterns, ocean cur-

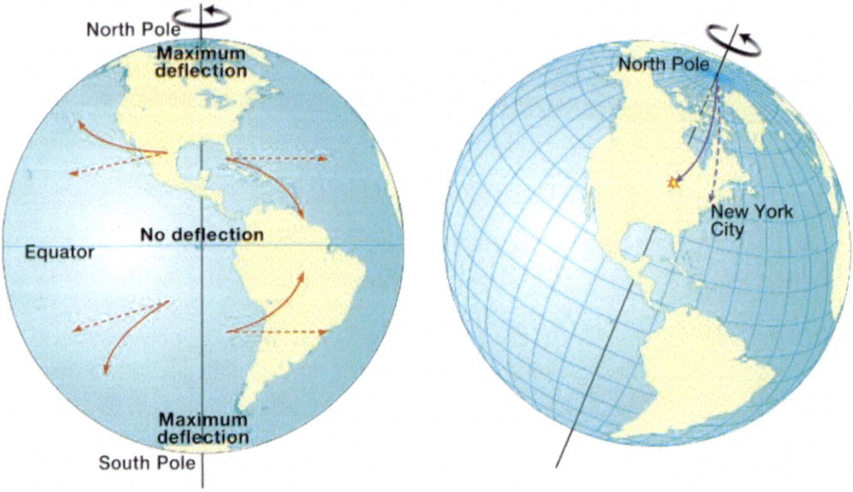

Fig. 2.14 Deflection of flow paths on individual air elements due to the Coriolis force

rents, and the general circulation of the atmosphere. The Coriolis effect plays a key role in the formation of cyclones, anticyclones, and prevailing wind patterns.

Ocean Currents: The Coriolis effect also impacts ocean currents, causing them to follow curved paths rather than straight lines. In the Northern Hemisphere, ocean currents are deflected to the right, while in the Southern Hemisphere, they are deflected to the left. This deflection helps establish major oceanic gyres, which are large systems of rotating ocean currents.

Weather Systems: The Earth's rotation affects the development and movement of weather systems. For instance, the rotation contributes to the formation of trade winds, westerlies, and polar easterlies, which are crucial for global climate patterns. Additionally, the rotation affects the path of storms and weather systems, influencing their direction and intensity.

Understanding these effects of Earth's rotation is essential for comprehending various natural processes, from the daily changes in weather to the broader patterns of ocean and atmospheric circulation.

The Earth's rotation brings every point on its surface into and out of the varying gravitational forces exerted by the Moon and the Sun, leading to the rise and fall of tides and water levels. Tides, the fluctuations in sea level, are caused by the combined effects of gravitational forces exerted by the Moon and the Earth–Moon orbit around each other (Fig. 2.15). Spring tides occur twice every lunar month throughout the year, regardless of the season. High tides and low tides, which also occur twice a month, happen when the Sun and the Moon are at right angles to each other.

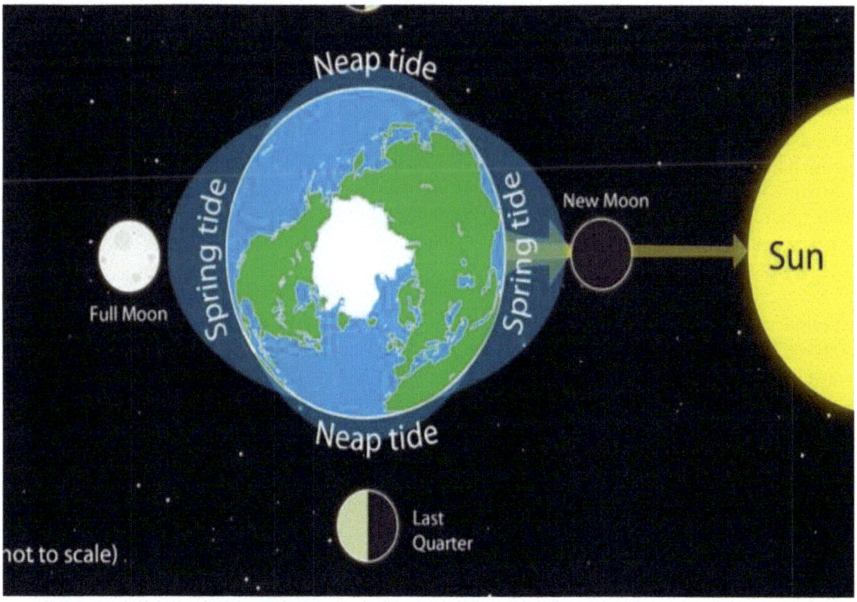

Fig. 2.15 Global view of tides

2.10 Earth's Tilt and Revolution Around the Sun

Earth's Tilt: The Earth's axis is inclined about its orbit around the Sun, a phenomenon known as "axial tilt" or "obliquity." The Earth's axis is tilted at an angle of approximately 23.5° relative to the plane of its orbit. This tilt is responsible for the changing seasons on Earth. When one hemisphere is tilted toward the Sun, it experiences summer, while the other hemisphere, tilted away from the Sun, experiences winter. The axial tilt also causes variations in the length of daylight over the year and influences the angle of sunlight, which affects climate patterns.

Earth's Revolution: Earth's revolution refers to its movement around the Sun in an elliptical orbit. It takes Earth approximately 365.25 days to complete one orbit around the Sun, forming the basis for our calendar year. The Earth's orbit is not a perfect circle but an ellipse, meaning that the distance between the Earth and the Sun varies slightly over a year. This variation has a minimal effect on the changing seasons. The plane in which the Earth orbits around the Sun is called the "ecliptic plane." The Earth's orbit is about 93 million miles (about 150 million kilometers) from the Sun at its closest point (perihelion) and slightly farther at its furthest point (aphelion). Together, the combination of the axial tilt of the Earth and its orbit around the Sun leads to seasonal changes and the annual cycle (Fig. 2.16).

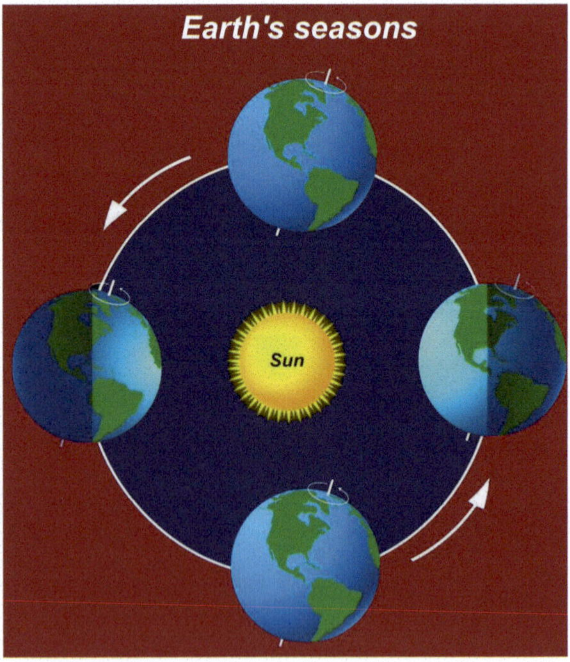

Fig. 2.16 Earth's revolution is the reason for the four seasons: summer, autumn, winter, and spring. However, the seasons are not determined by its orbit but by its axial tilt of 23.5°. Consequently, the Northern Hemisphere experiences winter when its pole tilts away from the Sun, while the Southern Hemisphere experiences summer. Conversely, when the North Pole is tilted toward the Sun, the Northern Hemisphere experiences summer and the Southern Hemisphere experiences winter

The axial tilt of the Earth ensures that different parts of the planet receive varying amounts of sunlight at different times of the year, leading to the well-known changes in climate and weather patterns that we experience.

Figure 2.16 illustrates the inclination of the Earth's axis and the orientation of its poles during its revolution. The Earth's revolution denotes its movement around the Sun. The path followed by the Earth during this revolution is called its orbit, which has an elliptical shape. On January 3rd, the Earth is at its Perihelion, positioned near the Sun at 147.5 Gm. On July 4th, the Earth is at its aphelion, farthest from the Sun in its orbit, at 152.5 Gm. The Earth moves in the opposite direction to the arrows on a clock and completes one full revolution in 3,652,422 days (a Tropical year).

2.10.1 Effects of Earth's Revolution

The Earth's revolution around the Sun is one of the fundamental processes that govern seasonal changes and the variation in daylight length across different latitudes. The revolution, combined with the axial tilt of approximately 23.5°, leads to the occurrence of equinoxes and solstices, which are key markers in the calendar year. These events have a profound effect on the distribution of sunlight, directly influencing the length of days and nights throughout the year.

(a) Equinoxes: Equal Day and Night

The equinoxes occur twice a year, around March 21st (vernal or spring equinox) and September 23rd (autumnal equinox). During these periods, the Earth's axial tilt is such that the Sun is positioned directly over the equator. This results in nearly equal amounts of daylight and darkness across the globe, with approximately 12 h of daylight and 12 h of night. The equinoxes mark the transition between seasons—spring and autumn—and are the only times of the year when every location on Earth experiences equal day and night.

(b) Solstices: The Extremes of Daylight

The solstices represent the points in the Earth's orbit when the axial tilt is most inclined toward or away from the Sun, resulting in the longest and shortest days of the year.

Summer solstice is another astronomical event that occurs when the Sun reaches its highest point on its trajectory across the sky about the Earth's equator (Fig. 2.17). This marks the northernmost point reached by the Sun and signals the transition between the periods of asymmetrical day and night. The summer solstice typically occurs around June 21st in the Northern Hemisphere. Now of the summer solstice, the North Pole of the Earth is most inclined toward the Sun. It is the longest day of the year in the Northern Hemisphere, indicating a long day and a short night. Following the summer solstice, as the days gradually shorten and the nights lengthen, temperatures start to decrease in the Northern Hemisphere.

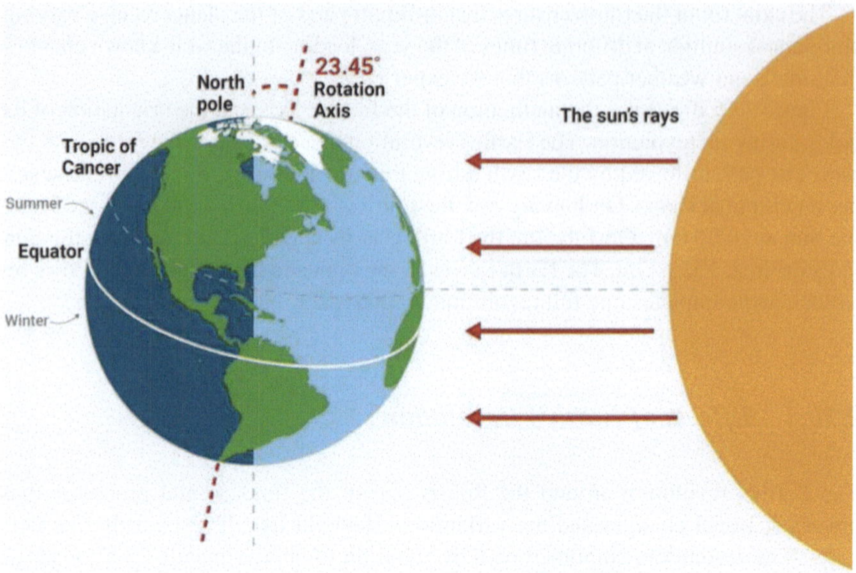

Fig. 2.17 The summer solstice is when the Sun is directly over the Tropic of Cancer, an imaginary line that circles the globe. This marks the beginning of astronomical summer in the Northern Hemisphere. June 21 has the most daylight

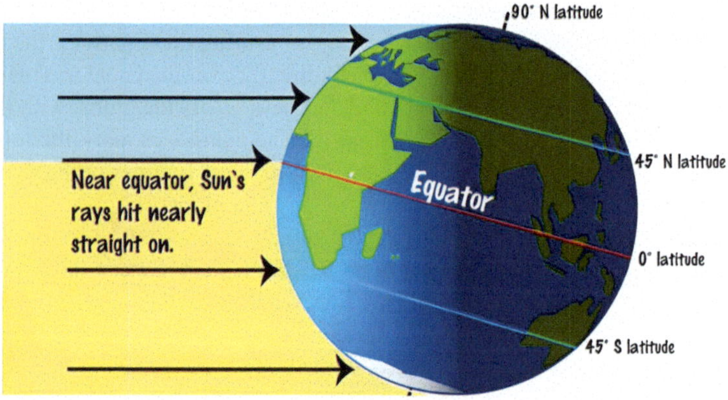

Fig. 2.18 Earth at the winter solstice (~December 21). At high latitudes, Sun's rays hit Earth at an angle. The North Pole is dark all day on December 21. Near the equator, the Sun's rays hit nearly straight on. South Pole has light all day on December 21

Winter solstice marks an astronomical event that occurs when the Sun reaches its lowest point on its trajectory across the sky about the Earth's equator (Fig. 2.18). This moment signifies the southernmost point reached by the Sun and marks the transition between the periods of asymmetrical day and night. It also denotes the shortest day and the longest night of the year in the Northern

Hemisphere. The winter solstice usually occurs around December 21st in the Northern Hemisphere, signaling the beginning of winter in that part of the world. At this juncture, the North Pole of the Earth is most inclined away from the Sun. With the winter solstice, the days begin to lengthen, and the gradual ascent of the Sun in the sky marks the onset of the period during which days gradually lengthen and become brighter in the Northern Hemisphere.

(c) Polar Day and Night

In the polar regions, the effects of Earth's axial tilt and revolution are most extreme.

Polar Day: During the summer solstice in the respective hemisphere, locations above the polar circles experience Polar Day, where the Sun does not set for an extended period. This continuous sunlight can last from a few days to several months, depending on how far north or south the location is. For instance, at the North Pole, the Sun remains above the horizon for 6 months, from the spring equinox to the autumn equinox.

Polar Night: Conversely, during the winter solstice, the same regions experience Polar Night, where the Sun does not rise above the horizon for weeks or months. The North Pole, for example, remains in darkness from the autumn equinox until the spring equinox.

(d) Impact on Climate and Ecology

The variations in the length of daylight and the angle at which sunlight strikes the Earth throughout the year play a crucial role in defining climate zones and ecological patterns. Longer days during the summer months result in more solar energy reaching the surface, warming the atmosphere and promoting photosynthesis in plants. This leads to distinct growing seasons in temperate and polar regions. In contrast, shorter days in winter reduce solar energy, contributing to cooler temperatures and dormancy in many plant species.

In tropical regions, where the day length remains relatively constant throughout the year, seasonal changes are less pronounced, and climate patterns are primarily driven by the movement of the Intertropical Convergence Zone (ITCZ) and other atmospheric circulation systems rather than the length of daylight.

Spring Equinox: The vernal equinox is an astronomical event that takes place when the Sun crosses the celestial equator, resulting in approximately equal lengths of day and night on Earth. This usually occurs on March 20th or 21st in the Northern Hemisphere. Now of the vernal equinox, the Earth's equator aligns with the Sun, ensuring that sunlight is evenly distributed across both poles of the Earth. This transition marks the onset of longer days and shorter nights in the Northern Hemisphere, symbolizing the beginning of calendar spring.

Autumn Equinox: The autumnal equinox, another astronomical phenomenon, occurs when the Sun crosses the celestial equator, once again resulting in equal durations of day and night on Earth. This marks the onset of autumn in the Northern Hemisphere, typically happening around September 22nd or 23rd. During the autumnal equinox, the Earth's equator is in direct alignment with the

Sun, leading to equal distribution of sunlight on both sides of the equator. Because of the autumnal equinox, days in the Northern Hemisphere become shorter while nights lengthen, signifying the transition from the warm summer months to the cooler days of autumn. This moment holds significant cultural and traditional value in many societies, often marked by various customs and celebrations.

References

Canup, R. M., & Asphaug, E. (2001). Origin of the moon in a giant impact near the end of the Earth's formation. *Nature, 412*(6848), 708–712.

Christopherson, R. W. (2012). *Geosystems: An introduction to physical geography*. Pearson Prentice Hall.

Hartmann, L., Cassen, P., & Kamp, I. (1998). Accretion of the early Earth and the origin of the Moon. *Icarus, 135*(1), 15–30.

Hutchison, R. (2004). *Meteorites: Petrologic, chemical and isotopic synthesis*. Cambridge University Press.

Jacob, D. J. (1999). *Introduction to atmospheric chemistry*. Princeton University Press.

Johnston, R. J., Gregory, D., Pratt, G., & Watts, M. (2000). *The dictionary of human geography*. Blackwell Publishers.

Koyré, A. (1957). *From the closed world to the infinite universe*. Harper & Row.

Krupp, N., & Dougherty, M. K. (2001). The magnetosphere of Jupiter's Moon Io. *Science, 291*(5503), 1073–1076.

Lamy, P., Toth, I., Fernandez, Y. R., & Weaver, H. A. (2004). The sizes, shapes, albedos, and colors of cometary nuclei. In M. Festou, H. U. Keller, & H. A. Weaver (Eds.), *Comets II* (pp. 223–264). University of Arizona Press.

Lewis, J. S. (1997). *Mining the sky: Untold riches from the asteroids, comets, and planets*. Addison-Wesley.

Lovelock, J. E. (2000). *Gaia: A new look at life on Earth*. Oxford University Press.

Lynden-Bell, D., & Pringle, J. E. (1974). The origin of the solar system. *Monthly Notices of the Royal Astronomical Society, 168*(3), 603–618.

McKnight, T. L., & Hess, D. (2014). *Physical geography: Landscape appreciation*. Prentice Hall.

Moore, P., Rees, N., & Sparrow, G. (2012). *The solar system*. Firefly Books.

NASA. (2021). *Europa clipper*. NASA website

Peixoto, J. P., & Oort, A. H. (1992). *Physics of climate*. American Institute of Physics.

Porco, C. C., et al. (2006). Cassini observations of the active south pole of Enceladus. *Science, 311*(5766), 1393–1401.

Rothery, D. A. (1999). *Satellites of the outer planets: Worlds in their own right*. Oxford University Press.

Smithson, P., Addison, K., & Atkinson, K. (2008). *Fundamentals of the physical environment*. Routledge.

Snyder, J. P. (1987). Map projections—A working manual

Sparrow, G. (2006). *Cosmos: A field guide*. Firefly Books.

Steffen, W., Sanderson, A., Tyson, P. D., Jäger, J., Matson, P. A., Moore, B., III, et al. (2004). *Global change and the Earth system: A planet under pressure*. Springer.

Strahler, A. N. (2013). *Introducing physical geography*. Wiley.

Weidenschilling, S. J. (1977). The distribution of mass in the planetary system and the formation of planetary cores. *Icarus, 30*(3), 289–304.

Chapter 3
Beneath the Surface: The Geosphere's Impact on the Planet

Global Ocean-Atmosphere Circulation explores the interconnected dynamics shaping Earth's climate system. From its outer crust to the deepest layers of the mantle and core, the geosphere plays a crucial role in shaping Earth's surface and influencing the conditions in which life exists. The interactions between the geosphere and other Earth systems—such as the atmosphere, hydrosphere, and biosphere—create complex feedback loops that drive geological and ecological phenomena. Understanding the structure, composition, and behavior of the geosphere is essential for recognizing the processes that have shaped the Earth over billions of years and continue to influence its evolution today. In this chapter, we will examine the different layers of the Earth, from the crust to the core, and explore their

© The Author(s), under exclusive license to Springer Nature Switzerland AG 2025
V. Spiridonov et al., *Atmospheric Perspectives*,
https://doi.org/10.1007/978-3-031-86757-6_3

composition and function. Additionally, we will discuss key processes such as plate tectonics, volcanism, and the geothermal gradient, which govern the Earth's internal dynamics. The study of the geosphere not only helps explain the planet's geological features but also informs us of our understanding of natural hazards, resource distribution, and the interconnectedness of Earth's systems. Through this exploration, we gain a deeper insight into the forces that shape the planet beneath our feet.

3.1 Geosphere

The geosphere encompasses the solid Earth in its entirety, including all rocks, minerals, and geological formations. It extends from the Earth's surface to its core, incorporating the crust, mantle, and core (Fig. 3.1). In simpler terms, the geosphere includes everything from the outermost layer of the Earth to its deepest interior. The geosphere is not just a collection of rocks and minerals; it is the foundation for many of Earth's dynamic processes. It plays a critical role in plate tectonics, volcanism, and the formation of mountains.

The Earth's crust, the thin outermost layer, is primarily composed of silicate minerals. The thickness of the crust varies: it ranges from 5 km to 10 km beneath the oceans and 35 km to 70 km beneath the continents (Press & Siever, 2000). Beneath the crust lies the mantle, a layer about 2900 km thick, made of solid rock that flows very slowly. The mantle's silicate minerals are denser than those in the crust. Deeper still is the core, which has a radius of about 3500 km and consists mostly of iron and nickel. The core is divided into two parts: the outer core, which is in a liquid state, and the inner core, which is solid (Kump et al., 2010).

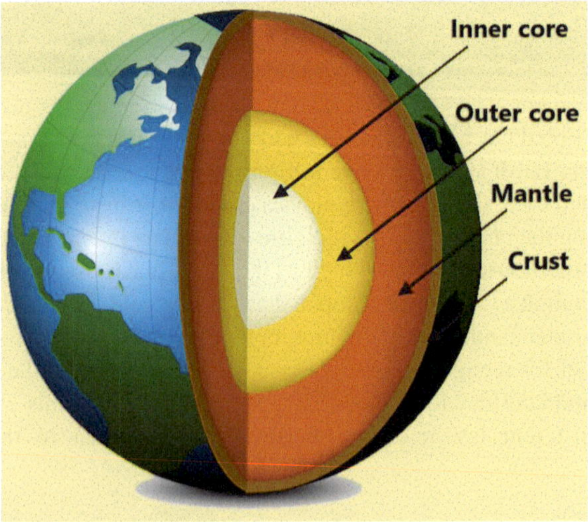

Fig. 3.1 Structure and composition of the geosphere

The Earth's formation began around 4.6 billion years ago from a massive cloud of dust and gas. As this cloud cooled and contracted under gravity, the Earth solidified. Heavier materials, like iron and nickel, sank to form the core, while lighter, rocky materials floated upward to form the crust and mantle. This differentiation is why the Earth is made up of distinct layers. The geosphere can be classified in two ways: chemically, into the crust, mantle, and core, or functionally, particularly in the case of the outer layers, the lithosphere, and asthenosphere.

3.2 Geothermal Gradient

The temperature within the Earth's geosphere changes in depth, and this change is known as the geothermal gradient. The geothermal gradient refers to the rate at which temperature increases as you go deeper below the Earth's surface (Fig. 3.2). Typically, the deeper you go, the hotter it gets. This gradient is often expressed at degrees Celsius per kilometer (°C/km) or degrees Fahrenheit per mile.

On average, the geothermal gradient is between 25 and 30 °C/km in the Earth's crust. However, temperatures in the mantle and core can reach thousands of degrees Celsius. This increase in temperature is primarily due to the Earth's internal heat, much of which is generated by the decay of radioactive isotopes in the mantle and crust.

The geothermal gradient is a crucial concept in geophysics and geology, as it influences the movement of heat within the Earth, the formation of geothermal resources, and the behavior of rocks and fluids at various depths.

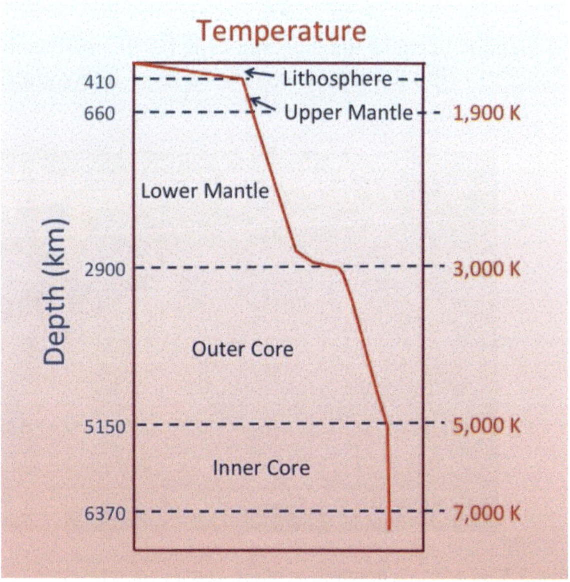

Fig. 3.2 Temperature profile of the Earth's layers. (Note: Temperature increases with depth)

3.3 The Chemical Composition of the Geosphere

The geosphere encompasses all the Earth's rocks and minerals, from the molten rock and heavy metals deep within the planet to the sand on beaches and the peaks of mountains. Its chemical composition is diverse, comprising a wide range of elements and compounds. The Earth's crust is primarily made up of the following elements: oxygen (O), silicon (Si), aluminum (Al), iron (Fe), calcium (Ca), sodium (Na), potassium (K), magnesium (Mg), and various trace elements (Fig. 3.3).

In essence, the crust is mostly composed of silicon oxide and aluminum oxide. Silicate minerals, which contain silicon and oxygen as their main components, dominate the Earth's crust. The most abundant elements in the crust are oxygen (about 46.6%) and silicon (about 27.7%) (Rudnick & Gao, 2014). Other common elements include aluminum, iron, calcium, sodium, potassium, and magnesium. The mantle, which makes up the largest volume of the Earth's layers, contains more iron and magnesium and less aluminum and silicon compared to the crust (Hofmann, 1997). The Earth's innermost layer, the core, is about 3500 km thick and primarily composed of iron and nickel (McDonough, 2016). The lithosphere, which includes both the crust and the upper mantle, behaves as a brittle, rigid solid. Beneath it lies the asthenosphere, a partially molten layer of the upper mantle that flows like a plastic material (Anderson, 1989).

The Earth's outer surface is its crust, a cold, thin, and brittle shell made of rock. Although the crust is very thin relative to the planet's radius, it is mainly composed of silicon, oxygen, and aluminum. There are two distinct types of crust, each with unique physical and chemical properties, which are summarized in Table 3.1. An oceanic crust forms from mafic magma that erupts on the seafloor, creating basaltic lava flows or cooling deeper to form intrusive igneous rock, such as gabbro (Herzberg, 2011). At the Earth's center lies the dense metallic core. It is believed to be metallic because the density of the Earth's surface layers is much lower than the planet's overall density, as calculated from its rotation (Birch, 1964). If the surface

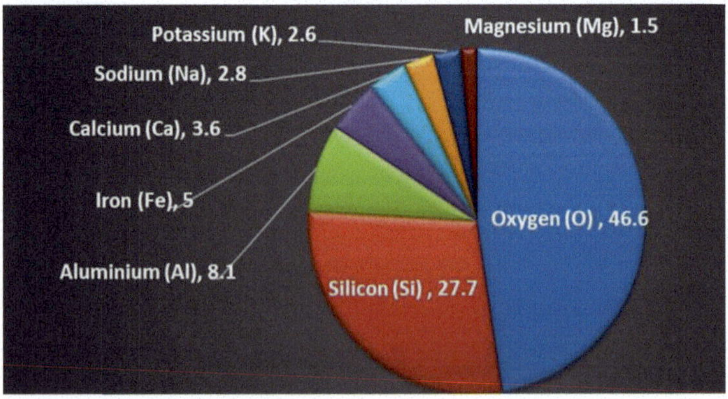

Fig. 3.3 Chemical composition of the Earth's geosphere

Table 3.1 Physical and chemical properties of oceanic and continental crust

Crust	Thickness	Density	Composition	Rock type
Oceanic	(5 – 12) km	$3.0\ g \cdot cm^{-3}$	Mafic	Basalt and gabbro
Continental	~35 km	$2.7\ g \cdot cm^{-3}$	Felsic	Various rock types

layers are less dense, the interior must be denser to balance the average. Calculations suggest that the core consists of about 85% iron and 15% nickel. Metallic meteorites, thought to represent the core, also exhibit this 85% iron/15% nickel composition (McDonough, 2016).

3.4 Lithosphere

The terms "lithosphere" and "geosphere" are closely related in Earth science, but they refer to different components of the Earth. The lithosphere is the Earth's rigid outer layer, encompassing the crust and the uppermost part of the mantle (Mason & Moore, 1982). This layer is divided into tectonic plates, which are large segments of the lithosphere that move and interact with each other. The lithosphere forms the Earth's uppermost surface layer, where life, both simple and complex, exists. It is a rigid shell that covers the entire planet and is home to plants and many animal species (Kardong, 2018). The term "lithosphere" is derived from the Greek words lithos, meaning rock or stone, and sphaíra, meaning sphere. As part of the geosphere, the lithosphere is one of the four terrestrial subsystems, alongside the hydrosphere, atmosphere, and biosphere (Fig. 3.4) (Tarbuck & Lutgens, 2016). It rests on the asthenosphere, a more ductile layer of the mantle beneath it (Turcotte & Schubert, 2002). The lithosphere is made of solid, rigid material and is divided into tectonic plates that move, causing various geological processes. This layer contains all the geological diversity on the planet and is crucial for ecosystems, which occur exclusively within this segment of the Earth. The lithosphere also holds essential resources such as gold, aluminum, iron, and other minerals, enabling humanity to create tools, products, and technologies that facilitate work and enhance various aspects of life (Condie, 2011).

3.4.1 Vertical Structure of the Lithosphere

The lithosphere is known for its rigid and brittle nature, playing a crucial role in Earth's geological processes, including plate tectonics (Bird, 2003). It consists of two main components: the outer layer, known as the crust, and the uppermost part of the mantle. The lithosphere is distinguished from the underlying mantle by a boundary called the Moho discontinuity, which lies at a depth of around 2500 km (Moho, 1964). Beneath this, the mantle transitions to a more ductile state, while the

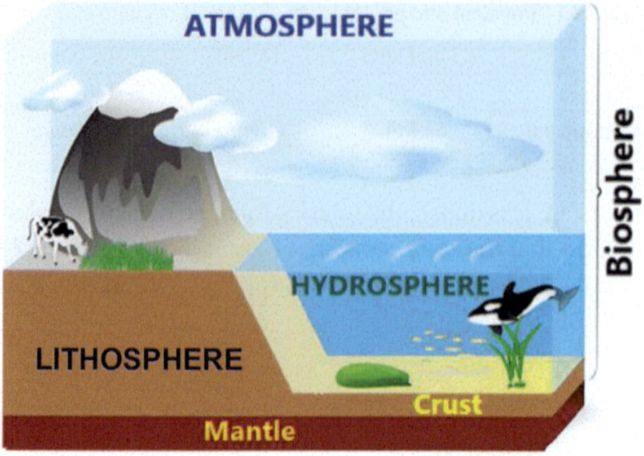

Fig. 3.4 The lithosphere as the outer layer of the geosphere: an integral component of the Earth system

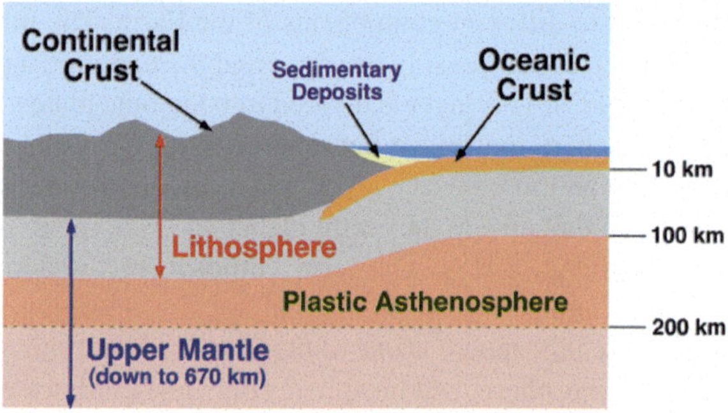

Fig. 3.5 Vertical structure of the lithosphere

core, located deeper, has a distinct outer layer extending over 2000 km (Lay & Wallace, 1995). Within the lithosphere, there are 12 major tectonic plates. These plates are rigid and move independently of one another (Davies, 1999). The vertical structure of the lithosphere is essential for understanding Earth's geology, tectonics, and the processes that shape its surface (Fig. 3.5). The interactions between these lithospheric plates are key drivers of geological phenomena, such as earthquakes, volcanic activity, and the formation of mountain ranges (Press & Siever, 2001).

3.4.2 *Composition of the Lithosphere*

The lithosphere is composed of the Earth's crust and the uppermost part of the mantle, extending from the surface to a depth of about 100 km (Taylor & McLennan, 2009). This layer includes both solid and semi-molten rocks, which vary in composition. The crust, the outermost part of the lithosphere, consists of hard rocks such as basalt and various minerals (Snow & Suskin, 2004). In the upper mantle, iron and magnesium silicates are prevalent, while the lower part of the lithosphere contains a

Fig. 3.6 San Andreas fault: Tectonic plate boundaries and their geological impact. (Note: Tectonic plate boundaries, such as the San Andreas Fault shown here, are sites where mountain-building events, volcanoes, or the formation of valleys and fissures can occur. Source: National Geographic. Photo Credit: Georg Gerster)

mix of magnesium, iron, and silicon oxides (Hofmann, 1988). Temperature variations in certain areas can lead to the formation of both solid and semi-molten rocks. Beneath the lithosphere, the core, composed primarily of iron and nickel, is divided into an upper and a lower part (Allegre, 1982). Temperatures in the lower core can exceed 3000 °C (Karato & Wu, 1993). The lithosphere's most notable feature is its tectonic activity. Large plates of the lithosphere, known as tectonic plates, interact and move over the more ductile asthenosphere below (Fig. 3.6). The plate tectonic theory explains the dynamic nature of the Earth's surface, suggesting that these plates are constantly shifting and interacting with the underlying asthenosphere (Jaupart & Mareschal, 2003). The asthenosphere, located approximately 100–700 km below the Earth's surface, is characterized by its partially molten or ductile state, which allows it to flow over geological timescales (Fig. 3.7) (McKenzie, 1984).

3.4.3 Types of Lithospheres

Continental Lithosphere: This lithosphere includes the outer part of the Earth's mantle and the continental crust. It is roughly 120 km thick and predominantly composed of granite rock. The continental lithosphere forms the continents and major mountain systems.

The oceanic lithosphere consists of the Earth's outer shell and oceanic crust. It is thinner than the continental lithosphere, with a thickness of about 60 km. Predominantly composed of basalt, it features mountain ridges on the ocean floor that can rise to 7 km. Over time, the oceanic lithosphere becomes denser as it cools and interacts with the asthenosphere.

Thermal Lithosphere: In this type, the focus is on the part of the mantle that conducts heat. It plays a crucial role in the thermal dynamics of the Earth.

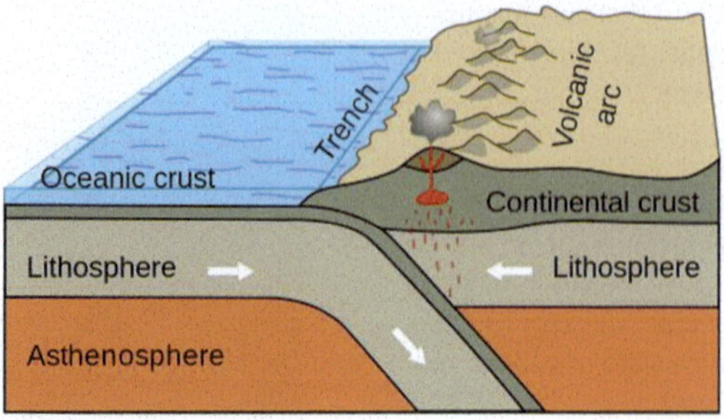

Fig. 3.7 Asthenosphere as the base on which the lithosphere rests. (Credit: http://www.differencebetween.net/)

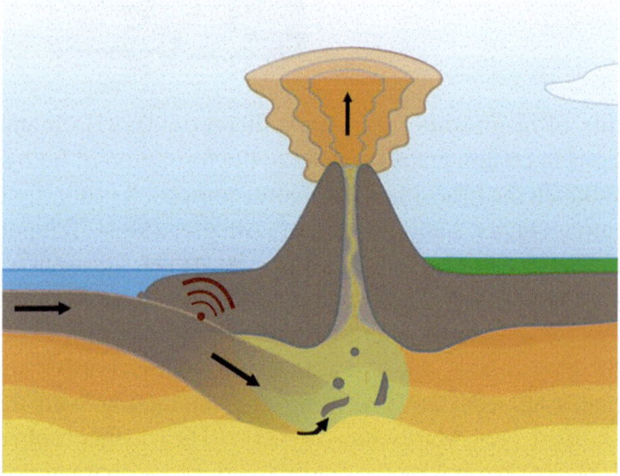

Fig. 3.8 Subduction resulting from lateral movements in the oceans. (Source: https://www.nps.gov/subjects/geology)

Seismic Lithosphere: This type is characterized by a reduction in the speed of seismic waves. It reflects the areas where seismic activity and ground motion are measured.

Elastic Lithosphere: Also known as the bending lithosphere, this type refers to the portion where tectonic plate movements occur. It is where the deformation and flexing of the lithosphere take place, influencing geological processes and plate tectonics.

3.4.4 Convergent Boundaries

Convergent boundaries are regions where tectonic plates collide, leading to lateral movements that create folds and wrinkles in the Earth's crust. This process forms mountain ranges, such as the Himalayas, where Mount Everest is located, and the Andes in South America. Similar interactions occur in oceanic plates through a process known as subduction (Fig. 3.8). In geological terms, subduction is the process where one tectonic plate is forced beneath another at a convergent boundary.

As the subducting plate sinks into the Earth's mantle, gravity pulls it downward. Although generally solid, the mantle is hotter than the crust and allows the subducting plate to descend at angles between 25° and 45°. Oceanic plates are unique in experiencing subduction due to their density and cooler temperatures compared to continental plates. When two oceanic plates converge, the older, denser plate is forced beneath the younger, less dense plate. Initially, oceanic plates are thin and hot when they form at mid-ocean ridges. As they move away from these ridges, their temperature decreases, causing the material on the underside to solidify. This solidification results in increased density, which causes the older plate to sink beneath younger or continental plates. While the movement of tectonic plates generally drives this process, some geologists speculate that very old plates, due to their significantly higher density, may sink more rapidly and unpredictably.

3.4.5 Definition and Key Features

A volcano is a rupture or opening in the Earth's crust through which magma from the mantle escapes to the surface, erupting as lava. These volcanic eruptions result from the pressure that builds up beneath the Earth's surface, driven by the movement of tectonic plates. Most volcanoes form along tectonic plate boundaries, particularly in a geologically active region called the Pacific Ring of Fire, which encircles the Pacific Ocean and is home to about 75% of the world's active volcanoes. Volcanoes can also form in areas away from plate boundaries, in locations called "hot spots," where mantle plumes rise to the surface. Volcanoes are often classified by the intensity and frequency of their eruptions. Non-explosive eruptions typically occur at divergent boundaries, where tectonic plates move apart, allowing magma to rise steadily and flow smoothly to the surface. These eruptions are characterized by the production of fluid basaltic lava, which spreads over large areas, forming broad, gently sloping cones known as shield volcanoes. The Hawaiian Islands, for example, are a prominent example of shield volcanoes created by non-explosive eruptions. In contrast, more explosive eruptions occur at convergent boundaries, where tectonic plates collide, causing subduction and the creation of thicker, more viscous lava. These eruptions can be violent, involving the ejection of ash, gas, and pyroclastic flows—fast-moving currents of hot gas and volcanic debris. Stratovolcanoes, or composite volcanoes, are commonly formed in these environments, with steep, symmetrical cones created by alternating layers of hardened lava flows and pyroclastic material. Famous examples of explosive volcanoes include Mount St. Helens in the United States and Mount Vesuvius in Italy. The vertical structure of a volcano, including its key components like the magma chamber, vent, and crater, is depicted in Fig. 3.9.

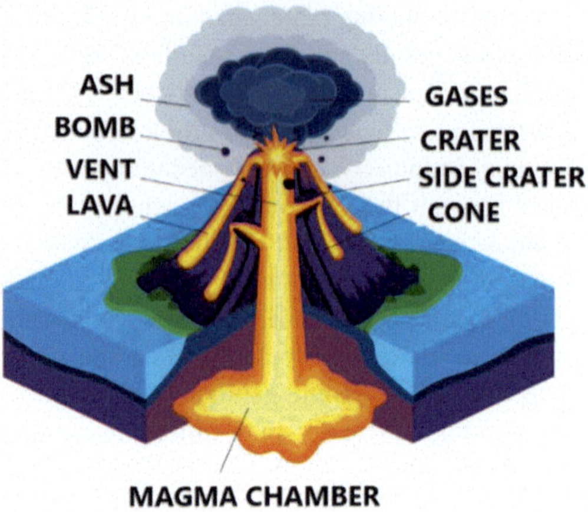

Fig. 3.9 Vertical structure of a volcano. The key components of a volcano, including the magma chamber, vent, crater, and volcanic cone, showing how magma moves from the Earth's interior to the surface during an eruption. (Credit: https://www.vectorstock.com)

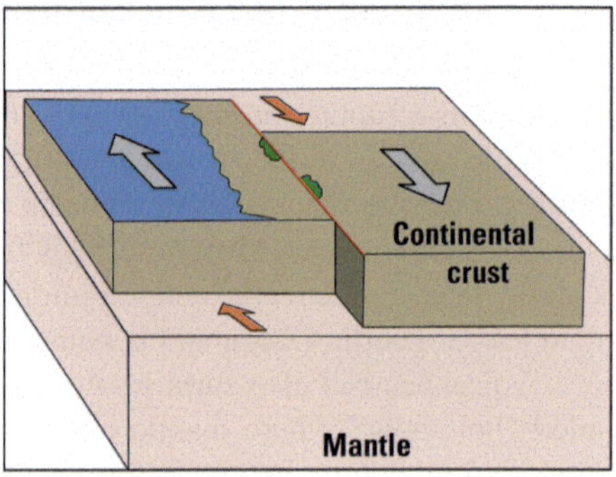

Fig. 3.10 Vertical profile of transform boundaries. (Source: https://www.usgs.gov/media/images)

3.4.6 *Transform Boundaries*

At transform boundaries, two tectonic plates slide past each other along strike-slip faults (Fig. 3.10). Unlike divergent or convergent boundaries, transform boundaries do not create significant oceanic or mountain formations. Instead, these boundaries are primarily associated with the horizontal movement of the plates. The movement is often not smooth, leading to the accumulation of stress along the fault lines. When this stress is released, it can result in large earthquakes. The San Andreas Fault in California is one of the most well-known examples of a transform boundary. While transform boundaries do not directly produce volcanic activity, the seismic activity along these faults can have significant impacts on nearby ecosystems and human infrastructure.

3.5 Seismic Waves

Seismic waves are elastic waves that travel through the Earth's interior, providing crucial information about the geosphere (Kennett, 2001). They are generated by various sources, including earthquakes, volcanic activity, explosions, and

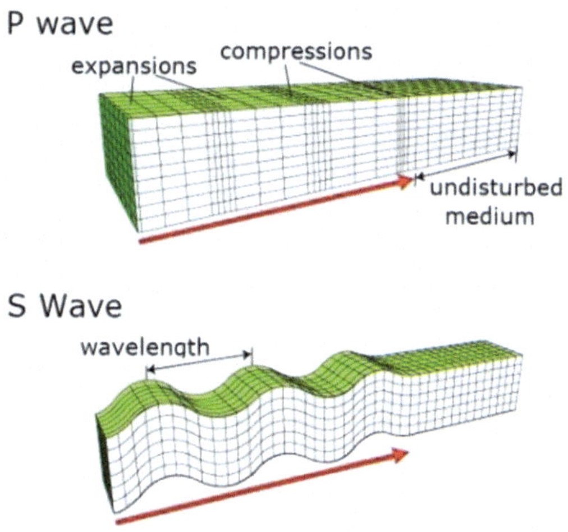

Fig. 3.11 P-wave and S-wave arrival times: P-waves arrive first due to their higher speed, followed by slower S-waves, which cause particle oscillations

man-made seismic sources. By studying the behavior of seismic waves, scientists can gain insights into the Earth's composition, structure, and dynamics. There are two main types of seismic waves: body waves and surface waves (Fig. 3.11).

Body Waves

- **P-Waves (Primary Waves):** These are the fastest seismic waves and can travel through solids, liquids, and gases. They are compressional waves, causing particles in the material to move in the same direction as the wave. P-waves are the first to be detected at a seismic station following an earthquake.
- **S-Waves (Secondary Waves):** Slower than P-waves, S-waves can only travel through solids. They are shear waves, meaning they cause particles to move perpendicularly in the direction of the wave. S-waves follow P-waves in an earthquake sequence and are primarily responsible for the shaking and damage associated with earthquakes.

P-waves and S-waves are often referred to collectively as body waves.

Surface Waves

Surface waves are seismic waves that travel along the Earth's surface, and they typically cause the most significant ground movement during an earthquake. Unlike body waves (P-waves and S-waves) that move through the Earth's interior, surface waves are slower but can have larger amplitudes, leading to more severe shaking and damage (Fig. 3.12).

- **Love Waves:** These waves move horizontally, causing the ground to move side to side. They are responsible for much of the horizontal shaking during an earthquake and can cause significant damage.
- **Rayleigh Waves:** These waves exhibit both horizontal and vertical motion, producing an elliptical rolling effect. Rayleigh waves cause rolling and swaying movements during an earthquake.

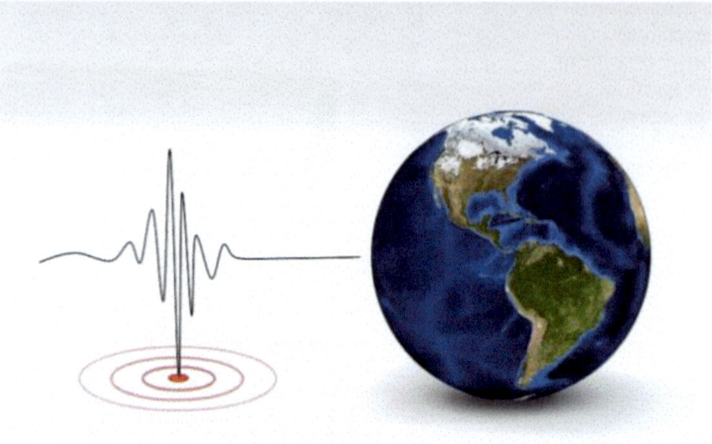

Fig. 3.12 Surface waves: Love and Rayleigh waves traveling along the Earth's surface, generated by various seismic sources. Surface waves, including Rayleigh waves with rolling motion and Love waves with horizontal shearing, arrive afterward and contribute to strong ground shaking

Fig. 3.13 A depiction of an earthquake causing massive damage. (Source: https://www.nationalgeographic.com/environment/article/earthquake-safety-tips)

3.5.1 Earthquakes

An earthquake is a natural phenomenon caused by the movement of tectonic plates and the shifting of the Earth's crust, which releases a significant amount of energy and results in the shaking of the ground (Fig. 3.13) (Kanamori & Brodsky, 2004; Stein & Wysession, 2009). The strength of an earthquake is influenced by several factors, including the amount of energy released, the depth of the hypocenter (the point within the Earth where the earthquake originates), the distance from the epicenter (the point on the Earth's surface directly above the hypocenter), and the composition of the Earth's crust. Earthquakes are typically characterized by the shaking or dislocation of the Earth's surface.

3.5.2 *Movements in the Earth's Crust*

While the Earth may seem static, it is constantly in motion in various ways. The geosphere, which includes the Earth's crust and upper mantle, plays a crucial role in the Earth's systems, affecting everything from the atmosphere and oceans to hydrological and biogeochemical cycles. The Earth rotates on its axis approximately every 24 h, creating the cycle of day and night. Additionally, the Earth orbits the Sun, leading to the changing seasons. Although we do not directly perceive the Earth's rotation and orbit, another type of movement within the Earth's crust is less noticeable but significant. This crustal movement occurs slowly, typically at a rate of a few millimeters per year. Over time, the accumulated stress from this gradual movement can result in sudden shifts that we experience as earthquakes. These seismic events, along with volcanic activity and other geological processes, demonstrate that the Earth's crust is an active and dynamic system with both horizontal and vertical movements. The Earth's crust contains large tectonic plates separated by faults and cracks, and tectonic movements cause disturbances and deformations in rock masses, leading to various geological phenomena.

3.6 Fundamentals of Plate Tectonics

Plate tectonics is a scientific theory that describes the Earth's lithosphere as being divided into several large and small plates that float on the semi-liquid asthenosphere beneath them (McKenzie et al., 1974). The Earth's crust is not a uniform layer; instead, it is segmented into distinct tectonic plates separated by fractures or faults. These plates move along the surface of the underlying, more fluid mantle. The movement of these plates can have significant effects on the Earth's surface. The mantle, located beneath the crust, is warm and flexible, allowing the tectonic plates to move slowly. Much of Earth's tectonic activity occurs at the boundaries between these plates, which can lead to various geological phenomena such as earthquakes, volcanic eruptions, and the formation of mountain ranges (Jordan, 2007).

Tectonic plate boundaries are categorized into three types:

- **Transform Boundaries**: At these boundaries, plates slide past each other horizontally. The friction between the plates prevents them from moving smoothly, which can lead to significant earthquakes (Bird, 2003). Transform boundaries do not typically result in the formation of oceans or mountain ranges but are known for generating large earthquakes.

- **Divergent Boundaries**: These boundaries occur when plates move away from each other, creating gaps that are filled with magma from below. This process forms a new crust and often takes place under the sea, such as at mid-ocean ridges. For example, the Pacific Ocean is expanding by about 18 cm per year due to divergent plate boundaries (Tolstoy et al., 2006).
- **Convergent Boundaries**: At convergent boundaries, plates move toward each other and collide. This interaction can lead to the creation of mountain ranges, such as the Southern Alps formed by the collision between the Australian plate and the Pacific plate (Smith & Briggs, 2004). When oceanic plates converge, one plate may be forced beneath the other in a process called subduction, which can lead to the formation of volcanic island arcs.

Understanding these boundaries and the movement of tectonic plates is essential for comprehending the dynamic nature of Earth's geology and the various processes that shape our planet's surface.

3.6.1 *Major Tectonic Plates*

Seven major tectonic plates cover nearly 95% of the Earth's surface (Fig. 3.14). These plates are key players in Earth's geological processes, including seismic and volcanic activity. Here are the most important ones:

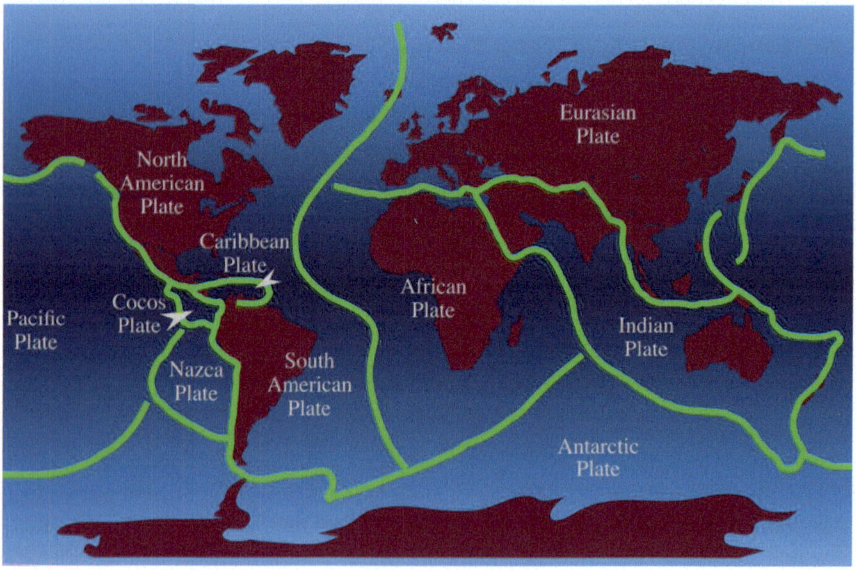

Fig. 3.14 Major tectonic plates covering most of the Earth's surface. (Credit: https://commons.wikimedia.org/wiki/File:Plattengrenzen_hg.png)

- ***Eurasian Plate:*** *This plate encompasses Europe and much of Asia, including Japan. It extends over the sea floor east of the Atlantic Ridge and is a significant zone of tectonic activity, contributing to the volcanic activity in the Pacific Ring of Fire.*
- ***Pacific Plate:*** *One of the largest oceanic plates, it forms the majority of the Pacific Ocean floor and is a major component of the Pacific Ring of Fire. It is surrounded by eight other tectonic plates, making it a crucial area for geological activity.*
- ***South American Plate:*** *This plate features a convergent boundary along its western edge, leading to high seismic activity and notable volcanoes. It is involved in complex interactions with neighboring plates, contributing to its dynamic geological features.*
- ***North American Plate:*** *Part of the Pacific Ring of Fire, this plate connects with the Pacific Plate along its western side. It is a major player in seismic and volcanic activities in the region.*
- ***African Plate:*** *This plate exhibits diverse geological characteristics. Its northern boundary, where it collides with the Eurasian Plate, is responsible for the formation of the Alps and the Mediterranean Sea. To the west, the Atlantic Ocean is expanding, and geological processes suggest that a rift could eventually split the continent.*
- ***Arabian Plate:*** *A relatively small plate, it is situated to the east of Africa. The western boundary is characterized by the ongoing rifting of the Red Sea, which is considered a young sea body and marks significant geological activity.*

Understanding these major tectonic plates and their interactions helps explain many of the Earth's geological phenomena, including earthquakes, volcanic eruptions, and mountain formation.

3.7 Geological Phases

The lithosphere plays a crucial role in sustaining life on Earth by insulating the planet's warm interior and providing a habitat for flora and fauna. This outer shell of the Earth acts as a mediator between the biosphere and deeper geological processes, facilitating the exchange of organic elements. These elements, buried in the crust, decompose over time and contribute to the formation of valuable resources such as gas, oil, and coal, which are essential for various industries. The interplay between the lithosphere, hydrosphere, and atmosphere creates a continuous nutrient cycle, supporting biological functions and maintaining ecosystem balance through food chains. Tectonic movements within the lithosphere drive geological processes that shape the Earth's surface. These movements, driven by the transfer of thermal energy from the Earth's core to the crust, generate convective currents in the mantle. These currents contribute to the formation of mountain ranges, as well as earthquakes and volcanic eruptions, which, despite their immediate destructive impacts, play a key role in long-term geological and ecological changes.

Over time, these geological processes lead to the creation of new habitats, promote plant growth, and stimulate adaptation processes in various species. The lithosphere is also a repository for many natural resources, including metals and precious stones, which form because of its composition and the dynamic interactions within the geosphere.

3.8 Gravity, Magnetism, and Earth's Heat

Gravity is the force that governs the motion of celestial bodies, keeping the planets in orbit around the Sun and the Moon around the Earth. It also influences ocean tides by pulling the seas toward the Moon. This fundamental force is responsible for the formation of stars and planets by attracting and aggregating mass. On Earth, gravity interacts with centrifugal force, which acts outward due to the planet's rotation. These forces are balanced to shape the Earth's surface and maintain its atmosphere, allowing life to thrive. While gravity is a universal force, its strength varies slightly across the planet due to the Earth's uneven distribution of mass, including subterranean structures and variations in relief. This variability can cause minor deviations in gravitational measurements.

The Earth's magnetic field, or geomagnetic field, resembles a magnetic dipole with its magnetic South Pole near the geographic North Pole and its magnetic North Pole near the geographic South Pole (Fig. 3.15). This magnetic field is essential for

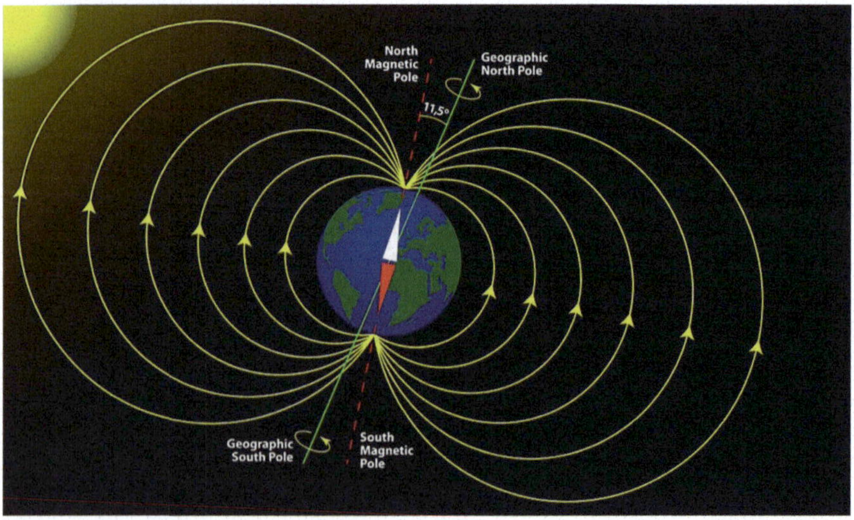

Fig. 3.15 Earth's magnetic field, magnetic poles, and geographic poles. The magnetic field resembles a magnetic dipole with the magnetic South Pole near the geographic North Pole and the magnetic North Pole near the geographic South Pole. (Credit: https://www.worldatlas.com/)

navigation, making compasses reliable tools for orientation. Extending from the Earth's core into space, the magnetic field interacts with the solar wind and protects the planet from harmful solar radiation. Although the Earth's magnetic field is relatively weak (roughly 0.05 mT), certain minerals in the crust exhibit magnetization. This magnetization results from the alignment of magnetic dipoles in minerals with the magnetic lines of force from the south to the north magnetic pole, cutting through the Earth's lithosphere.

Our planet's heat comes from several sources: the Sun's external heat, internal heat from the Earth's geosphere, and heat generated by radioactive processes within the Earth's core. The Sun provides essential warmth to Earth's surface, influencing weather and climate. Internal heat, due to the Earth's molten state and heated gases, is crucial for maintaining habitable conditions. Additionally, radioactive decay in the Earth's interior contributes significantly to this heat.

Radioactive elements such as uranium, thorium, and potassium decay over time, releasing heat and acting as a heat reservoir, which slows the Earth's cooling process and supports life. This radioactive decay contributes to about half of the heat necessary for our survival. The other half of the Earth's heat is a result of secular cooling, the slow cooling of the Earth since its formation.

The total heat energy released from the Earth is estimated at 46 TW, with 2 TW coming from volcanic eruptions and earthquakes. Most geothermal energy is continuously released through geothermal flow. Of this, between 15 TW and 25 TW are attributed to the steady release of heat from radioactive decay. These heat sources collectively influence the Earth's temperature regime.

The temperature change with depth is described by the geothermal gradient, which is defined as the increase in temperature with depth in the Earth. The typical value of the geothermal gradient close to the surface is 25.0 ° C/km. This gradient helps define the geothermal degree, indicating the depth at which the temperature rises by 1.0 ° C/km.

References

Allegre, C. J. (1982). High-pressure experiments and the composition of the Earth's core. *Philosophical Transactions of the Royal Society A, 305*(1481), 125–143.

Anderson, D. L. (1989). *Theory of the Earth*. Blackwell Scientific Publications.

Birch, F. (1964). Density and composition of mantle and core. *Journal of Geophysical Research, 69*(20), 4377–4388.

Bird, P. (2003). An updated digital model of plate boundaries. *Geochemistry, Geophysics, Geosystems, 4*(3), 1027.

Condie, K. C. (2011). *Earth as an evolving planetary system*. Academic Press.

Davies, G. F. (1999). *Dynamic Earth: Plates, plumes, and planetary processes*. Cambridge University Press.

Herzberg, C. (2011). Basalts, picrites, and komatiites: The mantle melts that drive plate tectonics. *Earth-Science Reviews, 106*(3–4), 37–52.

Hofmann, A. W. (1988). Chemical differentiation of the Earth's mantle. *Chemical Geology, 67*(3), 143–147.

Hofmann, A. W. (1997). Mantle geochemistry: The message from oceanic volcanism. *Nature, 385*(6613), 219–229.

Jaupart, C., & Mareschal, J. C. (2003). Heat flow and thermal structure of the lithosphere. In *Treatise on geophysics* (Vol. 1, pp. 239–256).

Jordan, T. H. (2007). The tectonic evolution of the Earth's lithosphere. *Annual Review of Earth and Planetary Sciences, 35*, 53–87.

Kanamori, H., & Brodsky, E. E. (2004). The physics of earthquakes. *Reports on Progress in Physics, 67*(8), 1429–1496.

Karato, S., & Wu, P. (1993). Rheology of the upper mantle: A synthesis. *Science, 260*(5109), 771–778.

Kardong, K. (2018). *Vertebrates: Comparative anatomy, function, evolution*. McGraw-Hill Education.

Kennett, B. L. N. (2001). *Seismic tomography: Theory and practice*. Cambridge University Press.

Kump, L. R., Kasting, J. F., & Crane, R. G. (2010). *The earth system* (3rd ed.). Pearson.

Lay, T., & Wallace, T. C. (1995). *Modern global seismology*. Academic Press.

Mason, B., & Moore, C. B. (1982). *Principles of geochemistry*. Wiley.

McDonough, W. F. (2016). The composition of the Earth. In *Treatise on geochemistry* (Vol. 3, 2nd ed., pp. 7–9).

McKenzie, D. (1984). The generation and differentiation of magma beneath Island arcs. *Journal of Petrology, 25*(3), 711–765.

McKenzie, D., Roberts, C., & Stone, P. (1974). Surface heat flow, plate boundaries, and the thermal state of the lithosphere. *Geophysical Journal International, 39*(1), 133–153.

Moho, A. (1964). *The Mohorovičić discontinuity*. International Geophysical Year.

Press, F., & Siever, R. (2000). *Understanding Earth*. W.H. Freeman and Company.

Press, F., & Siever, R. (2001). *Understanding Earth*. W. H. Freeman.

Rudnick, R. L., & Gao, S. (2014). Composition of the continental crust. In *Treatise on geochemistry* (Vol. 4, 2nd ed., pp. 1–51).

Smith, D., & Briggs, R. (2004). The formation and evolution of mountain belts. *Journal of Structural Geology, 26*(11), 2109–2124.

Snow, J. K., & Suskin, J. (2004). *Basaltic magmatism and plate tectonics*. Wiley.

Stein, S., & Wysession, M. (2009). *An introduction to seismology, earthquakes, and Earth structure*. Blackwell Publishing.

Tarbuck, E. J., & Lutgens, F. K. (2016). *Earth: An introduction to physical geology*. Pearson.

Taylor, S. R., & McLennan, S. M. (2009). *The continental crust: Its composition and evolution*. Blackwell Scientific Publications.

Tolstoy, M., et al. (2006). Seismic and hydroacoustic evidence for a new type of ocean ridge magmatism. *Nature, 442*(7101), 1040–1043.

Turcotte, D. L., & Schubert, G. (2002). *Geodynamics*. Cambridge University Press.

Chapter 4
Biosphere: Ecosystem Diversity and Environmental Change

The biosphere represents the zone of Earth where life exists, encompassing all living organisms and their interactions with one another and the physical environment (Margulis & Sagan, 1995). It is a dynamic and interconnected system that supports life, from microscopic bacteria in deep oceanic trenches to towering trees in rainforests and animals at high altitudes.

This life-supporting layer is unique to Earth, distinguishing it from other planets in the solar system. The biosphere overlaps with Earth's other spheres—the atmosphere, hydrosphere, and geosphere—forming a complex web of interactions that sustain life. For instance, the biosphere relies on the atmosphere for oxygen and carbon dioxide, the hydrosphere for water, and the geosphere for nutrients and physical support.

© The Author(s), under exclusive license to Springer Nature Switzerland AG 2025

V. Spiridonov et al., *Atmospheric Perspectives*,
https://doi.org/10.1007/978-3-031-86757-6_4

These interactions form the foundation of ecosystems, enabling life to thrive in diverse environments. Although the biosphere spans roughly 20 km in thickness, most life is concentrated within a much narrower range, from 500 m below the ocean's surface to about 6 km above sea level. This narrow band hosts a remarkable diversity of ecosystems and species, demonstrating the adaptability of life in varying conditions.

The biosphere plays a critical role in regulating Earth's climate and maintaining the balance of essential cycles, such as the carbon and nitrogen cycles. It not only sustains life but also supports human civilization by providing resources like food, water, and raw materials. Understanding the biosphere is therefore crucial for preserving biodiversity and addressing global environmental challenges.

4.1 The Biosphere

The term "biosphere" refers to the zone of Earth where life thrives. It encompasses all living organisms and their interactions with each other and the physical environment (Margulis & Sagan, 1995). The biosphere is a dynamic, interconnected system that includes ecosystems, communities, populations, and individual organisms (Fig. 4.1).

The biosphere comprises the regions on Earth that make life possible. It stretches from the deepest ocean waters and the roots of plants and trees to the highest mountain peaks. The biosphere is vital for all life on Earth, including humans and every other living organism. While Earth has three other spheres—the atmosphere, geosphere, and hydrosphere—but not all are home to living beings. The specific regions where life is found are collectively known as the biosphere (Vernadsky, 1998). The word "biosphere" comes from the Greek words "bios, "meaning "life," and "sphaira," meaning "sphere" or "globe." The biosphere represents the sum of all Earth's ecosystems. This habitable zone sets our planet apart from others in the solar

Fig. 4.1 The biosphere is an essential component of our planet

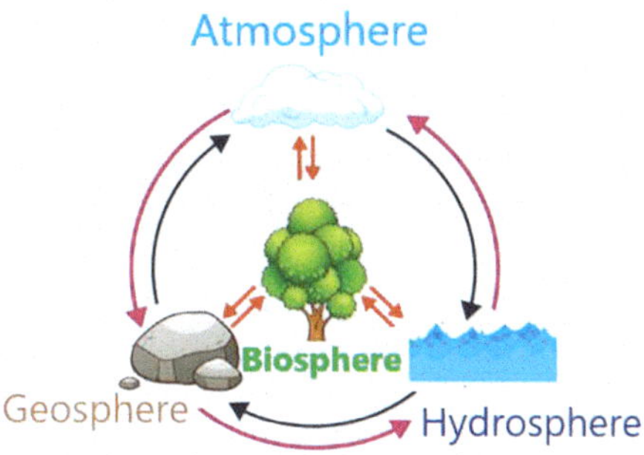

Fig. 4.2 "The Biosphere: Earth's Life Support System," life research. (Credit: https://byjus.com/chemistry/what-is-biosphere /)

system. Because life exists on land, in the air, and in water, the biosphere overlaps with all these other spheres (Fig. 4.2).

Although the biosphere spans approximately 20 km in thickness, nearly all life is concentrated within a narrow band, from about 500 m below the ocean's surface to around 6 km above sea level. The biosphere is a crucial component of our planet, playing a fundamental role in supporting and sustaining life. It encompasses all living organisms on Earth and the environments in which they exist (Smil, 2003; Clarke, 2008).

4.1.1 Definition of the Biosphere

The biosphere is the part of Earth where life exists, encompassing the planet's surface, oceans, and lower atmosphere where organisms thrive (Moore, 2015). It represents the sum of all ecosystems on Earth, forming a narrow zone where soil, water, and air combine to sustain life (Kump et al., 2010). This zone supports a diverse range of organisms, from microscopic fungi and bacteria to large animals. The biosphere is distinguished by its capacity to contain all living organisms and the products of their activities, making it essential for ecosystem maintenance and species survival (Chapin et al., 2011). Furthermore, the biosphere plays a critical role in regulating the Earth's climate.

4.1.2 The Emergence and Evolution of the Biosphere

The biosphere, Earth's life-supporting layer, has existed for approximately 3.5 billion years. The earliest life forms, known as prokaryotes, thrived in an oxygen-free environment. These primitive single-celled organisms, including bacteria and archaea, were the pioneers of life. Over time, some prokaryotes developed the ability to harness light, water, and carbon dioxide to produce energy-rich compounds, a process known as photosynthesis. This innovation not only allowed them to sustain themselves but also resulted in the release of oxygen as a byproduct, dramatically altering Earth's atmosphere.

As oxygen levels in the atmosphere increased, it paved the way for the evolution of more complex life forms. Autotrophs, from simple unicellular algae to multicellular vascular plants, played a crucial role in this transformation by utilizing atmospheric carbon dioxide and contributing oxygen. The rise in oxygen concentration led to greater biodiversity, supporting the emergence of aerobic organisms that could thrive in oxygen-rich environments. This shift enabled the biosphere to sustain a wide variety of complex life forms, including plants, animals, and eventually humans.

A key question that arises is: How is biodiversity connected to the biosphere? The answer lies in the concept of ecological niches. A "healthy" biosphere is rich in biodiversity, where various species occupy different ecological niches. An ecological niche refers to the role a species plays in its environment, including how it obtains food, interacts with other species, and contributes to the ecosystem. In the biosphere, autotrophs serve as the primary producers, forming the base of the food chain. Heterotrophs, unable to produce their food, depend on other organisms for sustenance and fulfill roles as consumers and natural biological controls. Herbivores feed exclusively on plants, carnivores on animal flesh, and omnivores on both. Another vital niche is occupied by decomposers, such as fungi, which break down dead organic matter. These organisms convert complex organic materials into simpler compounds, recycling nutrients back into the ecosystem and sustaining the cycle of life.

4.1.3 Constituent Elements of the Biosphere

The constituent elements of the biosphere are the living organisms that inhabit it, ranging from the smallest microorganisms, like bacteria, to the largest animals, such as whales and elephants. These diverse forms of life are the foundational components of the biosphere, which encompasses all living organisms on Earth and the environments they occupy.

The biosphere and ecosystems are interrelated concepts that represent different levels of organization in the study of life on Earth. The biosphere is the global sum of all ecosystems, including the entirety of life along with the atmosphere, lithosphere (Earth's crust and upper mantle), and hydrosphere (water in all its forms) where life exists (Fig. 4.3). It is a dynamic system, characterized by continuous interactions between living organisms and their physical environment.

In contrast, an ecosystem is a more localized and specific unit within the biosphere (Fig. 4.4). It consists of living organisms (biotic components) and their physical environment (abiotic components) interacting as a system. Ecosystems can range in size from small ponds to vast forests, and they can be found in both terrestrial and aquatic environments. Key components of ecosystems include producers, such as plants and algae, which generate energy through photosynthesis; consumers, including animals that feed on producers or other organisms; decomposers, like bacteria and fungi, which break down dead organic matter; and the physical environment that supports these life forms.

Fig. 4.3 The biosphere: the part of Earth where life exists, encompassing all living organisms and their interactions with each other and the environment

Fig. 4.4 Ecosystems: different environments or habitats where organisms live, such as forests, deserts, oceans, and grasslands

Thus, the biosphere is composed of a vast network of ecosystems, each playing a vital role in sustaining the balance and diversity of life on Earth.

4.1.4 Components of the Biosphere

The biosphere consists of both abiotic and biotic components. Abiotic components include the lithosphere, hydrosphere, and atmosphere—parts of Earth that do not contain living organisms and where life is not sustained (Odum & Barrett, 2005). These components encompass elements like air, soil, sunlight, and water. For example, the part of the sky where birds fly is included in the biosphere, while the upper layers of the atmosphere, where sustainable life is absent, are not considered part of the biosphere (Pidwirny, 2019).

Biotic components, however, include all living organisms, such as plants, animals, and microorganisms. These biological elements are essential for the functioning and maintenance of ecosystems, serving as the primary source of nutrition within these systems (Chapin et al., 2011) (Fig. 4.5).

Plants are primary producers, known as autotrophs because they create their food through photosynthesis, a process that involves chlorophyll assimilation. In addition to producing food, plants play a crucial role in recycling waste materials. They serve as the foundational energy source for all living organisms, including animals and humans.

Animals are consumers, also referred to as heterotrophs, as they cannot produce their food from inorganic sources. Instead, they rely on plants or other animals for nourishment. The food animals consume is metabolized to release energy, which is stored for future use and is essential for their growth and development.

Microorganisms form a significant part of ecosystems and include fungi, algae, bacteria, and viruses. These organisms often act as decomposers, breaking down waste and dead organic matter. This decomposition process not only recycles nutrients back into the environment but also serves as a food source for the microorganisms themselves.

Fig. 4.5 Biotic and abiotic factors: biotic factors include living components, such as plants, animals, and fungi, while abiotic factors consist of non-living components like air, water, soil, and climate, all of which interact within the biosphere

4.2 Biodiversity

Biodiversity, or biological diversity, refers to the vast variety and variability of life forms on Earth. It includes species diversity, genetic diversity, and ecosystem diversity, encompassing everything from the smallest microorganisms to the largest ecosystems (Wilson, 1999). Biodiversity is more than just a scientific concept; it is a crucial component of the health and resilience of our planet. The conservation of biodiversity is essential for the well-being of both ecosystems and human society (Cardinale et al., 2012). Healthy biodiversity ensures ecosystem stability, providing services, such as clean air, water purification, pollination, and climate regulation (Millennium Ecosystem Assessment (MEA), 2005) (Fig. 4.6).

Fig. 4.6 Biodiversity conservation: essential across ecological, economic, social, and ethical dimensions. (Credit: https://www.epa.gov/)

4.2.1 Diversity of Species

Species Richness: This dimension of biodiversity focuses on the number of different species present in a given area. A higher number of species generally indicates greater biodiversity, reflecting a more varied and complex ecosystem.

Species Evenness: This aspect describes the distribution of individuals among different species within an area. High species evenness means that no single species overwhelmingly dominates, promoting a balanced and stable ecosystem.

4.2.2 Genetic Diversity

Genetic Variation: Genetic diversity refers to the range of genetic information within a population or species. It is essential for the adaptation and survival of organisms, as it enables them to evolve and adjust to changing environmental conditions.

Genetic Resilience: Populations with higher genetic diversity tend to be more resilient to diseases, environmental changes, and other threats. This resilience enhances their ability to survive and maintain stability in the face of challenges.

4.2.3 Ecosystem Diversity

Habitat Diversity: Ecosystem diversity encompasses a range of different ecosystems, such as forests, wetlands, grasslands, and coral reefs. Each ecosystem supports distinct species and fulfills specific ecological roles, contributing to the overall biodiversity of the biosphere.

Ecological Processes: Every ecosystem is characterized by its own set of ecological processes, including nutrient cycling, energy flow, and species interactions. Biodiversity plays a crucial role in maintaining the functioning and stability of these processes, ensuring the health and resilience of ecosystems.

The Biosphere as a Complex System: The biosphere is a multifaceted and interconnected system where diverse ecosystems interact and influence each other, highlighting the importance of preserving ecosystem diversity for overall planetary health.

4.2.4 Functional Diversity

Functional Roles: Functional diversity refers to the variety of ecological roles and functions that different species perform within an ecosystem. These roles include nutrient cycling, pollination, seed dispersal, predation, and habitat formation, among others. Each species contributes uniquely to ecosystem processes, and the interactions among species help regulate ecosystem dynamics. The loss of certain species can disrupt these essential functions, leading to imbalances that may affect ecosystem stability, productivity, and resilience. In many cases, the disappearance of key functional species, such as top predators or primary producers, can trigger cascading effects throughout the ecosystem, reducing its ability to recover from disturbances, such as climate change, natural disasters, or human activities. Functional diversity, therefore, plays a critical role in maintaining ecosystem health and the services it provides to human societies.

4.2.5 The Importance of Biodiversity

Biodiversity refers to the variety of life on Earth, encompassing species, ecosystems, and genetic diversity within species. Its significance is far-reaching and integral to both ecological and human systems. Diverse ecosystems are inherently more resilient and stable, capable of adapting to environmental changes, such as climate fluctuations or the introduction of new species. This resilience ensures the continued functionality of ecosystems in the face of challenges.

For humanity, biodiversity provides essential resources. Industries such as agriculture, forestry, and fisheries rely heavily on biological diversity, while ecosystems serve as potential sources for new medicines, materials, and technological innovations. The balance within ecosystems is maintained by the unique role's species play; predators regulate prey populations, preventing overgrazing and preserving equilibrium. Every species, no matter how small, contributes to the intricate web of life.

Biodiversity is also vital for human health. Many medical advancements derive from plants, animals, and microorganisms, and healthy ecosystems provide essential services like clean air and water. Genetic diversity within species offers the raw material for adaptation, enabling organisms to respond to environmental changes, diseases, and other threats. Additionally, plants contribute to climate regulation by absorbing carbon dioxide through photosynthesis, a critical process in mitigating climate change.

In essence, biodiversity is fundamental not only for sustaining ecological balance and ecosystem health but also for supporting human survival, health, and well-being. Its preservation is vital for the stability and future of life on Earth.

In summary, biodiversity is essential not only for maintaining ecological balance and resilience but also for providing resources and services critical to human health and survival.

4.2.6 Threats to Biodiversity

Biodiversity, encompassing the incredible variety of life on Earth, is under serious threat, largely due to human activities. The loss of biodiversity has profound implications for ecosystems, economies, and human well-being, as the interconnected web of life supports essential ecological functions and services.

One of the most pressing challenges is habitat loss, driven by urban expansion, agricultural development, and deforestation, which destroy and fragment ecosystems, leaving species with fewer resources and shrinking territories. Pollution further exacerbates the problem; contaminants in the air, water, and soil disrupt the delicate balance of ecosystems, harming both individual species and entire communities.

Climate change, with its shifting temperature and precipitation patterns, compounds these issues by altering ecosystems and forcing species to migrate to new areas, often with mixed success. The introduction of invasive species adds another layer of complexity, as these non-native organisms frequently outcompete indigenous species, upsetting local ecological dynamics. Overexploitation, including the unsustainable harvesting of wildlife and natural resources, accelerates species declines, with some driven to extinction.

Efforts to mitigate these threats emphasize a multifaceted approach. Establishing protected areas remains a cornerstone strategy for conserving critical habitats. Habitat restoration projects aim to repair damaged ecosystems, fostering biodiversity recovery and improved ecological function. The adoption of sustainable practices in agriculture, fishing, and forestry helps balance human needs with environmental preservation. Additionally, raising awareness and educating communities about the value of biodiversity are vital steps in promoting conservation and responsible stewardship.

By addressing these challenges through coordinated, science-based actions, humanity can help safeguard the rich diversity of life and ensure the resilience and health of the planet's ecosystems for generations to come.

4.3 The Biosphere as a Complex System

The biosphere is a highly intricate and interconnected system essential for the survival and sustainability of all life on Earth. It operates across multiple levels of organization, each playing a unique role in maintaining the dynamic equilibrium of the planet's life-supporting systems.

Biomes represent vast regions defined by their distinctive vegetation, climate, soil, and wildlife. Examples include tundra landscapes with permanently frozen subsoil, grasslands characterized by moderate rainfall, forests rich in biodiversity, arid deserts with sparse vegetation, and aquatic systems, such as rivers, lakes, and oceans teeming with life. These large-scale zones provide the foundation for diverse ecosystems.

Ecosystems encompass biological communities and their interactions with the physical environment. These systems integrate living organisms (biotic factors) with non-living elements (abiotic factors), such as water, soil, and air. Ecosystems can be terrestrial, like grasslands and forests, aquatic, such as freshwater rivers and marine oceans, or even artificial environments like terrariums created by human intervention.

Within ecosystems, species coexist and interact, forming communities that depend on specific environmental conditions, such as temperature, pH, and nutrient availability. These communities are composed of population groups of individuals of the same species inhabiting a shared habitat. Populations can fluctuate in size, with overpopulation potentially straining ecological resources and declines posing risks to species' survival.

At the most fundamental level are organisms, the living entities that make up the biosphere. Organisms exhibit cellular organization and biological processes governed by their genetic material. They range from eukaryotes, such as humans, plants, and animals with complex cellular structures, to simpler prokaryotes like bacteria, which lack organelles.

This hierarchical structure demonstrates the complexity and interdependence of the biosphere. Each level—from the vast expanse of biomes to individual organisms—contributes to the overall stability and functionality of Earth's life-sustaining systems.

4.3.1 Factors Affecting the Biosphere

The biosphere is continuously influenced and altered by both living and non-living factors. These factors impact ecosystems and the organisms within them in various ways. Key factors affecting the biosphere include:

1. The Tilt of the Earth: The Earth's axial tilt significantly affects the biosphere by creating seasonal temperature variations. This tilt causes one hemisphere to experience warmer temperatures while the other cools, influencing the types of species that can thrive in different regions throughout the year. Seasonal changes play a crucial role in determining species distribution and ecosystem dynamics.
2. Natural Disasters: Natural disasters such as volcanic eruptions, earthquakes, and floods can have profound and long-lasting impacts on the biosphere. These events can drastically alter landscapes, destroy habitats, and disrupt ecosystems. The resulting changes to the environment can affect the survival and distribution of species.
3. Minor Factors: Various minor factors also affect the biosphere, including climate change, water availability, and soil erosion. These factors can lead to gradual or sudden disruptions in ecosystems, influencing species' life cycles and interactions.

Human activities are pivotal in shaping the biosphere's future. Actions such as deforestation, fossil fuel combustion, and pollution contribute to changes in atmospheric composition, such as fluctuations in oxygen and carbon dioxide levels. These activities can disrupt ecosystems and impact species' survival. Thus, understanding and mitigating human impacts is crucial for the health and sustainability of the biosphere.

4.4 The Biogeochemical Cycle

The biogeochemical cycle describes the continuous movement of nutrients and elements between living organisms (biotic factors) and their physical environment (abiotic factors) (Schlesinger & Bernhardt, 2013). The term "biogeochemical" combines "bio" (biosphere), "geo" (geological components), and "chemical" (cycling elements), reflecting the interconnected nature of these systems.

On Earth, matter exists in the form of atoms and is neither created nor destroyed but is recycled through various forms. The Sun provides energy to the Earth, which is eventually radiated back into the atmosphere as heat. Key elements involved in these cycles include:

- Carbon
- Hydrogen
- Nitrogen
- Oxygen
- Phosphorus
- Sulfur

These elements cycle through the biosphere, hydrosphere, and lithosphere, interacting with both biotic and abiotic components of ecosystems (Smith & Smith, 2012).

Understanding these cycles is crucial for comprehending how nutrients are exchanged and how ecosystems maintain their balance and functionality. These cycles illustrate how energy is transferred and utilized within ecosystems, ensuring the movement of essential elements necessary for life. They are vital for recycling and storing elements, as well as regulating essential nutrients through physical processes (Falkowski & Raven, 2007). The interconnection between living and nonliving components within these cycles is crucial for the continued survival of ecosystems. However, some human activities disrupt these natural cycles, leading to adverse impacts on ecosystems. By closely examining these mechanisms, we can mitigate and potentially halt their harmful influences.

4.4.1 Types of Biogeochemical Cycles

Biogeochemical cycles are categorized into two main types:

- Gas Cycles: These include cycles for carbon, oxygen, nitrogen, and water (Fig. 4.7). They involve the movement of gases between the atmosphere and living organisms, as well as their interactions with other components of the Earth system.
- Sedimentary Cycles: These involve elements, such as sulfur, phosphorus, and the rock cycle. These cycles are primarily concerned with the movement of elements through the Earth's crust and sediments.

Fig. 4.7 Ecosystems are fundamental components of the natural world, representing intricate networks of interactions between living organisms and their physical environments. They vary widely in size and type, ranging from small microecosystems to expansive biomes

Here's a brief overview of some key biogeochemical cycles crucial for sustaining life on Earth:

Hydrological Cycle: The hydrological cycle describes the movement of water through the Earth's system. Water evaporates from bodies of water, cools and condenses into clouds, and returns to Earth as precipitation. This cycle is essential for regulating weather patterns and climate. Evapotranspiration, the process of water vapor being released from plants (via leaves) and the soil into the atmosphere, plays a key role in this cycle.

Carbon Cycle: The carbon cycle involves the exchange of carbon among the biosphere, geosphere, hydrosphere, atmosphere, and pedosphere. Plants absorb carbon dioxide from the atmosphere for photosynthesis, storing carbon in their tissues. When plants decompose, they contribute to soil formation and the creation of fossil fuels. Burning fossil fuels releases carbon dioxide back into the atmosphere. Animals consume plants and obtain carbon, which is returned to the atmosphere through respiration and decomposition. The carbon cycle is crucial for regulating atmospheric carbon dioxide levels and supporting life processes.

Nitrogen Cycle: The nitrogen cycle is a biogeochemical process through which nitrogen is transformed into various forms as it moves from the atmosphere to the soil, into organisms, and back to the atmosphere. Nitrogen, which constitutes about 78% of the atmosphere, is essential for life and circulates through terrestrial and marine ecosystems.

The cycle includes several key processes: nitrogen fixation, nitrification, denitrification, decomposition, and putrefaction. In nitrogen fixation, bacteria in plant roots convert atmospheric nitrogen into ammonia, a form usable by plants. Ammonia, often supplied as fertilizer, is further transformed into nitrites and nitrates through nitrification. These nitrates can be used by plants or returned to the atmosphere as nitrogen gas through denitrification, performed by specific bacteria. This ongoing process ensures the continuous availability of nitrogen for living organisms and maintains atmospheric balance.

Oxygen Cycle: The oxygen cycle involves the movement of oxygen through the atmosphere, lithosphere, and biosphere. Oxygen is a prevalent element on Earth, constituting about 21% of the atmosphere. Plants release oxygen during photosynthesis, which is essential for the respiration of humans and other animals. These organisms inhale oxygen and exhale carbon dioxide, which is then reabsorbed by plants. The plants use this carbon dioxide in photosynthesis to produce more oxygen, thus continuing the cycle.

The oxygen cycle is a continuous process that maintains the balance of oxygen in the atmosphere, supports life through respiration and photosynthesis, and influences various ecological and atmospheric processes.

4.4.2 Significance of Biogeochemical Cycles

Biogeochemical cycles are essential for maintaining the balance of key elements in ecosystems. They facilitate the movement of crucial elements—such as carbon, nitrogen, and phosphorus—through ecosystems, ensuring that these elements are continuously available to support the growth and survival of living organisms.

These cycles prevent the depletion of essential nutrients by recycling elements. They store and transform these elements into various forms, making them accessible for future use and helping to maintain ecosystem stability. Additionally, biogeochemical cycles help regulate the concentrations of important elements like carbon dioxide and oxygen in the environment, influencing climate and weather patterns. They also illustrate the intricate connections between living organisms (biotic factors) and their physical environment (abiotic factors), which are crucial for ecosystem health and functionality. However, human activities, such as deforestation, pollution, and industrial processes, can disrupt these natural cycles.

Understanding how these cycles operate is key to recognizing how human actions impact ecosystems and to developing strategies for mitigating their effects.

Overall, biogeochemical cycles are fundamental to the sustainability of life on Earth. Understanding their dynamics is crucial for protecting and preserving natural systems, ensuring their resilience and continued functionality.

4.5 Ecosystems and Global Biogeography

Ecosystems are fundamental units in the study of nature, representing communities of interacting living organisms within a specific environment. They range from small-scale systems, like those involving individual organisms, such as insects or plants, to larger systems, such as lakes, mountain ranges, or entire forests.

An ecosystem is a dynamic and interconnected system where living organisms (biotic components) interact with their physical environment (abiotic components) to form a functional unit (Fig. 4.7). These components are interdependent and continuously influence each other. An ecosystem consists of two inseparable components:

Abiotic Components: These are the physical aspects of the environment, such as climate, temperature, humidity, nutrient concentration, and pH.

Biotic Components: These include all living organisms, such as animals, plants, and microorganisms, which interact and depend on one another.

These biotic and abiotic components are linked through nutrient cycles and energy flows. Natural ecosystems are considered "balanced" systems due to the stable interactions between different organisms. However, ecosystems are continuously evolving due to dynamic processes and environmental changes, such as climate shifts or natural phenomena. As a result, the ecological community (biocenosis) must adapt to these new conditions, which can lead to gradual changes or, in extreme cases, the complete transformation or disappearance of the ecosystem.

According to the law of conservation of energy, energy cannot be created or destroyed, only transformed or transferred. This principle is crucial in ecosystems, where energy flow is vital for sustaining life. Almost all energy in Earth's ecosystems originates from the Sun. Once this solar energy reaches Earth, it is distributed among ecosystems in complex ways. Analyzing this distribution is often done through food chains or food webs, which illustrate how energy moves through different trophic levels in an ecosystem.

All organisms, whether living or decaying, have the potential to transfer energy within an ecosystem. However, matter is not extensively consumed in ecosystems. The situation with energy is different. According to the second law of thermodynamics, not all energy can be fully utilized. Throughout the food chain, energy is converted into useful work, but it always returns as heat. A food chain tracks the movement of energy and nutrients from one organism to another.

The food chain consists of various levels known as trophic levels, beginning with producers that absorb sunlight. Energy flows to organisms that consume or decompose these producers, continuing up to top predators, which are those that generally do not consume other animals. Each trophic level contains a specific amount of biomass, which is transferred from one level to the next. However, this transfer is not highly efficient due to the second law of thermodynamics.

Typically, only 2–40% of the consumed energy is converted into usable biomass at each level, a process known as environmental efficiency. As a result, energy loss increases with each successive trophic level, which explains the abundance of decomposers and insects compared to the relatively few top predators like tigers, sharks, and eagles. This inefficiency also makes top predators more vulnerable to extinction when their ecosystems are disrupted.

Ecosystems function by capturing and transferring energy, with sunlight being the primary source. Plants convert sunlight into chemical energy through photosynthesis, which is then transferred through the food chain as organisms consume one another. Nutrients such as carbon, nitrogen, and phosphorus cycle through ecosystems, with decomposers breaking down dead organisms and organic matter to return essential nutrients to the environment. In ecosystems, different trophic levels represent the positions of organisms in the food chain. Primary producers (plants) occupy the bottom level, followed by herbivores, carnivores, and decomposers. The term "trophic level" refers to the various steps in the food chain or ecological pyramid where food (or energy) is transferred from one organism to another.

4.5.1 Types of Ecosystems

Ecosystems are complex networks of living organisms interacting with their physical environment, and they can be classified based on their geographical location, climatic conditions, and dominant forms of vegetation and animal life. Understanding the different types of ecosystems is essential for studying biodiversity, ecological processes, and the impact of human activities on the natural world. Ecosystems vary in scale, ranging from small localized systems to vast regions like biomes that span continents. They play crucial roles in sustaining life, regulating climate, and providing resources and services essential for human well-being. Below are some of the major types of ecosystems found on Earth:

- Terrestrial Ecosystems: These ecosystems occur on land and include forests, grasslands, deserts, and tundra.
- Aquatic Ecosystems: These water-based ecosystems encompass oceans, lakes, rivers, and wetlands.
- Artificial Ecosystems: Man-made environments such as urban areas, agricultural fields, and gardens are also considered ecosystems.
- Biomes: These are large-scale ecosystems characterized by distinct climates, vegetation, and wildlife. Examples include tropical rainforests, deserts, grasslands, and tundra.
- Marine Ecosystems: Part of the broader category of aquatic ecosystems, marine ecosystems cover more than 70% of the Earth's surface and are distinguished by their high salt content. They include various systems, such as the surface ocean, deep sea, pelagic zones, and the seafloor.

Figure 4.8 provides an overview of these ecosystem types, highlighting their structural and functional diversity. Terrestrial ecosystems, such as forests and

Fig. 4.8 Overview of ecosystem types: terrestrial, aquatic, and artificial

grasslands, primarily operate on land, while aquatic ecosystems encompass marine and freshwater environments. Artificial ecosystems, shaped by human intervention, include urban landscapes and agricultural zones. This classification underscores the complexity and interconnectedness of ecosystems across the globe.

Additionally, nearshore marine ecosystems include coral reefs, mangroves, and seagrass meadows. Marine ecosystems are characterized by both abiotic and biotic components. Biotic components include the organisms and species within these systems, such as predators, parasites, and competitors. Abiotic components involve factors like nutrient concentration, temperature, sunlight, turbulence, salinity, and density.

4.6 Biomes

Biomes are large-scale ecological communities characterized by distinct climate patterns, vegetation types, and animal life. Each biome represents a unique combination of temperature, precipitation, and seasonal variations, which together create specific environments that support distinct ecosystems. Understanding these biomes is crucial for appreciating the diversity of life on Earth and the intricate ways in which organisms have adapted to their surroundings. They play a crucial role in shaping Earth's ecosystems and are vital for maintaining biodiversity (Barbour & Billings, 1988; Whittaker, 1975). Table 4.1 lists the six primary biomes on Earth,

Table 4.1 The main biomes and their specific abiotic factors

Biom	Characteristic abiotic factor
Tundra	Characterized by extremely cold temperatures, short growing seasons, and low biodiversity. It has permafrost, limiting plant growth to hardy species like lichens and mosses. Animal life is adapted to the harsh climate with insulation adaptations like thick fur or fat
Taiga (boreal forest)	Dominated by coniferous forests, this biome experiences cold winters and moderate precipitation. Wildlife such as moose and wolves are well adapted to the cold, while the vegetation consists of evergreen trees like spruce and pine
Temperate forest	Marked by four distinct seasons, temperate forests support a diverse range of deciduous trees, such as oak and maple, along with rich wildlife. The moderate climate provides a balanced environment for numerous plant and animal species
Grassland (savanna or prairie)	Grasslands vary from temperate prairies to tropical savannas. These biomes are characterized by vast open spaces dominated by grass, with few trees. Seasonal precipitation and drought play major roles, supporting large herbivores like bison or zebras, along with their predators
Desert	Defined by extremely low precipitation, deserts host specialized plants like cacti and animals adapted to conserve water. Temperature extremes, both hot and cold, create challenging living conditions, yet desert ecosystems are rich in adapted biodiversity
Tropical rainforest	These biomes are found near the equator, experiencing high temperatures and heavy rainfall year-round. Tropical rainforests are the most biodiverse biomes, with layered canopies that host a vast array of plant and animal species, from large mammals to insects and amphibians

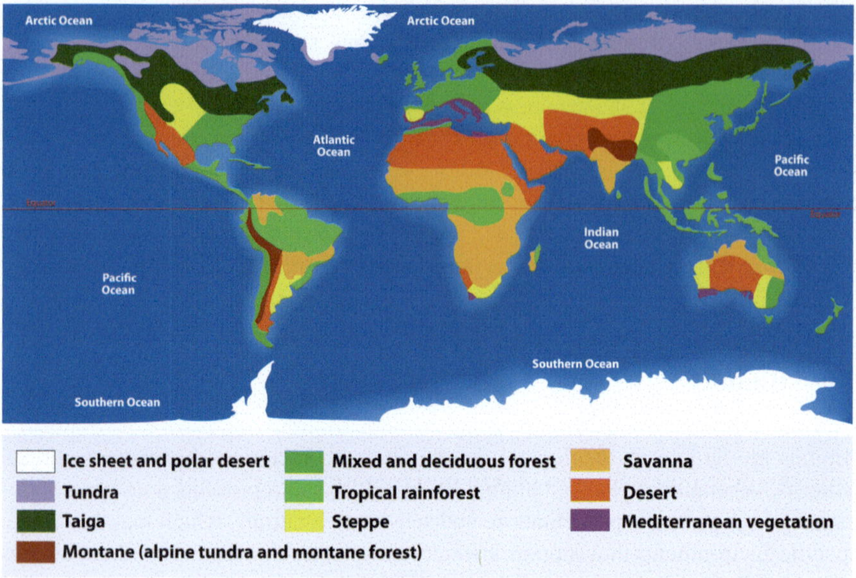

Fig. 4.9 The major biomes of the globe. A depiction of the Earth's major biomes, including tropical rainforests, deserts, tundra, grasslands, and temperate forests, each characterized by distinct climates, vegetation, and wildlife. (Credit: https://www.internetgeography.net/)

each characterized by distinct abiotic factors that shape their associated ecosystems, as depicted in Fig. 4.9. Additionally, Table 4.2 provides a more comprehensive description of each biome, examining the influence of various climatic conditions, including monthly averaged rainfall and temperature, which play a critical role in defining the biome's structure and function.

Understanding these biomes provides valuable insight into the mechanisms driving the distribution of life on Earth. The diversity within each biome reflects the adaptation strategies of organisms to their respective environments, while the study of biomes highlights the ecological interdependencies that sustain life on the planet (Begon et al., 2006).

The following section provides an overview of several key biomes, each with its own characteristic climate and ecological features. By examining the tropical rainforest, temperate grassland, desert, temperate forest, taiga, and tundra biomes, we gain insight into the diverse living conditions and adaptations that define each environment. The following descriptions highlight the major climatic factors and typical vegetation for each biome, illustrated with monthly temperature and rainfall distributions that reflect the average annual climate conditions. These graphics offer a visual representation of how climate influences biome characteristics and support the varied plant and animal life within these ecosystems (Ricklefs, 2008).

Biomes contain fauna and flora (animals and plants) that have adapted to their specific environments. They are often defined by abiotic (non-living) factors, such as climate, topography, geology, soils, and vegetation.

Table 4.2 A detailed description of the six primary biomes, including monthly averaged rainfall and temperature, highlighting the climatic conditions that influence each biome's structure, vegetation, and ecosystem dynamics

Tropical rainforest biome: Characterized by warm temperatures and abundant rainfall throughout the year, the tropical rainforest biome spans between 23.5° north and 23.5° south of the equator. It is hot and humid year-round, with approximately 12 h of sunlight daily. This biome is rich in biodiversity, teeming with a vast array of plant and animal species Nearly all plants in tropical rainforests are evergreen, growing rapidly and adapting to capture maximum sunlight. The dense vegetation offers abundant food sources and habitats, supporting a wide variety of animal life. Despite the nutrient-poor soils, the rapid breakdown of leaves and organic matter ensures a continuous supply of nutrients, sustaining the lush ecosystem	  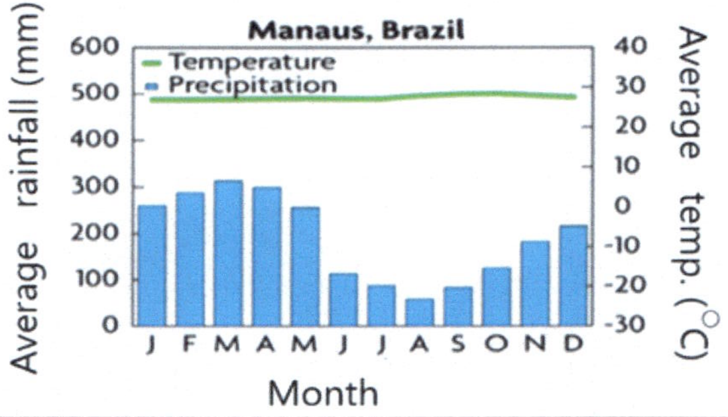

(continued)

Table 4.2 (continued)

Temperate grassland biome: Dominated by grass as the primary vegetation, the temperate grassland biome occurs under a range of climatic conditions between 40° and 60° latitude, both north and south of the equator. This biome experiences distinct seasons, with hot summers reaching up to 40 °C and very cold winters plunging to −40 °C Precipitation is moderate, ranging from 250 to 500 mm annually, primarily falling in late spring and early summer. The amount of light varies throughout the year, and the precipitation is generally insufficient to support large trees, resulting in mainly herbaceous vegetation Temperate grasslands are home to fewer animal species compared to tropical grasslands. Common mammals include bison, wild horses, and rodents like moles and rats. The warm summer temperatures facilitate rapid decomposition, contributing to relatively thick, nutrient-rich soils	  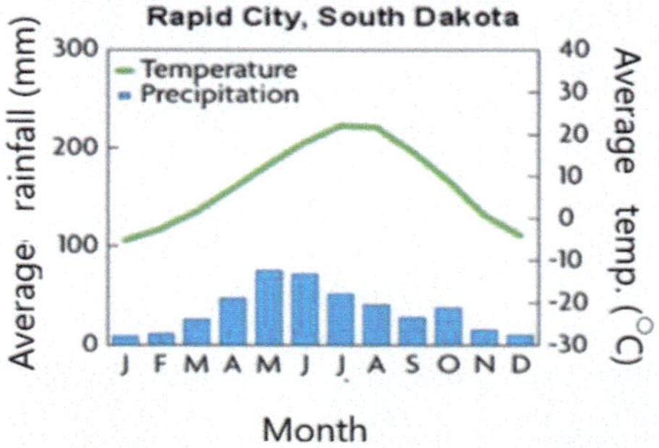

Desert biome: Characterized by an arid environment with minimal rainfall, the desert biome encompasses four distinct types: hot, semi-arid, coastal, and cold deserts. Deserts are typically located between 5° and 30° latitude, both north and south of the equator, within a belt of high pressure and low precipitation. Annual rainfall is extremely low, often less than 250 mm

Deserts experience extreme temperature variations, with very cold nights (around 0 °C) and hot days that can reach up to 45 °C. The lack of cloud covers allows for abundant sunshine during the day. Plant life is sparse due to limited rainfall, with vegetation primarily consisting of cacti and thorny plants. Many desert plants have short life cycles, emerging only when it rains

Animal species in the desert are relatively few, but those that inhabit this biome are highly adapted to its harsh conditions. Common desert dwellers include scorpions, lizards, snakes, and various insects

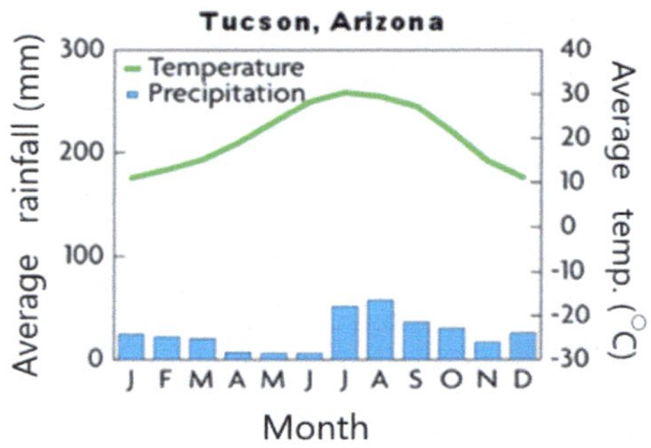

(continued)

Table 4.2 (continued)

Temperate forest biome: This biome includes both deciduous forests and temperate rainforests. Temperate deciduous forests are characterized by hot summers and cold winters, with deciduous trees being the dominant plant species. Located between 40° and 60° latitude, both north and south of the equator, this biome experiences high rainfall and mild temperatures typical of mid-latitudes. It also features four distinct seasons with varying daylight hours—shorter days in winter and longer days in summer Annual rainfall can reach up to 1500 mm, occurring throughout the year. Though fewer in number compared to tropical rainforests, plant species in temperate forests are rich and diverse, predominantly composed of broad-leaved trees like oaks and shrubs, which shed their leaves in autumn. This seasonal leaf fall contributes to the nutrient-rich, dense soils, as the leaf litter breaks down quickly The mild climate and abundant plant life provide ample food and habitats for a variety of mammals, including foxes, squirrels, and mice, as well as numerous bird and insect species	  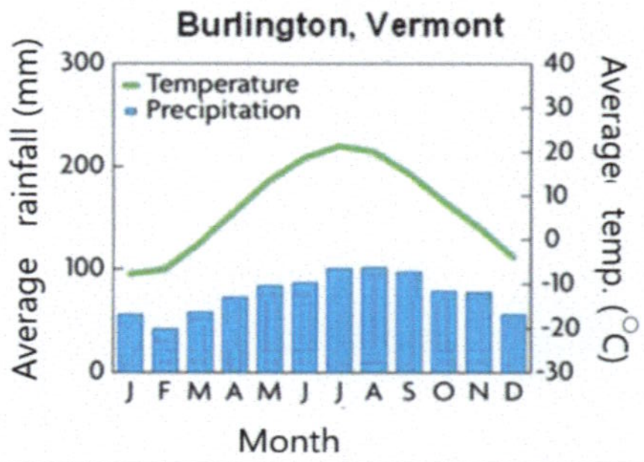

Taiga: This biome is found in cold northern climates, typically at high latitudes (around 60° north of the equator) and in mountainous regions. It is characterized by long, cold winters with temperatures dropping to −20 °C and short, mild summers with temperatures around −10 °C. Rainfall is limited, with less than 500 mm per year, most of which falls as snow. The region experiences extended daylight during summer months, but little to no daylight during winter. Clear skies provide ample sunlight during the day The vegetation is dominated by evergreen trees, particularly coniferous species like pine and fir, which can continue growing whenever there is enough sunlight. Lichens and low-growing mosses are also common. Due to the harsh conditions and limited food sources, the taiga supports relatively few animal species. Notable animals include black bears, wolves, and deer. The needles from the coniferous trees decompose slowly due to the low temperatures, resulting in thin, nutrient-poor, and acidic soils. The ground remains frozen for most of the year, further limiting plant growth	  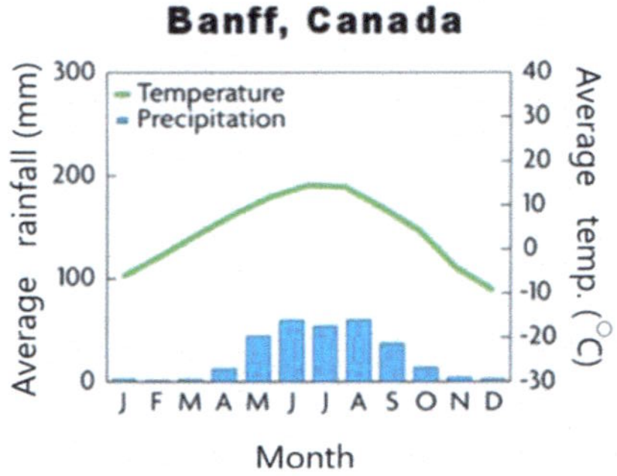

(continued)

Table 4.2 (continued)

Tundra: Far northern latitudes with long winters (10 months) limited precipitation, and eternal frost. Far north, above 60° N in Northern Europe, Alaska, and Russia. Below zero most of the year. Summer (5–10) °C, winter −30 °C. Little precipitation—less than 250 mm per year, most of which falls as snow. Almost continuous sunlight during the summer and little or no sunlight during the winter. In summer, there is more cloudiness. Few trees due to the lack of light in winter. Vegetation includes mosses, grasses, and low shrubs

Relatively few species of animals. Examples include arctic hares, arctic foxes, and birds

Some animals migrate south during the winter. Sparse vegetation produces little leaf litter, and organic matter slowly decomposes due to low temperatures. The soil is thin and poor in nutrients

Below the soil surface is a layer of permanently frozen land (permafrost)

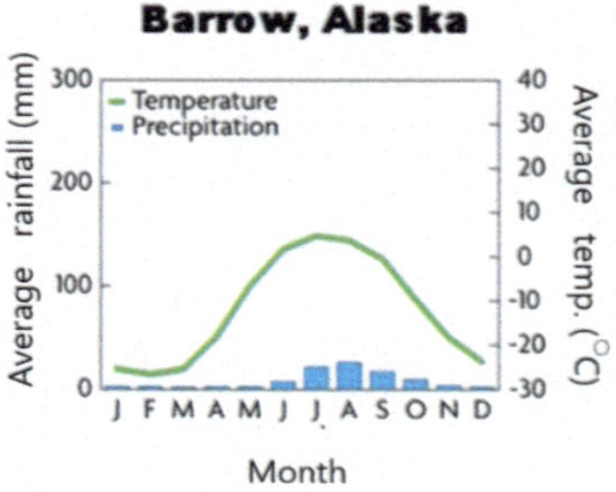

4.7 Factors Affecting the Biosphere

The biosphere, encompassing all life on Earth, is continually shaped and influenced by a range of living (biotic) and non-living (abiotic) factors. These elements interact in complex ways, affecting ecosystems and the species within them. Key factors include:

The biosphere, encompassing all life on Earth, is constantly shaped by a range of living (biotic) and non-living (abiotic) factors. These elements interact in intricate ways, influencing ecosystems and the species within them. Key factors include:

Earth's Tilt: The tilt of the Earth on its axis plays a critical role in regulating the biosphere by driving seasonal temperature variations. This axial tilt creates distinct seasons, influencing migration patterns, reproduction cycles, and plant growth. In temperate zones, many animals hibernate or migrate to escape harsh winters, while plants enter a dormant phase. The distribution of solar energy also shapes ecosystems, with tropical rainforests experiencing minimal seasonal change and polar regions requiring adaptation to extreme cold.

Natural Disasters: Catastrophic events like volcanic eruptions, earthquakes, floods, hurricanes, and wildfires can have both immediate and long-term effects on the biosphere. These events may lead to habitat destruction, species displacement, and biodiversity loss. For example, volcanic eruptions can blanket ecosystems with ash, altering soil composition, while floods can reshape landscapes by eroding soil and redistributing nutrients. However, some ecosystems rely on periodic disturbances—such as forest fires—to maintain ecological balance, and disasters can sometimes create opportunities for regeneration.

Climate Change: One of the most pressing factors affecting the biosphere today is climate change. Rising global temperatures, shifting precipitation patterns, and more frequent extreme weather events are causing widespread disruptions. Species are being forced to adapt, migrate to more favorable habitats, or face extinction. Examples include coral bleaching, polar ice melting, and the shifting of biomes, which are reshaping ecosystems around the globe.

Water Availability: Access to water is a fundamental factor for life in the biosphere. The distribution of freshwater resources influences the types of ecosystems that can thrive, ranging from lush wetlands to arid deserts. Changes in water availability—due to drought, pollution, or overuse—can lead to ecosystem degradation and reduced biodiversity. Aquatic species are vulnerable to changes in water quality, and human activities, such as dam construction, can disrupt natural water cycles, further impacting ecosystems.

Soil Erosion and Degradation: Soil is essential for supporting plant growth and providing a habitat for many microorganisms. Human activities, such as deforestation, overgrazing, and unsustainable agricultural practices, contribute to soil erosion and degradation, leading to desertification and the loss of arable land. Healthy soil is crucial for nutrient cycling, water retention, and carbon sequestration, making its preservation vital for maintaining biodiversity and ecosystem services.

Pollution: Human-induced pollution, including plastic waste, chemical runoff, air pollution, and excessive nutrient inputs, has severe effects on the biosphere. Pollution can lead to habitat destruction, biodiversity loss, and the disruption of ecosystem services. For example, plastic pollution threatens marine life, while agricultural runoff can create nutrient overloads that lead to dead zones in aquatic ecosystems, reducing oxygen levels and killing aquatic species.

Human Activities: Deforestation, urbanization, industrialization, and agriculture all exert significant pressure on the biosphere. Habitat fragmentation caused by human development isolates species, reduces genetic diversity, and limits migration routes, making it harder for organisms to adapt to environmental changes. Furthermore, overexploitation of natural resources—such as overfishing, logging, and mining—impedes the ability of ecosystems to recover and maintain balance.

In summary, the biosphere is dynamic and constantly subject to various natural and anthropogenic influences. The interplay of these factors determines the resilience and adaptability of ecosystems and the long-term sustainability of life on Earth. Understanding and mitigating the human impact on the biosphere is essential to preserving biodiversity and ensuring the health of our planet.

4.8 Life and the Future of the Biosphere

The future of the biosphere is deeply interconnected with human activities and the choices we make regarding environmental stewardship. As our global population grows and industrialization continues to expand, the pressure on natural ecosystems is intensifying. This has triggered significant transformations in Earth's life systems, threatening the delicate balance that sustains all species. Two critical dimensions shaping the biosphere's future are the shifts in atmospheric composition and the degradation of ecosystems, but several other factors also play a pivotal role.

Changes in atmospheric gases, particularly carbon dioxide, methane, and oxygen, are having profound impacts on the biosphere. Carbon dioxide, primarily released through the burning of fossil fuels, has been a central driver of global climate change (IPCC, 2021). This shift is manifesting in rising global temperatures, altered weather patterns, and an increase in the frequency of extreme climate events, such as droughts, floods, and storms.

These disruptions challenge the survival of numerous species, from polar bears in the Arctic to coral reefs suffering bleaching due to ocean acidification. Not only do ecosystems struggle to adapt to these rapid changes, but human societies also face reduced food security, freshwater shortages, and economic losses tied to agricultural and ecological instability (Sala et al., 2000).

The loss of biodiversity is another pressing issue. Human activities such as deforestation, urban sprawl, and overexploitation of resources are leading to the extinction of species at unprecedented rates. Ecosystem resilience—nature's ability to recover from disturbances—is diminished when biodiversity declines. Key species, such as pollinators or apex predators, are essential for maintaining ecosystem balance. Without them, food chains collapse, and the services ecosystems provide, such as pollination, water filtration, and carbon sequestration, are compromised (Cardinale et al., 2012). In the end, human well-being is intrinsically linked to the health of these systems, as our survival depends on the services nature provides for free.

Deforestation is a driving force behind these changes. Forests are not only home to countless species but are also vital carbon sinks, absorbing vast amounts of carbon dioxide from the atmosphere. When forests are cleared for agriculture, urban development, or logging, this stored carbon is released, exacerbating climate change. Moreover, the loss of forest ecosystems leads to soil degradation, disrupted water cycles, and the displacement of indigenous species, further amplifying ecological imbalances (Lewis et al., 2015).

Pollution in all forms, water and soil, presents a significant risk to the biosphere's future. Industrial emissions, agricultural runoff, and plastic waste are contaminating ecosystems worldwide, particularly in marine environments. Oceans, which are crucial to regulating global climate and supporting biodiversity, are now facing multiple threats, from oil spills and plastic debris to the silent but devastating impacts of ocean acidification. These pollutants not only harm marine life but also disrupt entire food chains, with consequences that extend to human populations reliant on the oceans for food and livelihoods (Jambeck et al., 2015).

Climate change continues to exert cascading effects on the planet's ecosystems. Rising sea levels threaten coastal ecosystems and communities, while changes in ocean currents and atmospheric circulation patterns could reshape entire climate zones. The Arctic is warming at twice the global average, leading to the melting of ice sheets and permafrost, which further accelerates climate feedback loops. As habitats are lost or altered, species must either adapt, migrate, or face extinction—challenges compounded by the already rapid pace of environmental change (IPCC, 2021). Despite these daunting challenges, hope remains. Technological innovations and scientific research are providing us with powerful tools to better understand and protect the biosphere. Satellite technologies allow for real-time monitoring of deforestation, pollution, and wildlife populations. Advances in conservation biology are helping to restore ecosystems and reintroduce endangered species to their native habitats.

Furthermore, cutting-edge fields like synthetic biology and geoengineering hold the promise of mitigating some of humanity's negative impacts, though these technologies must be approached with caution due to their potential risks.

The global push toward sustainability and conservation is also gaining momentum. Initiatives such as reforestation, ecosystem restoration, and the establishment of protected areas are crucial to preserving biodiversity and maintaining the ecological functions upon which all life depends. The transition to renewable energy sources is key to reducing the carbon emissions that drive climate change. At the same time, international agreements like the Paris Agreement represent vital steps toward global cooperation in addressing environmental challenges (UNEP, 2020).

The future of the biosphere hinges on our ability to make informed, responsible decisions. Human activities have already placed immense stress on natural systems, but by reducing our environmental footprint, conserving critical ecosystems, and embracing sustainable development, we can still steer the biosphere toward a more stable and thriving future. The well-being of all species, including our own, depends on our commitment to ensuring the long-term health of the planet.

4.9 Estuary and Freshwater Ecosystems

An estuary is a partially enclosed coastal body of brackish water, where freshwater from rivers and streams mixes with salt water from the sea. These areas are dynamic transition zones between rivers and oceans, characterized by their mixture of fresh and salt water. Estuaries are among the most productive ecosystems, supporting a diverse array of plant and animal species.

Key characteristics of estuaries include

- A unique blend of fresh salt water, creating brackish conditions.
- Rivers flowing into estuaries carry a rich supply of nutrients, supporting high biological productivity.
- The combination of fresh and salt water, along with abundant nutrients, allows for a high diversity of species to thrive.
- Estuaries provide essential shelter and spawning grounds for many species, offering protection and an environment conducive to growth. Unfortunately, over 80% of estuaries have been lost due to land development and human activities, leading to significant ecological impacts.

Estuaries are critical ecosystems that play a vital role in supporting biodiversity and maintaining the health of marine and freshwater environments (Fig. 4.10).

Estuaries are sensitive ecosystems that can be significantly impacted by human activities, such as pollution, habitat destruction, and changes in water flow. Protecting estuaries is crucial for maintaining their ecological balance and ensuring their continued health and productivity.

Fig. 4.10 An estuary is a partially enclosed body of water formed where a river flows into the sea or ocean. Estuaries are of great importance for biodiversity

Fig. 4.11 Freshwater ecosystems

4.9.1 Freshwater Ecosystems

Freshwater ecosystems encompass a range of environments where the water has low salinity or is entirely freshwater (Fig. 4.11). These ecosystems can be classified into several types, including rivers, lakes, ponds, wetlands, and streams. Each type has unique characteristics and plays a vital role in supporting life.

Rivers and Streams: Rivers and streams are the two primary types of watercourses on Earth, each playing a key role in the water cycle and hydrologic systems. Rivers are typically larger watercourses with significant volume and discharge, crossing vast areas and often serving as routes for transportation and sources of water for various purposes. They have multiple sources, such as floods, snowmelt, or groundwater, and flow from their source to their mouth. Rivers are divided into different sections, including headwaters, midwaters, and estuaries. Their shape and width can vary considerably, supporting diverse ecosystems with rich biodiversity.

Streams are smaller watercourses compared to rivers. They can be wide or narrow, with a smaller volume of water. Streams usually have a shorter and faster course and originate from various sources. While they support a variety of life forms, their biodiversity may be lower compared to rivers. Both rivers and streams are crucial to the water cycle, provide essential resources, and play a significant role in shaping the landscape.

Lakes and Ponds: Lakes and ponds are freshwater bodies with varying surface areas and depths (Fig. 4.12). Lakes are larger bodies of water that can be natural or artificial, created by dams or other human interventions. They have considerable surface area and depth and support diverse ecological communities. Ponds are smaller compared to lakes and can also be natural or artificial. Despite their size, they are vital for supporting various life forms and providing ecological services.

Wetlands: Wetlands are characterized by their periodic or permanent water presence, which can be either fresh or brackish. These ecosystems are crucial for maintaining ecological balance and supporting biodiversity. Wetlands are areas where

Fig. 4.12 (**a**) Lakes are generally larger and deeper than small ponds. (**b**) A pond or small lake ecosystem includes biotic (living) plants, animals, and microorganisms, as well as abiotic (nonliving) physical and chemical interactions. Both support a range of plants and animals

water accumulates on the Earth's surface, often covered with grasses, plants, and aquatic vegetation. They typically form in areas with poor drainage, allowing water to be retained. Wetlands provide essential habitats for waterfowl, fish, and other aquatic organisms, contribute to water balance, and play a significant role in flood control and water regulation.

References

Barbour, M. G., & Billings, W. D. (1988). *North American terrestrial vegetation*. Cambridge University Press.

Begon, M., Townsend, C. R., & Harper, J. L. (2006). *Ecology: From individuals to ecosystems*. Blackwell Publishing.

Cardinale, B. J., Duffy, J. E., Gonzalez, A., et al. (2012). Biodiversity loss and its impact on humanity. *Nature, 486*, 59–67.

Chapin, F. S., Matson, P. A., & Vitousek, P. M. (2011). *Principles of terrestrial ecosystem ecology*. Springer.

Clarke, A. (2008). *Life in cold environments*. Oxford University Press.

Falkowski, P. G., & Raven, J. A. (2007). *Aquatic photosynthesis* (2nd ed.). Princeton University Press.

IPCC. (2021). *Climate Change 2021: The physical science basis*. Contribution of Working Group I to the Sixth Assessment Report of the Intergovernmental Panel on Climate Change.

Jambeck, J. R., Geyer, R., Wilcox, C., et al. (2015). Plastic waste inputs from land into the ocean. *Science, 347*(6223), 768–771.

Kump, L. R., Kasting, J. F., & Crane, R. G. (2010). *The Earth system*. Pearson.

Lewis, S. L., Edwards, D. P., & Galbraith, D. (2015). Increasing human dominance of tropical forests. *Science, 349*(6250), 827–832.

Margulis, L., & Sagan, D. (1995). *What is life?* University of California Press.

Millennium Ecosystem Assessment (MEA). (2005). *Ecosystems and human well-being: Biodiversity synthesis*. World Resources Institute.
Moore, J. (2015). *The biosphere: Protecting our global environment*. Oxford University Press.
Odum, E. P., & Barrett, G. W. (2005). *Fundamentals of ecology* (5th ed.). Brooks/Cole.
Pidwirny, M. (2019). *The atmosphere and the biosphere*. PhysicalGeography.net.
Ricklefs, R. E. (2008). *The economy of nature* (6th ed.). W. H. Freeman and Company.
Sala, O. E., Chapin, F. S., Armesto, J. J., et al. (2000). Global biodiversity scenarios for the year 2100. *Science, 287*(5459), 1770–1774.
Schlesinger, W. H., & Bernhardt, E. S. (2013). *Biogeochemistry: An analysis of global change* (3rd ed.). Academic Press.
Smil, V. (2003). *The Earth's biosphere: Evolution, dynamics, and change*. MIT Press.
Smith, T. M., & Smith, R. L. (2012). *Elements of ecology* (8th ed.). Pearson.
UNEP. (2020). *Emissions gap report 2020*. United Nations Environment Programme.
Vernadsky, V. I. (1998). *The biosphere*. Copernicus Books.
Whittaker, R. H. (1975). *Communities and ecosystems*. Macmillan Publishing.
Wilson, E. O. (1999). *The diversity of life*. W. W. Norton & Company.

Chapter 5
Hydrosphere and Cryosphere: Key Challenges

The hydrosphere encompasses all forms of water found on Earth's surface, subsurface, and atmosphere. This includes vast oceans, seas, lakes, rivers, groundwater reservoirs, glaciers, and even water vapor in the atmosphere. Representing about 71% of Earth's surface, the hydrosphere plays a central role in regulating the planet's energy balance, supporting life, and driving key processes such as the hydrological cycle. This cycle involves the continuous movement of water through evaporation, condensation, precipitation, and runoff, linking the hydrosphere with the atmosphere, biosphere, and geosphere. The hydrosphere's influence extends to weather systems, climate patterns, and sustaining ecosystems and human activities like agriculture, industry, and freshwater supply.

© The Author(s), under exclusive license to Springer Nature Switzerland AG 2025
V. Spiridonov et al., *Atmospheric Perspectives*,
https://doi.org/10.1007/978-3-031-86757-6_5

The cryosphere, a subcomponent of the hydrosphere, refers to Earth's frozen water systems. This includes ice caps, glaciers, permafrost, snow cover, sea ice, and frozen lakes and rivers. Found primarily in polar regions and high altitudes, the cryosphere is vital for Earth's climate regulation. It reflects a significant portion of solar radiation (albedo effect), helping to stabilize global temperatures. Changes in the cryosphere, such as melting glaciers and shrinking sea ice, are key indicators of climate change, influencing sea-level rise and altering ecosystems.

Together, the hydrosphere and cryosphere are intricately connected. For example, melting glaciers contribute to rising sea levels, affecting coastal hydrospheric systems. Their interaction plays a critical role in Earth's climate system, driving ocean circulation, redistributing heat, and sustaining biodiversity. Understanding these systems is essential for addressing global challenges, including freshwater scarcity, climate change, and habitat conservation.

5.1 The Hydrosphere: Earth's Water Systems

The hydrosphere encompasses all the water on Earth's surface, including oceans, seas, lakes, rivers, groundwater, glaciers, and even atmospheric water vapor. It plays a crucial role in supporting life on Earth and is a key component of the Earth system, participating actively in the hydrological cycle and influencing both weather patterns and climate. The distribution of water within the hydrosphere is vital not only for sustaining ecosystems but also for human activities such as agriculture, industry, and drinking water supply. Moreover, the hydrosphere interacts with other Earth systems, including the atmosphere, biosphere, and geosphere, to regulate environmental conditions across the planet (Gleick, 1993; Lenton & Watson, 2011).

5.1.1 Definition of Hydrosphere

The hydrosphere encompasses all the water present on Earth, including oceans, lakes, rivers, streams, groundwater, glaciers, and water vapor in the atmosphere. This vast system exists in three states: liquid, solid (ice), and gaseous, highlighting its central role in Earth's physical and biological processes (Pidwirny, 2006). Covering approximately 70% of Earth's surface, the hydrosphere is dominated by oceans. As illustrated in Fig. 5.1, about 97.5% of Earth's water is saline, primarily contained within oceans and seas, while the remaining 2.5% is freshwater. However, most of this freshwater is locked in glaciers and ice caps, with only a small portion accessible in rivers, lakes, and groundwater. This limited freshwater availability makes it an essential resource for sustaining terrestrial life (Shiklomanov, 1993).

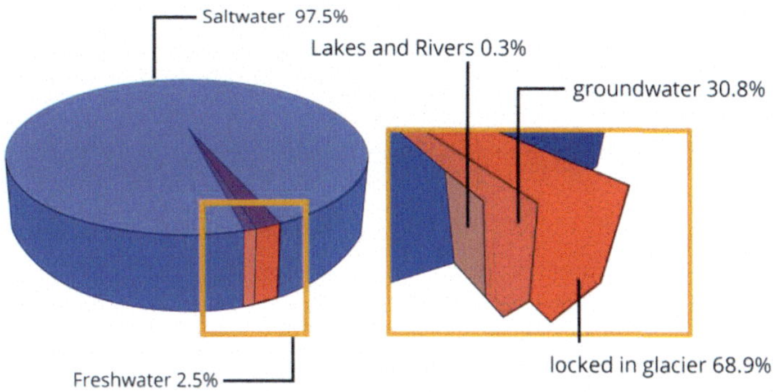

Fig. 5.1 Pie chart of water distribution on Earth. (Credit: https://www.vedantu.com/evs/water-distribution-on-earth)

5.2 Gaseous State of Water

Water vapor is present in the atmosphere primarily due to the evaporation of water from the Earth's surface, including oceans, seas, lakes, rivers, and soil. Water continuously circulates through the atmosphere: it evaporates from the surface, rises in warm updrafts, condenses into clouds, is carried by the wind, and then falls back to Earth as rain or snow. This cycle is a crucial mechanism for transferring heat and energy from the Earth's surface to the atmosphere and redistributing it across the planet (Fig. 5.2).

Water vapor is also the most significant greenhouse gas in the atmosphere. Heat radiated from the Earth's surface is absorbed by water vapor molecules in the lower atmosphere. These molecules then radiate heat in all directions, with some of the heat returning to Earth's surface. Consequently, water vapor is a major source of heat for the Earth's surface, in addition to sunlight.

Figure 5.3 illustrates the global average amount of water vapor in a vertical column of the atmosphere for September 2023. Water vapor in the atmosphere is distributed through convection, turbulent mixing, and diffusion. As water vapor rises from the Earth's surface, its content decreases rapidly with altitude. In the free atmosphere, water vapor falls more quickly than over mountain slopes, where it is in direct contact with evaporating soil. During temperature inversions, water vapor content may increase with height. Air moisture, driven by water vapor, is a crucial meteorological element. The content of water vapor in the atmosphere is quantified by measures that determine the amount of water vapor in the air and those that indicate the degree of saturation.

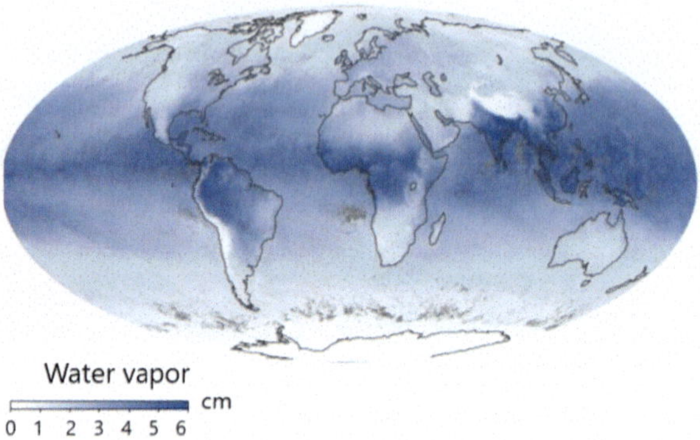

Fig. 5.2 Global distribution of water vapor for September 2023. (Source: NASA)

5.3 Hydrological Cycle

The hydrological cycle, also known as the water cycle, is essential for maintaining the Earth's water balance and supporting life. It involves the continuous movement of water within the Earth's atmosphere, surface, and subsurface environments, driven primarily by solar radiation. The cycle includes key processes such as atmospheric boundary layer interactions, evaporation, condensation, cloud formation, and precipitation, all of which contribute to the regulation of the Earth's energy balance (Oki & Kanae, 2006).

At its core, the hydrological cycle represents the journey of water as it changes states between liquid, solid, and gas and moves between the oceans, atmosphere, and land (see Fig. 5.3). Water evaporates from oceans, lakes, and other bodies of water, entering the atmosphere as water vapor. Plants also contribute through transpiration, releasing water vapor during photosynthesis. This combined process is referred to as evapotranspiration. As water vapor rises, it cools and condenses, which is essential in the formation of clouds and precipitation. This precipitation, in the form of rain, snow, or hail, falls back to the Earth's surface, where it either infiltrates the soil to become groundwater or flows over the land as runoff, eventually returning to rivers, lakes, and oceans (Chahine, 1992).

The quantitative aspect of the hydrological cycle is referred to as the water balance. This balance measures the amount of water vapor in the atmosphere and its fluctuations. Two critical processes influencing the water balance are evaporation and condensation. Evaporation occurs when liquid water converts to vapor, while condensation happens when water vapor cools and changes back to liquid form. Both processes are influenced by temperature, atmospheric pressure, and solar radiation. Humidity, which describes the amount of water vapor in the air, is closely linked to these processes and plays a key role in weather phenomena, such as cloud formation and storm development (Schneider et al., 2017).

The hydrological cycle is not only crucial for climate regulation but also for the distribution of freshwater resources, which are vital for ecosystems, agriculture, and human consumption. As climate change intensifies, the hydrological cycle is being

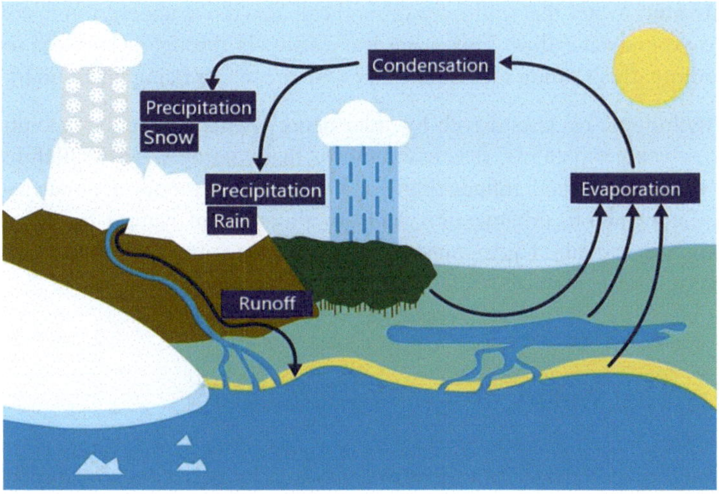

Fig. 5.3 The hydrological cycle and its main components, including evaporation/transpiration, advection, precipitation, infiltration, direct runoff, recharge of watercourses, surface water, and catchment areas

altered, leading to changes in precipitation patterns, more extreme weather events, and disruptions in water availability, highlighting the need for further research and monitoring (Huntington, 2006).

The key processes involved in the hydrologic cycle include evaporation, condensation, precipitation, transpiration, infiltration, surface runoff, and groundwater flow.

Evaporation: This process involves the conversion of liquid water into water vapor, primarily from the Earth's surface, such as oceans, rivers, lakes, and moist soil.

Transpiration: This is the release of water vapor from plants into the atmosphere. Plants absorb water from the soil through their roots, and this water eventually evaporates from the leaves.

Condensation: Water vapor in the atmosphere cools and condenses into tiny water droplets or ice crystals, forming clouds. This process occurs when air masses with different temperatures meet.

Precipitation: Water droplets in clouds combine and grow until they are heavy enough to fall to Earth's surface as precipitation. This can manifest in the form of rain, snow, sleet, or hail, depending on temperature and atmospheric conditions.

Infiltration: Precipitation that falls on the ground can be absorbed into the soil in a process known as infiltration. Some of this water is taken up by plant roots, while the rest moves deeper into the soil, eventually reaching groundwater.

Surface Runoff: Excess water that cannot be absorbed by the soil becomes surface runoff, flowing over the land and collecting in rivers, lakes, and oceans. Surface runoff is critical as it recharges water bodies and shapes the landscape.

Groundwater: Water that infiltrates the soil can also move underground, contributing to groundwater flow. This water travels slowly through permeable rock layers, eventually draining into rivers, lakes, or oceans, or being taken up by plants.

The hydrologic cycle is driven by solar energy, which powers evaporation and influences weather patterns. It is essential for maintaining water availability, supporting ecosystems, and influencing Earth's climate patterns. Human activities, such as deforestation, urbanization, and the alteration of natural landscapes, can impact the water cycle. Understanding the hydrologic cycle is crucial for sustainable water resource management.

5.4 Internal Waters

Inland waters encompass various freshwater bodies located within Earth's continental boundaries. These bodies of water play essential roles in supporting ecosystems, providing resources for human activities, and influencing local climates. Here is an overview of key types of inland waterways:

- Lakes: Large bodies of water surrounded by land, which can vary greatly in size, depth, and ecological characteristics. Lakes are classified into different types based on their origin: tectonic, volcanic, glacial, and artificial. Tectonic lakes, like Lake Baikal, are formed by geological processes such as the movement of Earth's crust, while artificial lakes are created by human activities such as dam construction (Smith, 2002).
- Rivers and Streams: Flowing bodies of water that transport water from higher to lower elevations. They are crucial for the water cycle, transporting nutrients, sediments, and organic matter. Rivers and streams also provide habitats for diverse aquatic species and play a vital role in shaping landscapes through erosion and sediment deposition (Miller & Smith, 1986).
- Reservoirs: Artificial lakes created by damming rivers. Reservoirs serve multiple purposes, including water supply, flood control, irrigation, and hydroelectric power generation. However, they can also have significant ecological impacts, such as altering natural flow regimes and affecting aquatic habitats (Petts, 1984).
- Ponds: Smaller, shallow bodies of water that can be either natural or man-made. Ponds are often found in both urban and rural areas and support diverse plant and animal life. They play an important role in local ecosystems, providing breeding grounds for amphibians and serving as habitats for various insects and plants (Wetzel, 2001).

5.4.1 Catchment Area

A catchment area, also known as a watershed or drainage basin, is a geographic region where water, sediments, and dissolved materials drain into a common outlet, such as a river, lake, or ocean (see Fig. 5.4a). Catchment areas are crucial for

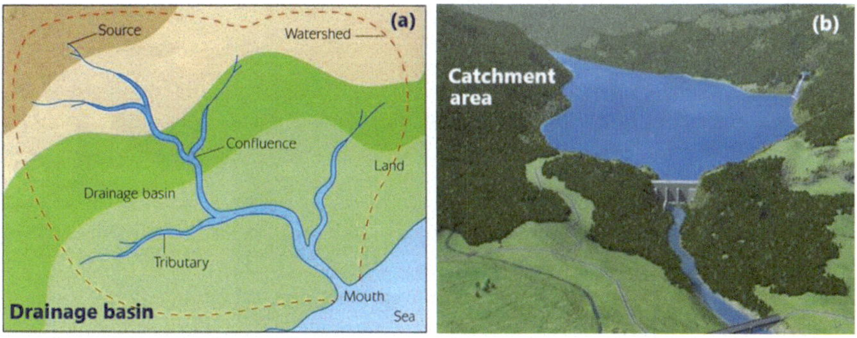

Fig. 5.4 (**a**) Detailed view of a drainage basin: This wider area includes the headwaters of a river and its tributaries. Its boundaries are defined by watersheds that separate it from other drainage basins. All surface water in this area drains through rivers, gullies, watercourses, and tributaries into the mainstream, eventually flowing into the sea or lake. (**b**) Catchment area: An area drained by smaller rivers, gullies, and watercourses. Several such catchment areas combine to form a larger drainage basin for a single river

understanding the hydrologic cycle and are fundamental for water resource management, environmental conservation, and land-use planning. In a catchment area, all internal waters, including rivers, streams, lakes, and ponds, flow into a larger body of water, which is referred to as the catchment or drainage basin (see Fig. 5.4b). The size and shape of a catchment area can influence the flow and distribution of water. Several factors affect water flow within a catchment area, including:

- Topography: The shape and slope of the land surface influence how water flows and accumulates. Steeper slopes lead to faster runoff and less infiltration, while flatter areas allow for greater water absorption and slower runoff (Niemi et al., 1990).
- Land Use: Human activities such as agriculture, urban development, and deforestation can impact water flow and quality within a catchment area. For example, impervious surfaces like roads and buildings increase runoff and reduce groundwater recharge (Dunne & Leopold, 1978).
- Soil Type: Soil composition affects the infiltration and storage of water. Sandy soil typically has higher infiltration rates compared to clay soils, which can affect the volume and timing of runoff (Hillel, 1998).
- Vegetation: Plant cover influences water retention and erosion control. Vegetation helps slow runoff, increase infiltration, and stabilize soil, reducing the risk of erosion and sedimentation (Aber et al., 1995).
- Climate: Precipitation, wind, and temperature impact water flow and distribution.
- Human Activities: Structures such as dams can alter natural water flow patterns.

5.5 Ocean Waters

Ocean waters cover approximately 71% of the Earth's surface, making them a fundamental component of the Earth's climate system and supporting a diverse range of life forms. The five major oceans are:

- The Pacific Ocean: The largest and deepest of the world's oceans, the Pacific Ocean spans more than 63 million miles^2 (165 million km^2). Its vast expanse plays a pivotal role in global weather and climate. The Pacific is known for the El Niño-Southern Oscillation (ENSO), a climate phenomenon originating here that has significant impacts on global weather patterns, including altering monsoon cycles, influencing the frequency and intensity of tropical cyclones, and contributing to extreme weather events such as droughts and floods (Cane, 2005).
- The Atlantic Ocean: The second largest ocean, stretching between the Americas, Europe, and Africa, the Atlantic Ocean is integral to global climate regulation. It is home to the Gulf Stream, a powerful current that transports warm water from the Gulf of Mexico toward Europe, moderating the climate of Western Europe and contributing to the region's relatively mild winters (Saunders & Jones, 2009). Additionally, the Atlantic Ocean plays a crucial role in the Atlantic Meridional Overturning Circulation (AMOC), a key component of the global thermohaline circulation that regulates heat distribution and influences climate patterns worldwide (Buckley & Marshall, 2016).
- The Arctic Ocean: The smallest and shallowest of the oceans, the Arctic Ocean is largely covered by sea ice, which plays a critical role in regulating the Earth's temperature by reflecting sunlight. However, due to climate change, the Arctic ice cover is diminishing rapidly, leading to the opening of new shipping routes and increased access to natural resources (Stroeve et al., 2012). The reduction in ice cover also accelerates global warming through a feedback loop known as the ice-albedo effect, where decreased ice reduces reflectivity and increases heat absorption (Screen & Simmonds, 2010).
- The Indian Ocean: Bordered by Asia, Africa, and Australia, the Indian Ocean is the third largest and is significant for its impact on regional climate systems. It influences the monsoon systems that affect South Asia and East Africa.
- The Indian Ocean Dipole (IOD) is a climate phenomenon that affects weather patterns, causing variations in rainfall and temperature across the Indian Ocean region (Saji et al., 1999).
- The Southern Ocean: Officially designated as a distinct ocean in 2000, the Southern Ocean encircles Antarctica and is defined by the Antarctic Circumpolar Current (ACC), the world's largest ocean current. This current plays a critical role in global ocean circulation and climate regulation by redistributing heat and nutrients. The Southern Ocean is also a significant site for carbon sequestration, as its cold waters absorb substantial amounts of carbon dioxide, influencing global carbon cycles and climate change mitigation efforts (Gordon et al., 2000).

Two essential parameters in the study of ocean waters are temperature and salinity, which significantly influence ocean currents, marine ecosystems, and the global climate system. These factors will be explored in more detail in the following sub-section.

5.5.1 *The States of Water on Earth*

Water on Earth exists in three fundamental states: liquid, solid, and gaseous. These states are central to Earth's hydrological cycle, climate system, and the dynamic interactions between the atmosphere, oceans, and land surfaces.

Liquid State: The Ocean's Role in Global Systems

The ocean represents most of the Earth's liquid water, covering approximately 71% of the globe. As a dynamic system, it plays a pivotal role in regulating climate by absorbing solar energy and redistributing heat across the planet. Ocean currents, both surface and deep, transport warm and cold water between the equator and the poles, influencing weather patterns, regional climates, and marine ecosystems. In addition to its heat capacity, the liquid state of water in oceans drives the global hydrological cycle through the process of evaporation. This evaporation is the largest single source of atmospheric moisture, contributing to the formation of clouds and precipitation. Water vapor, the gaseous state of water, fuels the development of weather systems such as tropical storms, hurricanes, and monsoons. The continuous exchange of moisture between the ocean and atmosphere is vital for sustaining freshwater resources on land.

Solid State: Ice and Its Climatic Influence

In polar regions and high-altitude environments, water exists primarily in its solid state, in the form of ice. Glaciers, ice sheets, and sea ice play critical roles in Earth's energy balance by reflecting incoming solar radiation, a process known as the albedo effect. Polar ice, particularly in the Arctic and Antarctic, acts as a thermal buffer, regulating the temperature of surrounding oceans and contributing to global climate stability. Seasonal variations in the extent of sea ice, particularly in the Arctic, have a profound influence on global weather patterns and ocean currents. The melting of polar ice due to global warming contributes to sea-level rise and alters the salinity of the oceans, which in turn affects ocean circulation systems. These changes are closely monitored as they have the potential to disrupt ecosystems, weather systems, and the livelihoods of coastal communities worldwide.

Gaseous State: Water Vapor and Climate Interaction

Water in its gaseous state, water vapor, is a potent greenhouse gas and a crucial component of the atmosphere. While water vapor is naturally occurring, its concentration increases through the evaporation of ocean waters, particularly in tropical and subtropical regions where solar heating is most intense.

Water vapor plays a significant role in trapping heat within the atmosphere, thus contributing to the greenhouse effect. Water vapor also has an important feedback mechanism in the climate system. As global temperatures rise, more water evaporates, increasing the amount of water vapor in the atmosphere, which in turn amplifies warming through enhanced heat retention. This positive feedback loop makes water vapor a key player in climate dynamics.

The transition between the three states of water—liquid, solid, and gaseous—has profound impacts on Earth's environment. Evaporation, condensation, freezing, and melting are processes that not only drive the hydrological cycle but also influence energy exchange between the ocean and atmosphere. For instance, when water evaporates, it absorbs heat from its surroundings, cooling the ocean surface. Conversely, when water vapor condenses into clouds, it releases heat into the atmosphere, affecting weather patterns and storm formation.

In polar regions, phase transitions between liquid and solid states are especially crucial. The formation and melting of sea ice influence ocean salinity, which affects global ocean circulation patterns, particularly the thermohaline circulation, a system that helps regulate the distribution of heat across the globe. These changes are essential for maintaining the delicate balance of Earth's climate system.

The polar regions, where solid and liquid states of water interact intensively, serve as critical zones for regulating Earth's climate. Sea ice reflects a significant portion of incoming solar radiation, reducing heat absorption by the ocean. This reflection, known as ice-albedo feedback, plays a crucial role in maintaining cooler global temperatures. However, as polar ice melts due to rising global temperatures, darker ocean waters are exposed, leading to increased heat absorption and further acceleration of ice melt—a worrying trend that contributes to the phenomenon of Arctic amplification.

5.5.2 Challenges and Advances in Oceanographic Research

The world's oceans are increasingly under pressure from the dual impacts of climate change and direct human activities such as pollution, overfishing, and habitat destruction. These challenges threaten marine ecosystems, disrupt climate-regulating processes, and jeopardize the livelihoods of millions who depend on ocean resources. Recent research underscores the urgency of addressing these issues through innovative solutions and global cooperation.

One of the most significant challenges is the impact of climate change, which drives ocean warming, sea-level rise, and acidification. The ocean absorbs over 90%

of the excess heat generated by global warming, resulting in rising sea temperatures. This warming has profound effects on marine biodiversity, particularly coral reefs, which experience widespread bleaching and ecosystem collapse under prolonged thermal stress. The consequences extend to marine food webs and fisheries, disrupting the livelihoods of coastal communities. Additionally, warmer ocean waters fuel stronger storms, intensifying extreme weather events such as hurricanes and typhoons, which in turn cause severe coastal flooding and infrastructure damage.

Sea-level rise, a result of thermal expansion and the melting of polar ice caps and glaciers, poses a direct threat to coastal communities and ecosystems. Low-lying regions and island nations face the prospect of displacement and land loss, while coastal wetlands, vital for natural flood defense and wildlife habitats, are at risk of submersion. These changes amplify the vulnerabilities of populations living in affected areas and demand urgent adaptation measures.

Ocean acidification presents another major concern. As the ocean absorbs increasing amounts of atmospheric carbon dioxide (CO_2), chemical reactions lower the water's pH, disrupting the ability of calcifying organisms such as mollusks and corals to build their shells and skeletons. This process undermines entire marine ecosystems and has cascading effects on biodiversity, fisheries, and coastal protection systems. The economic and ecological implications of this phenomenon are particularly severe for regions reliant on shellfish and coral reef tourism.

Marine pollution remains a pervasive problem. Vast quantities of plastic waste, including microplastics, enter the ocean each year, harming marine life through ingestion and entanglement while introducing toxic chemicals into food chains. Runoff from agricultural and industrial sources contributes to chemical contamination and nutrient overloading, which create oxygen-depleted "dead zones" in coastal areas, rendering them uninhabitable for marine species. Oil spills, though less frequent, cause devastating and long-lasting impacts on marine ecosystems.

Overfishing has led to the depletion of key fish stocks, threatening food security and ecosystem stability. Unsustainable practices such as bottom trawling and dynamite fishing exacerbate habitat destruction, particularly in sensitive environments like coral reefs and seagrass beds. While sustainable fishing initiatives and marine protected areas (MPAs) offer hope, the persistence of illegal, unreported, and unregulated (IUU) fishing undermines conservation efforts and global fisheries management.

The role of the ocean in regulating Earth's climate system is increasingly understood to be under threat. Large-scale ocean circulation systems, such as the Atlantic Meridional Overturning Circulation (AMOC), are critical for heat distribution between the tropics and higher latitudes. Emerging evidence suggests that melting ice sheets and freshwater input into the North Atlantic may weaken these systems, potentially triggering severe climate disruptions, particularly in Europe and North America.

Efforts to address these challenges are gaining momentum through international collaboration. Organizations such as the United Nations are promoting ocean conservation through initiatives like Sustainable Development Goal 14, which aims to protect marine resources and ecosystems. Technological advances, such as satellite monitoring, have enhanced the ability to track changes in sea level, ocean temperature, and circulation patterns, supporting more effective policy decisions. Projects like Seabed 2030 aim to map the ocean floor entirely, offering new insights into ocean dynamics and biodiversity.

Oceans also serve as the globe's largest carbon sink, absorbing roughly 30% of human-generated CO_2. Blue carbon ecosystems, including mangroves, seagrasses, and salt marshes, play a vital role in carbon sequestration, storing carbon at rates far exceeding terrestrial forests. Protecting and restoring these ecosystems is essential to mitigating climate change, as their degradation releases stored carbon, accelerating global warming.

While the challenges facing the oceans are daunting, they also present an opportunity to advance scientific understanding and foster global cooperation. By prioritizing marine conservation and sustainable practices, humanity can mitigate these threats, preserve the ocean's critical role in Earth's climate system, and ensure the resilience of marine ecosystems for future generations.

5.5.3 *Ocean Temperature*

Ocean temperature varies with depth, season, and latitude (see Fig. 5.5). In the tropics, the ocean surface remains warm year-round, with temperatures varying seasonally between 1 and 2 °C approximately. At mid-latitudes, mid-ocean temperatures fluctuate around 8 ° C throughout the year. At polar latitudes, the surface temperature remains close to the freezing point of seawater, approximately −1.9 ° C.

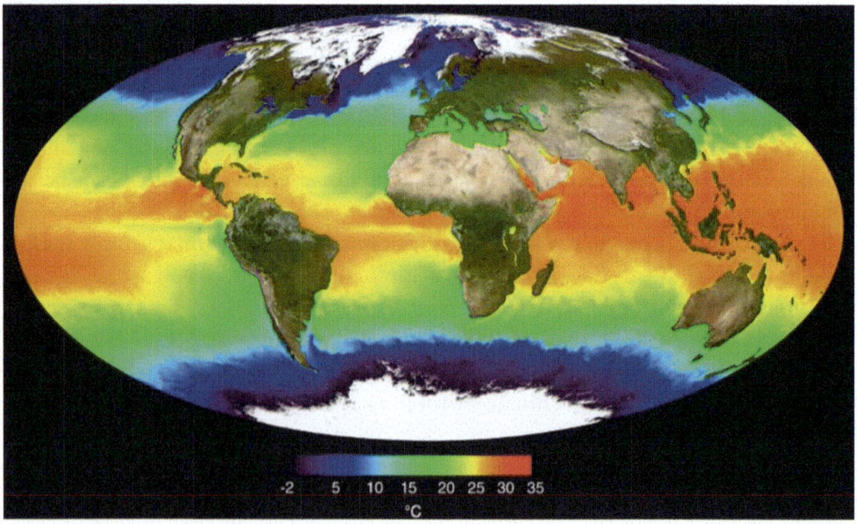

Fig. 5.5 Global distribution of sea surface temperature (SST). (Source: NASA)

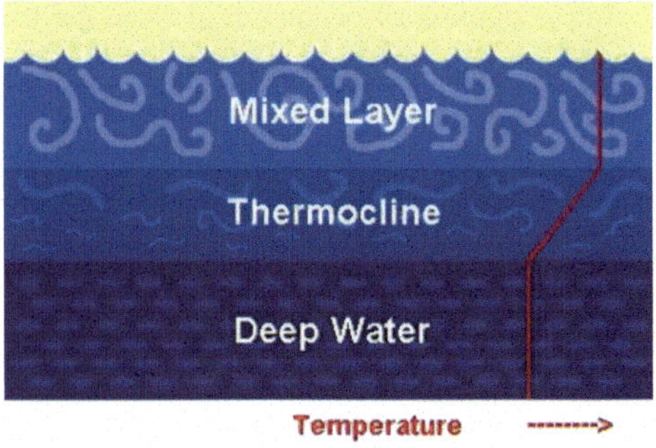

Fig. 5.6 Vertical structure of the ocean and distribution of temperature in the thermocline

Sunlight heats the upper layer of ocean water, known as the mixed layer. Below this layer, at depths greater than 200 m, the temperature decreases rapidly in a zone called the thermocline. The thermocline serves as the transition layer between the well-mixed surface layer and the colder, deeper layers of the ocean. In this zone, the temperature decreases sharply with increasing depth (see Fig. 5.6). The temperature gradient within the thermocline can vary based on location, season, and local environmental conditions.

The depth of the thermocline can range significantly. In some bodies of water, it may be only a few meters deep, while in others, it may extend to much greater depths. Its depth and intensity often change seasonally: during warmer months, the surface layer warms, and the thermocline can become more pronounced, whereas in colder months, the surface layer cools, and the thermocline may become less distinct.

The thermocline plays a crucial role in the distribution of heat and nutrients in aquatic environments, influencing the vertical movement of water and the distribution of marine life. For instance, certain fish species may prefer habitats above or below the thermocline. In the context of ocean circulation, the thermocline is a critical feature. It separates the warm surface layer from the cooler, deeper waters and affects the vertical movement of water masses. Understanding the thermocline is important for oceanography, limnology (the study of lakes), and fisheries management. Researchers study the thermocline to better grasp the dynamics of aquatic ecosystems and the impact of climate change on water temperatures.

Ocean temperatures exhibit seasonal variation: they warm slightly in the summer and cool slightly in the winter. These temperature changes are less pronounced than on land due to water's slower heat absorption and release compared to land. Ocean temperatures range from (25 – 28 ° C) at the equator, (12 – 17 ° C) in temperate zones, and are generally colder than (10 ° C) at extreme northern and southern latitudes.

5.5.4 Ocean Depth and Thermocline

The thermocline plays a crucial role in the distribution of heat and nutrients in aquatic environments, influencing the vertical movement of water and the distribution of marine life. For instance, certain fish species may prefer habitats above or below the thermocline.

In the context of ocean circulation, the thermocline is a critical feature. It separates the warm surface layer from the cooler, deeper waters and affects the vertical movement of water masses. Understanding the thermocline is important for oceanography, limnology (the study of lakes), and fisheries management. Researchers study the thermocline to better grasp the dynamics of aquatic ecosystems and the impact of climate change on water temperatures.

Ocean temperatures exhibit seasonal variation: they warm slightly in the summer and cool slightly in the winter. These temperature changes are less pronounced than on land due to water's slower heat absorption and release compared to land. Ocean temperatures range from 25 to 28 °C at the equator, 12–17 °C in temperate zones, and are generally colder than 10 °C at extreme northern and southern latitudes.

5.5.5 *Ocean Salinity*

Ocean salinity refers to the concentration of dissolved salts in seawater, typically expressed in parts per thousand (ppt) or practical salinity units (PSU). The average salinity of the world's oceans is approximately 3.5%, meaning that, on average, about 35 g of salt is dissolved in 1000 g (1 kg) of seawater.

Salinity is not uniform across the oceans and can vary depending on location (see Fig. 5.7). Factors influencing ocean salinity include precipitation, evaporation, river runoff, and melting ice. For instance, areas with high evaporation and low

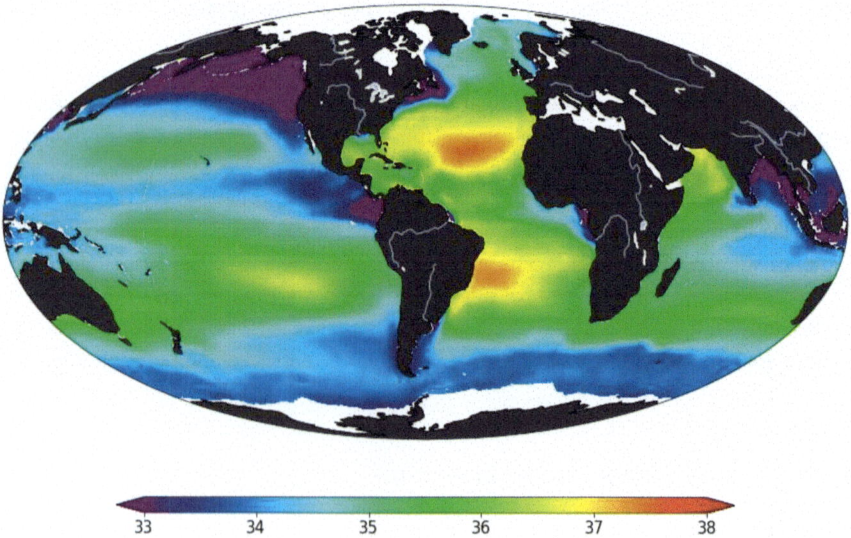

Fig. 5.7 Global distribution of ocean salinity: orange colors indicate regions with high salinity due to water evaporation leaving the salt behind. Areas of low salinity are shown in blue and purple, representing regions with high river runoff, significant rainfall, and melting ice

precipitation, such as the subtropics, generally have higher salinity. In contrast, regions with high precipitation and significant freshwater input, such as near the equator or at high latitudes, tend to have lower salinity.

Salinity plays a crucial role in ocean circulation, which in turn affects global climate. Salt dissolved in seawater influences currents that distribute heat and carbon around the globe. Thus, salinity not only impacts the health of marine environments but also contributes to climate regulation by driving global ocean circulation.

Ocean salinity is crucial in Earth's climate system and ocean circulation. Changes in salinity can affect the density of seawater, influencing ocean currents and the global distribution of heat. Scientists monitor ocean salinity to understand and predict shifts in climate and ocean circulation patterns. Salts from the lithosphere dissolve into the ocean as rivers flow over land and carry these salts to the sea. The salinity of the oceans typically ranges from 3.4% to 3.7%.

5.5.6 *Ocean Circulation*

Ocean circulation refers to the large-scale movement of seawater in Earth's oceans, playing a crucial role in regulating the planet's climate by redistributing heat, nutrients, and gases. Ocean currents can be warm or cold, depending on their origin and direction of flow. These currents are driven by wind patterns, the Earth's rotation (Coriolis effect), and variations in water temperature and salinity. Warm currents, originating in tropical regions, transport heat toward higher latitudes, significantly influencing coastal climates by raising temperatures. Notable examples include the Gulf Stream in the Atlantic Ocean and the Kuroshio Current in the Pacific Ocean.

In contrast, cold currents flow from polar regions toward the equator, cooling the air above them and often enhancing marine productivity through nutrient upwelling. Examples include the California Current and the Peru Current. The interplay of warm and cold currents regulates global temperatures, influences weather patterns, and supports marine ecosystems.

Ocean circulation is categorized into two main types: surface currents, primarily driven by wind, and deep currents, driven by density differences in seawater (thermohaline circulation) (Broecker, 1991; Ganachaud & Wunsch, 2000). Together, these systems illustrate the intricate connections between the oceans and the atmosphere, underscoring their importance in Earth's climate system.

- **Surface Currents**: Driven primarily by winds, these currents are influenced by the major wind belts—trade winds, westerlies, and polar easterlies. The Coriolis effect, due to Earth's rotation, also impacts surface currents. In the Northern Hemisphere, surface currents generally flow clockwise, while in the Southern Hemisphere, they flow counterclockwise. These currents move horizontally and are shaped by global wind patterns.

- **Deep Currents (Thermohaline Circulation)**: Deep currents, driven by differences in temperature (thermo) and salinity (haline), are a crucial component of ocean circulation. The primary driver of these currents is the sinking of dense, cold, and salty water in polar regions, which initiates a continuous global circulation pattern (see Fig. 5.8). In polar regions, the surface water becomes cold and salty, increasing its density. This dense water sinks into the deep ocean, flows along the ocean floor, and eventually rises to the surface in other regions. This process redistributes heat and nutrients globally, playing a vital role in regulating the Earth's climate.
- **Coastal Currents:** Local features such as coastlines, bays, and estuaries can influence the direction and strength of nearshore currents. Additionally, tides caused by the gravitational pull of the Moon and Sun create currents that move in and out of coastal areas.

Impact of Ocean Circulation: Ocean circulation profoundly affects climate, weather patterns, and marine ecosystems. It transports heat from the equator to the poles and influences regional climate variations. The redistribution of nutrients by ocean currents supports marine life and impacts ecosystems. Changes in ocean circulation patterns can significantly affect weather, sea level, and global ecosystems. Ocean currents are continuously moving seawater in specific directions. The combined effect of all these currents constitutes ocean circulation. Due to the Coriolis effect, ocean currents influenced by prevailing winds move clockwise in the

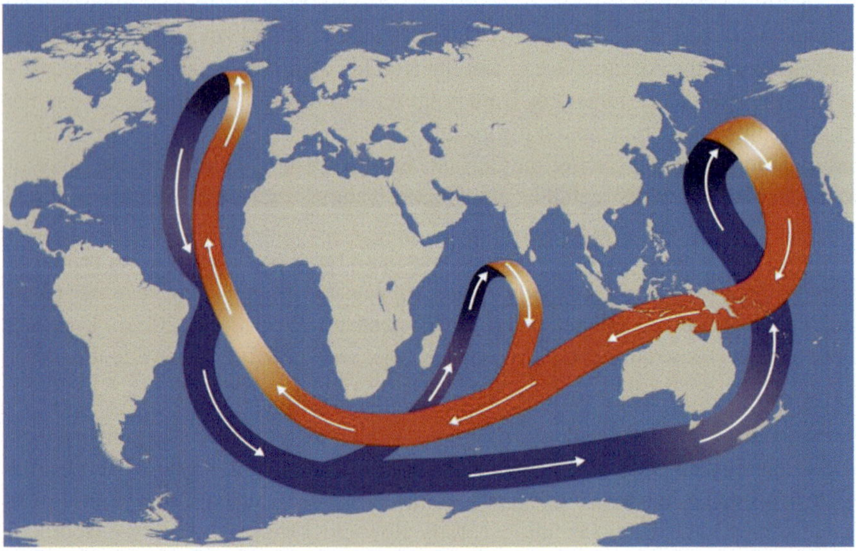

Fig. 5.8 Thermohaline circulation: a key component of Earth's climate system, the thermohaline circulation plays a significant role in regulating global climate by influencing the movement of heat and nutrients across the oceans

Northern Hemisphere and counterclockwise in the Southern Hemisphere. This creates large circular currents, or gyres, in ocean basins.

5.6 Cryosphere

The cryosphere encompasses all parts of the Earth's surface where water exists in its solid form. This includes both permanent and transient ice and snow, which play crucial roles in the Earth's climate system and various environmental processes.

Permanent Ice and Snow: This category includes glaciers, ice caps, and polar ice sheets that cover significant portions of the Earth's surface. The Antarctic and Greenland ice sheets are the largest examples, holding about 99% of the Earth's freshwater ice (Fretwell et al., 2013). These ice masses influence global sea levels and climate patterns through processes such as ice-albedo feedback and freshwater fluxes into the ocean.

Hydrometeors: These are transient forms of ice, including snow, sleet, hail, and frost, which result from meteorological processes. Hydrometeors contribute to the Earth's hydrological cycle by affecting precipitation patterns and water availability in different regions. Snow and ice accumulation in mountain regions, for instance, is crucial for sustaining river flows and replenishing groundwater resources (Mote et al., 2018).

The cryosphere plays a pivotal role in the Earth's climate system. It influences global sea levels, affects weather patterns, and impacts various ecosystems. By reflecting solar radiation through its high albedo, the cryosphere helps regulate the planet's temperature. Additionally, changes in the cryosphere, such as melting ice, influence ocean currents and atmospheric circulation, which can have far-reaching effects on global climate and weather patterns (Vaughan et al., 2013).

5.6.1 Key Components of the Cryosphere

Glaciers are large masses of crystalline ice, snow, rocks, sediments, and sometimes liquid water that originate from land and move downhill under the influence of their weight (see Fig. 5.9). They are found on every continent and serve as a primary source of freshwater for many rivers and lakes. Glaciers also contribute to sea-level changes through their advance and retreat.

Ice Sheets: Ice sheets are vast continental glaciers that cover large areas of the Earth, such as the Antarctic and Greenland ice sheets (see Fig. 5.10). They contain the majority of Earth's freshwater ice and have a significant impact on sea levels.

Ice Caps and Ice Areas: Ice caps are extensive masses of glacial ice that cover less than 50,000 km^2 of land area, making them smaller than ice sheets but larger than alpine glaciers (Fig. 5.11). These formations are typically found in polar and subpolar regions, such as Antarctica and Greenland, as well as in some mountainous

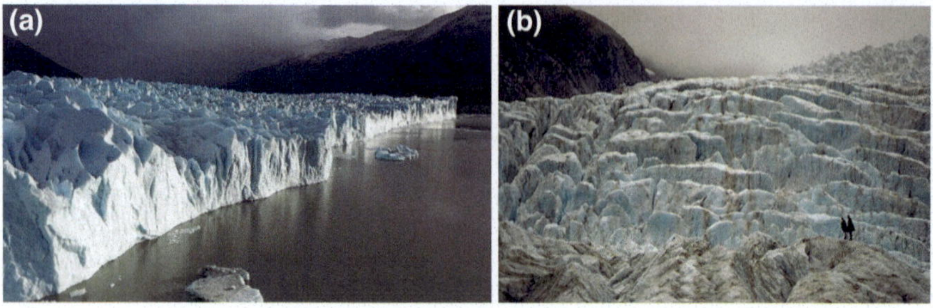

Fig. 5.9 (**a**) Argentina's melting glaciers symbolize the impact of climate change. (Credit: https://www.france24.com/); (**b**) photograph of a glacier. (Credit: James L. Amos/National Geographic Creative)

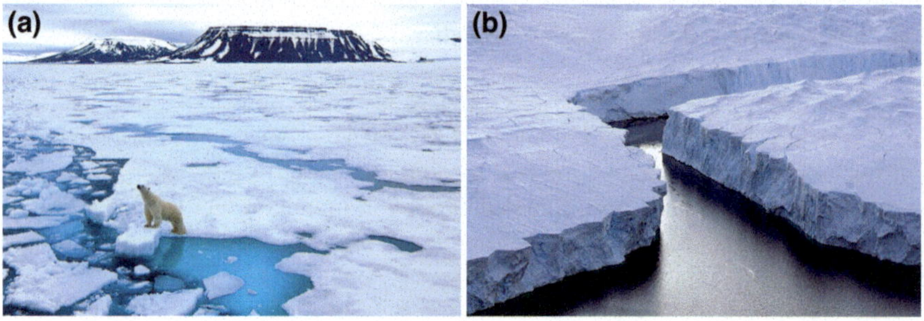

Fig. 5.10 (**a**) The Antarctic ice sheet, a critical component of Earth's climate system. (Credit: NASA); (**b**) the Greenland ice sheet, one of the largest sources of freshwater ice. (Credit: National Geographic)

Fig. 5.11 Ice cap. (Credit: National Park Service)

areas. Ice caps develop over thousands of years as snow accumulates and compacts into ice. The immense weight of the ice causes it to flow outward from the center, forming a dome-shaped mass.

Ice caps are primarily located in high-latitude regions, particularly near the poles. For instance, the Antarctic Ice Sheet dominates the continent of Antarctica, while the Greenland Ice Sheet covers the island of Greenland. Smaller ice caps are also

Fig. 5.12 Sea ice. (Credit: Copernicus)

present in other polar and mountainous regions worldwide. Like ice sheets, ice caps play a crucial role in regulating global sea levels. When ice caps lose mass due to melting or the calving of icebergs, it contributes to rising sea levels.

Changes in the size and mass of ice caps are often used as indicators of climate change. Monitoring the retreat or advance of ice caps provides valuable insights into temperature trends within specific regions.

Sea Ice: Sea ice refers to frozen seawater that forms and floats on the surface of the ocean. It plays a key role in the Earth's climate system and has various characteristics and functions. Sea ice forms on the surface of the ocean when seawater freezes (Fig. 5.12). As sea ice forms, it expels salt, creating pockets of brine in the ice. This process leads to the formation of relatively fresh ice. Sea ice typically undergoes seasonal changes, expanding during the winter and shrinking during the summer. In the Northern Hemisphere, sea ice reaches its maximum extent in late winter, while in the Southern Hemisphere, it peaks in late winter or early spring. Sea ice is most prominent in the polar regions, especially around Antarctica in the Southern Hemisphere and the Arctic Ocean in the Northern Hemisphere. These regions experience significant seasonal variations in sea ice coverage. It plays a role in regulating Earth's climate by reflecting sunlight and insulating the ocean.

Snow Cover: Snow cover refers to the extent of the Earth's surface covered with snow. The snowpack is a critical component of the Earth's climate system, with significant implications for ecosystems, weather patterns, and climate change. Satellites equipped with sensors, such as passive microwave and visible/infrared instruments, monitor snow cover globally. These observations provide information on the extent, duration, and thickness of snow cover. Snow depth, density, and water content are measured using instruments such as snow gauges, snow strips, battens, or solid metal rods.

Snow cover varies seasonally: In the Northern Hemisphere, it typically peaks in winter and is minimal in summer, while in the Southern Hemisphere, the pattern is reversed. The extent and duration of snow cover depend on factors such as latitude, altitude, and regional climate patterns. Freshly fallen snow has a high albedo, meaning it reflects a significant portion of incoming solar radiation (Fig. 5.13). This albedo effect contributes to regulating the Earth's temperature. When snow cover

Fig. 5.13 Fresh snow cover, in daylight during winter. (Credit: 123RF)

decreases and darker surfaces like bare ground or water are exposed, more solar energy is absorbed, leading to warming.

The snowpack acts as a natural reservoir, storing water in the form of snow during colder months and releasing it during warmer months, thereby contributing to river flow and water availability. Changes in snowpack can impact water resources for various uses, including agriculture, industry, and domestic consumption. Many plant and animal species are adapted to seasonal changes in snow cover. Variations in weather and the duration of snow cover can affect ecosystems, influencing the behavior, migration patterns, and survival of species.

Climate change has led to alterations in snow cover patterns. Warmer temperatures may reduce the extent and thickness of snowpacks, particularly at lower elevations, contributing to the global phenomenon of glacial retreat. In Fig. 5.14, the global distribution of snow cover for January 2023 is shown. During this month, which is summer in the Southern Hemisphere, snow cover is observed only in the extreme northern parts of Antarctica around the South Pole. Conversely, January represents winter in the Northern Hemisphere, characterized by extensive snow cover.

The timing and rate of snowmelt significantly impact the seasonal runoff of rivers and streams. Changes in snowmelt patterns can have profound implications for water availability downstream, potentially leading to issues such as flooding or drought. Scientists use indices such as snow cover extent (SCE) and snow water equivalent (SWE) to quantify and analyze snow cover. SCE measures the area covered by snow, while SWE represents the amount of water contained in the snowpack.

Understanding and monitoring snow cover is crucial for assessing the impacts of climate change, managing water resources, and studying the interconnectedness of Earth's systems. Ongoing research and advancements in technology continue to enhance our understanding of snowpack dynamics. Seasonal snowfall plays a significant role in the cryosphere, affecting temperature regulation, freshwater supply, and various human activities, including winter sports.

Permafrost: Permafrost is soil, sediment, or rock that remains at or below freezing temperatures for at least two consecutive years. It is commonly found in

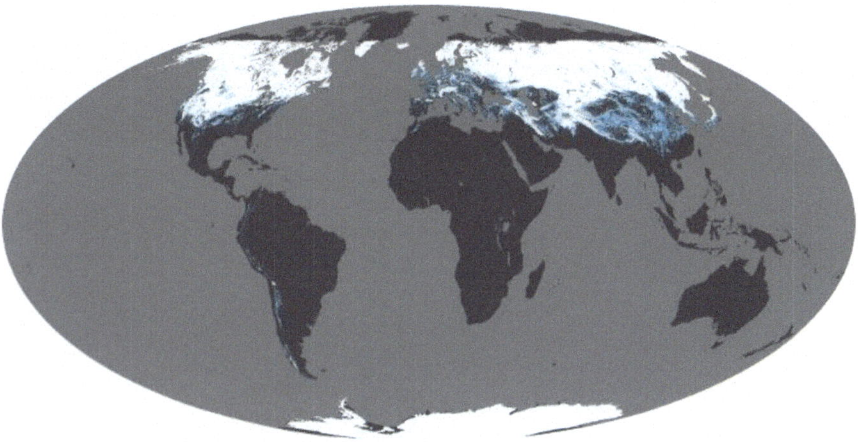

Fig. 5.14 Global snow cover for January 2023. (Source: NASA)

polar and high-altitude regions where temperatures remain consistently low. Permafrost is characterized by its temperature, not by the presence of snow or ice, and contains significant amounts of organic carbon. It comprises both ice and unfrozen water.

Permafrost typically has a layered structure. The top layer, known as the active layer, thaws and refreezes annually, while the deeper layer remains permanently frozen. Permafrost is present in both the Northern and Southern Hemispheres but is more extensive in the Northern Hemisphere.

It covers large areas of Alaska, Canada, Siberia, and Northern Europe (Fig. 5.15). Permafrost is predominantly found in high-latitude regions and is particularly sensitive to climate change.

Changes in permafrost due to rising temperatures can have significant impacts, including the release of stored carbon, alterations in hydrology, and impacts on ecosystems. The thawing of permafrost can contribute to global warming through the release of greenhouse gases and can affect infrastructure in permafrost regions.

5.6.2 The Cryosphere and the Climate System

The cryosphere plays a crucial role in the Earth's climate system by interacting with other components, including the atmosphere, oceans, and biosphere. It significantly influences weather patterns, ocean circulation, and global climate. Key interactions include:

Weather Patterns: The cryosphere impacts atmospheric conditions through its high albedo effect, which reflects sunlight and helps regulate temperatures. Changes in snow cover and ice extent can influence local and global weather patterns.

Ocean Circulation: Sea ice affects ocean circulation by influencing the exchange of heat and moisture between the ocean and the atmosphere. The formation and

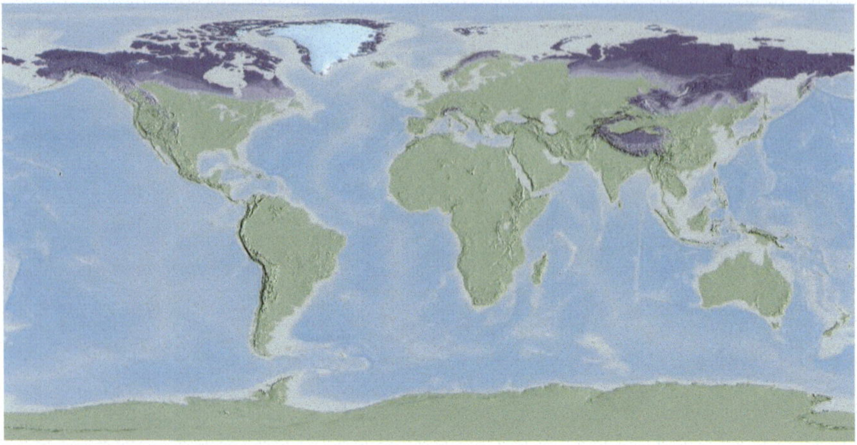

Fig. 5.15 Permafrost regions are depicted in darker shades of lavender over the light green continents, while the Greenland ice sheet is shown in light blue

melting of sea ice contribute to the thermohaline circulation, which redistributes heat around the globe.

Global Climate: Melting glaciers and ice sheets contribute to rising sea levels, which can lead to coastal erosion and flooding. Additionally, the loss of ice reduces the Earth's albedo, causing more solar energy to be absorbed by the oceans and land, further accelerating global warming.

A major concern related to the cryosphere is its response to climate change. Rising global temperatures are causing glaciers and ice sheets to melt at an accelerated rate. This contributes to rising sea levels and alters regional climate patterns. The melting of permafrost also releases stored carbon, further enhancing the greenhouse effect and potentially leading to more significant climate changes.

Understanding and monitoring the cryosphere are essential for predicting the impacts of climate change and making informed decisions on mitigation and adaptation strategies. Scientists utilize various tools and technologies, including remote sensing and field measurements, to study the cryosphere and its role in Earth's climate system.

References

Aber, J. D., Melillo, J. M., & Steudler, P. A. (1995). Soil and water impacts of land use changes in the northeast. *Ecological Applications, 5*(4), 1350–1360.

Broecker, W. S. (1991). The great ocean conveyor. *Oceanography, 4*(2), 79–89.

Buckley, M. A., & Marshall, J. (2016). Observations, simulations, and mechanisms of the Atlantic Meridional Overturning Circulation. *Annual Review of Marine Science, 8*, 69–94.

Cane, M. A. (2005). The evolution of El Niño, past and future. *Earth and Planetary Science Letters, 230*(3–4), 227–240.

Chahine, M. T. (1992). The hydrological cycle and its influence on climate. *Nature, 359*(6394), 373–380.

Dunne, T., & Leopold, L. B. (1978). *Water in environmental planning*. W. H. Freeman and Company.

Fretwell, P., et al. (2013). Bedmap2: Improved ice bed, surface, and thickness datasets for Antarctica. *The Cryosphere, 7*(1), 375–393.

Ganachaud, A., & Wunsch, C. (2000). Large-scale ocean circulation and the meridional overturning circulation. *Ocean Circulation and Climate*, 35–54.

Gleick, P. H. (1993). *Water in crisis: A guide to the world's fresh water resources*. Oxford University Press.

Gordon, A. L., et al. (2000). The Southern Ocean's role in the climate of the 21st century. *Nature, 406*(6794), 157–163.

Hillel, D. (1998). *Environmental soil physics*. Academic Press.

Huntington, T. G. (2006). Evidence for intensification of the global water cycle: Review and synthesis. *Journal of Hydrology, 319*(1–4), 83–95.

Lenton, T. M., & Watson, A. J. (2011). *Revolutions that made the Earth*. Oxford University Press.

Miller, J. R., & Smith, J. A. (1986). Fluvial processes and sediment transport. *Hydrological Processes, 1*(1), 87–95.

Mote, P. W., et al. (2018). Mountain snowpack in the western United States: Impacts and projections. *Wiley Interdisciplinary Reviews: Climate Change, 9*(2), e507.

Niemi, T., Makkonen, L., & Koivusalo, H. (1990). The influence of topography on runoff and erosion. *Geographical Annals: Series A, 72*(3), 175–186.

Oki, T., & Kanae, S. (2006). Global hydrological cycles and world water resources. *Science, 313*(5790), 1068–1072.

Petts, G. E. (1984). *Impounded rivers: Perspectives for management*. Wiley.

Pidwirny, M. (2006). The hydrosphere. In *Fundamentals of physical geography* (2nd ed.). University of British Columbia Okanagan. http://www.physicalgeography.net/fundamentals/8b.html

Saji, N. H., et al. (1999). A dipole mode in the tropical Indian Ocean. *Nature, 401*(6751), 360–363.

Saunders, P. M., & Jones, D. A. (2009). The Gulf Stream and the North Atlantic Oscillation. *Journal of Climate, 22*(14), 3720–3729.

Schneider, T., Kaul, C. M., & Pressel, K. G. (2017). Possible climate transitions from breakup of stratocumulus decks under greenhouse warming. *Nature Geoscience, 10*(8), 614–619.

Screen, J. A., & Simmonds, I. (2010). The central role of diminishing sea ice in recent Arctic temperature amplification. *Nature, 464*(7293), 1334–1337.

Shiklomanov, I. A. (1993). World fresh water resources. In P. H. Gleick (Ed.), *Water in crisis: A guide to the world's fresh water resources*. Oxford University Press.

Smith, R. L. (2002). *Freshwater ecosystems*. Cambridge University Press.

Stroeve, J. C., et al. (2012). The Arctic's rapidly shrinking sea ice cover: A research review. *Climatic Change, 110*(3–4), 1005–1027.

Vaughan, D. G., et al. (2013). Observations: Cryosphere. In *Climate change 2013: The physical science basis* (pp. 317–382). Cambridge University Press.

Wetzel, R. G. (2001). *Limnology: Lake and river ecosystems*. Academic Press.

Chapter 6
Atmosphere: The Vital Layer of the Globe

6.1 Origin of the Atmosphere

The Sun and the planets of the solar system are thought to have formed approximately 4.6 billion years ago. This estimation is derived from analyses of radioactive isotopes with long half-lives, which provide valuable insights into the age of these celestial bodies (Halliday, 2001). During this early stage, depicted in Fig. 6.1, which

© The Author(s), under exclusive license to Springer Nature Switzerland AG 2025
V. Spiridonov et al., *Atmospheric Perspectives*,
https://doi.org/10.1007/978-3-031-86757-6_6

Fig. 6.1 A conceptual illustration of a protoplanetary disk, the birthplace of planets, moons, and asteroids around a forming star. (Credit: https://en.wikipedia.org/wiki/Formation_and_evolution_of_the_Solar_System)

shows an artist's representation of the early stages of planetary formation around a young star, the solar system emerged from a rotating disk of gas and dust.

The Earth's atmosphere likely began forming during this period, as volatile compounds were released from its interior through volcanic activity and outgassing processes. These events played a crucial role in shaping the planet's early environment (Canup & Righter, 2000; Ćurić & Spiridonov, 2023).

6.1.1 Formation of the First Atmosphere

Earth's atmosphere is notably distinct from the other planets in the solar system, and certainly from the Sun. It is hypothesized that Earth formed through the accretion of solid matter fragments. Earth's first atmosphere depicted in Fig. 6.2, likely consisted mainly of hydrogen and helium, the primary gaseous components of the dusty envelope surrounding the Sun from which the planets, including Earth, originated (Lissauer & de Pater, 2013). These light gases, however, were lost to space early in Earth's history due to the weak gravitational pull and solar wind during this period (Lambeck, 1991).

However, the initial Earth and its surrounding atmosphere were extremely hot. In such a high-temperature environment, the hydrogen and helium molecules, being very light, moved rapidly and eventually escaped Earth's gravitational pull, drifting into outer space. Several factors contributed to the loss of this primordial atmosphere:

(a) *Insufficient Gravitational Force*: The relatively small Earth lacked the gravitational strength to retain such light gases, especially given the high temperatures of the early planet.
(b) *Cataclysmic Collision*: Approximately 30 million years after Earth's formation, it is believed that a large object, roughly the size of Mars, collided with Earth.

Fig. 6.2 The initial atmosphere. An illustration depicting Earth's first atmosphere, primarily composed of hydrogen and helium. The image captures the early stage of Earth's development, before the loss of these light gases into space. (Illustration by Peter Sawyer © Smithsonian Institution)

Fig. 6.3 Formation of the Moon. An illustration showing the Moon's formation from debris following a massive collision with early Earth

This impact is thought to have initiated the formation of the Moon and led to the loss of Earth's original hydrogen–helium atmosphere (Fig. 6.3).

Many scientists, drawing on their understanding of the gases present in the universe, support the thesis that Earth's first atmosphere was composed of helium, hydrogen, ammonia, and methane. However, another school of thought suggests that the first atmosphere likely consisted of gases released by volcanic activity from Earth's hot inner core. This group of gases included nitrogen (N_2), carbon dioxide (CO_2), hydrogen (H_2), carbon monoxide (CO), and water vapor (H_2O).

6.1.2 Second Atmosphere

During this period, Earth's surface was extremely hot, with widespread volcanic activity and constant bombardment by celestial objects of various sizes. As the planet gradually cooled, the second atmosphere began to take shape. Increased volcanic activity released additional gases into the atmosphere. This second, denser atmosphere consisted of gases expelled from the hot interior through volcanic eruptions or fissures in the Earth's crust (Fig. 6.4). The composition of the second atmosphere was predominantly water vapor (about 80–90%), carbon dioxide (6–12%), and sulfur dioxide (1–2%), with smaller concentrations of hydrogen, carbon monoxide, hydrogen sulfide, methane, and molecular nitrogen. The composition of gases released from Earth's interior today is believed to be similar to what it once was. However, in the past, hydrogen, carbon monoxide, methane, and other greenhouse gases were likely more abundant due to the lower oxidation state of Earth's mantle.

Water vapor played a crucial role during this era, as its condensation led to the formation of oceans on Earth's surface. The presence of liquid water enabled the absorption of carbon dioxide through a process known as carbonization, where CO_2 dissolved in water and precipitated as carbonate minerals. Over time, the second atmosphere evolved, leading to a composition more closely resembling what we see today, with nitrogen and oxygen becoming the dominant gases. The development of an oxygen-rich atmosphere is closely linked to the emergence of photosynthetic organisms, such as cyanobacteria, which began producing oxygen as a byproduct of photosynthesis. This process, along with volcanic outgassing and cooling, contributed to the formation of Earth's second atmosphere, setting the stage for the evolution of life as we know it. It is believed that Earth's second atmosphere resulted from heating and planetary differentiation. This atmosphere likely consisted of gases like those released by modern volcanoes: carbon dioxide, nitrogen, water vapor, hydrogen, and other trace gases. Planetary differentiation caused lighter elements to migrate to the upper layers of Earth, allowing these lighter gases to escape into space. Consequently, the second atmosphere contained little to no oxygen. In

Fig. 6.4 The second atmosphere. As a result of volcanic activity and the formation of oceans, the second atmosphere had a different composition than the initial atmosphere. It was primarily composed of nitrogen, carbon dioxide, water vapor, and traces of other gases

addition, the lack of an ozone layer meant that ultraviolet radiation reached the Earth's surface directly. However, with the influx of water vapor and the subsequent cooling of the planet, clouds and oceans began to form. This gradual transition from a volatile environment to one more hospitable to life marked a crucial step in the development of Earth's atmosphere.

6.1.3 Earth's Third Atmosphere

During the first two billion years of Earth's evolution, the atmosphere contained only trace amounts of oxygen, likely produced by the dissociation of water molecules (H_2O) due to solar radiation. Evidence of this early oxygen is found in the formation of iron oxides in some ancient rocks (Fig. 6.5). The presence of oxygen also led to the formation of the ozone layer, which helped reduce ultraviolet (UV) radiation reaching the Earth's surface. With the proliferation of photosynthetic organisms, such as cyanobacteria and early plants, oxygen levels began to rise significantly. This increase in oxygen was accompanied by the formation of more rust in rocks, indicating higher levels of oxygen. Between 2.5 billion years ago and about 500 million years ago, oxygen concentrations approached near-present levels. As oxygen increased, carbon dioxide (CO_2) levels decreased due to several factors:

Photosynthesis: Photosynthetic organisms used CO_2 to produce organic matter, some of which was removed from the system when it fell to the ocean floor or became part of sedimentary rocks. Chemical Weathering: This process involves the decomposition of rocks on Earth's surface through rainwater, extreme temperatures, and biological activity. Initially occurring without life, this process was later accelerated by living organisms. Marine organisms incorporated carbonate into their shells, which, upon their death, sank to the ocean floor, thereby sequestering CO_2 for extended periods. Additionally, sulfur compounds were removed from the

Fig. 6.5 Formation of rust in rocks. The presence of rust in ancient rocks indicates the existence of small amounts of oxygen in Earth's early atmosphere

atmosphere through acid rain and deposited on Earth's surface as sulfates. Nitrogen (N_2) levels increased slowly and steadily, as it is relatively inert and not easily removed from the atmosphere. During the period of the second atmosphere, the Sun was approximately 30% fainter than it is today. This raises a critical question: why didn't the Earth freeze under such conditions? The answer lies in the atmospheric composition and geophysical processes at the time. Volcanic activity continued to release large amounts of CO_2 into the atmosphere. With limited liquid water available, the chemical weathering process—which typically removes CO_2 from the atmosphere—was significantly reduced. As a result, CO_2 levels increased, amplifying the greenhouse effect until it became strong enough to warm the Earth. Once the warming began, it likely occurred rapidly. The significant accumulation of oxygen in Earth's atmosphere occurred later, driven by photosynthesis from early microscopic organisms like cyanobacteria. This marked the transition to the third atmosphere, characterized by the presence of oxygen. These processes unfolded over vast geological timescales, reflecting the complex interplay between geological and biological factors. Scientists rely on geological and fossil records to understand how Earth's atmosphere has evolved over billions of years.

The formation of today's atmosphere shown in Fig. 6.6. followed a similar process to that of the second atmosphere. However, to fully grasp this, the atmosphere should be viewed as part of an interconnected system that includes the hydrosphere (the total mass of water on Earth's surface), the biosphere (the realm of living organisms), and the lithosphere (the solid, sedimentary part of the Earth's crust). The estimated mass of the Earth is $5.198 \cdot 10^{24}$ kg, while the mass of the atmosphere is a million times less and about 300 times less than the mass of the hydrosphere. The total mass of volatile compounds on Earth's surface is only 0.0025% of Earth's total mass. The current atmosphere has been relatively stable for about a billion years.

Fig. 6.6 Atmosphere: Earth's protective covering. Sunset over the Indian Ocean as seen by astronauts aboard the International Space Station (ISS). The image offers a stunning view of Earth's atmosphere from orbit. The curvature of the Earth is visible along the horizon, or limb, which extends from the center-left to the lower right corner. Above Earth's shadowed surface, a brilliant spectrum of colors delineates the various layers of the atmosphere. (Credit: NASA)

6.2 Basic Characteristics of the Atmosphere

The Earth's atmosphere is a thin but vast gaseous envelope that surrounds the globe, held in place by gravity. It plays a critical role in regulating energy exchange between the Sun, the Earth's surface, and different regions, which is essential for maintaining thermal equilibrium and shaping the planet's climate (Wallace & Hobbs, 2006). Earth's atmosphere is unique in its ability to support life, thanks to complex interactions with the ocean and other processes that create a favorable environment for the biosphere (Hartmann, 1994). If the Earth were reduced to the size of an apple, the atmosphere's thickness would be less than the apple's skin. Nearly 99% of the atmosphere's total mass is concentrated within the first 30 km above the surface (Andrews, 2010), with no clearly defined upper boundary as it gradually thins out into space (Vallis, 2017). Without the atmosphere, life as we know it would not exist: the sky would be dark, the Sun would appear white, and there would be no clouds, precipitation, winds, or weather. In an atmosphereless world, temperatures would vary drastically between day and night, lakes and oceans would vanish, fire could not burn, and sound would not travel (Salby, 2012).

The atmosphere is a dynamic fluid medium, subject to movements ranging from small-scale turbulence to global circulations (Spiridonov & Ćurić, 2021). These air movements interact with key components such as water vapor, ozone, and clouds, which significantly influence the Earth's energy balance and climate system. The atmosphere's unique composition and dynamic behavior make it indispensable for sustaining life on Earth.

6.2.1 Main Factors Determining the Behavior of the Atmosphere

Gravity is the most important factor influencing the behavior of the atmosphere, as it holds the Earth's gaseous envelope in place despite the lack of a solid upper boundary. The Earth's gravitational field prevents the atmosphere from escaping into space and shapes many of its characteristics. One such characteristic is the atmosphere's geometry: about 75% of its mass is compressed into a layer just 10 km thick, which is less than 1% of the Earth's radius. This gravitational compression results in the stratification of the atmosphere, with its mass and properties arranged in layers by height.

Gravity also affects air movement, making large-scale motions predominantly quasi-horizontal, while vertical movements are much smaller by comparison. The stratification extends to atmospheric components like water vapor and ozone, which are distributed in horizontal layers, except during small-scale processes such as convection, where vertical motions become more comparable to horizontal ones. Air compressibility further complicates the atmosphere's behavior. Changes in pressure can cause air parcels to alter their volume, leading to shifts in the concentration of their constituents, even though the number of molecules remains constant. For instance, condensation reduces water vapor content in air passing through clouds, while photodissociation of oxygen molecules increases ozone levels in air exposed to sunlight. Various other processes similarly alter the characteristics of air as it moves through different atmospheric conditions. In addition to gravity, the Earth's rotation significantly influences atmospheric movement and the spatial distribution of air properties. As the Earth rotates, it behaves as a non-inertial system, requiring adjustments to the laws of mechanics to account for rotational accelerations. These rotational forces are critical for large-scale movements, such as cyclones and anti-cyclones, and they drive meridional circulation, transferring heat and other properties between the equator and the poles. While gravity primarily causes vertical stratification, Earth's rotation tends to organize air characteristics in the meridional direction. It's important to note that these processes do not act in isolation but interact in complex ways. For example, heat transfer via radiation creates the thermal structure of the atmosphere, which in turn drives circulation patterns that redistribute radiation-active components like water vapor, clouds, and ozone. Such interactions complicate the study of atmospheric behavior, as the combined effects are often as significant as those of individual factors.

6.3 Chemical Composition of the Atmosphere

The atmosphere is a complex mixture of solid, liquid, and gaseous constituents. Gases in the atmosphere are typically classified as either constant, where their concentrations remain relatively stable over time, or variable, where their concentrations fluctuate due to various processes (Schumann, 2012). The stable, or permanent, gases include nitrogen (78.1%), oxygen (20.9%), and argon (0.934%), which together comprise the bulk of the atmosphere (Jacob, 1999). Other stable gases, present in much smaller amounts, include neon, helium, hydrogen, and krypton (Brasseur & Solomon, 2005).

These gases play a crucial role in maintaining atmospheric stability and supporting life on Earth. The remaining atmospheric gases, known as trace gases, are found in much smaller concentrations but are essential for many of the Earth's processes (Fig. 6.7). Trace gases include carbon dioxide, ozone, methane, and nitrous oxide—each of which has a significant impact on the Earth's climate and atmospheric chemistry. For instance, carbon dioxide and methane are potent greenhouse gases that contribute to global warming, while ozone plays a critical role in protecting life by absorbing harmful ultraviolet radiation from the Sun.

Although some atmospheric gases like water vapor and ozone exist in small quantities, they play a crucial role in the energy balance of the atmosphere, which in turn influences its dynamics. These gases are not uniformly distributed throughout the atmosphere; rather, they are created and dissipated locally, leading to significant variability in their concentrations. Given their importance in atmospheric radiation and chemical processes, it is essential to quantify these variable

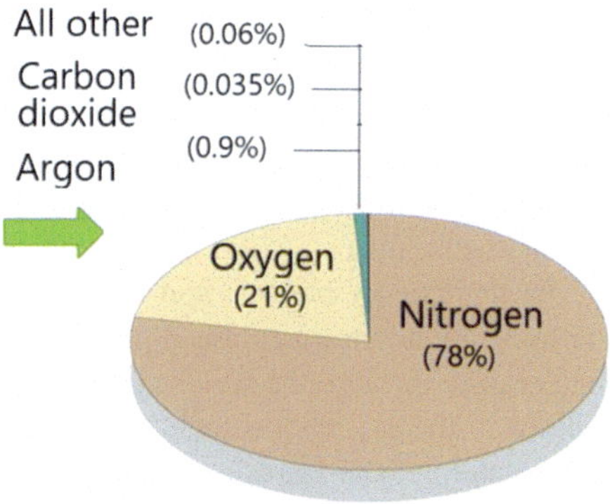

Fig. 6.7 The chemical composition of the Earth's atmosphere, showing the relative abundances of major and trace gases

components. This can be done in several ways. One method involves using the absolute measures of a given component, expressed through its density, ρ_i, or its number density (concentration), $[i]$, which is defined as:

$$[i] = \left(\frac{N_A}{M_i}\right)\rho_i \tag{6.1}$$

where N_A is Avogadro's number, and M_i is the molar mass. Other measures of absolute concentration include partial pressure p_i and partial volume V_i. However, due to the compressibility of air, absolute concentration can be ambiguous. In other words, the volume representation of a substance may not be expressed clearly. For this reason, relative measures—expressed about air or dry air—are often more useful. The relative concentration of the ith component is defined by the molar ratio N_i:

$$N_i = \frac{p_i}{p} = \frac{V_i}{V} \tag{6.2}$$

where p is the total pressure, and V is the corresponding volume. In meteorology, a common measure is the mixing ratio:

$$r_i = \frac{m_i}{m_s} \tag{6.3}$$

where the subscript s refers to dry air. If r_i represents water vapor in the troposphere, it is typically expressed in g/m^3, whereas stratospheric ozone is measured in parts per million (ppm) of dry air mass. In addition to gases, the atmosphere contains various aerosols—tiny particles of dust, soot, and ice—that can influence climate by reflecting sunlight or acting as nuclei for cloud formation. Sulfur compounds, fluorocarbons, and other particulates also contribute to atmospheric processes, often with complex and far-reaching effects on both climate and air quality. The Earth's atmosphere is not just a passive layer of gases but a dynamic system that plays a critical role in sustaining life and regulating the globe's climate. The primary gases, nitrogen and oxygen, are crucial for life and many processes occurring near the Earth's surface. Nitrogen is essential for building proteins in living organisms, while oxygen is vital for respiration. Alongside these primary gases, the atmosphere contains secondary gases that, though present in much smaller quantities, profoundly impact the Earth's system. These secondary gases include water vapor, carbon dioxide, methane, nitrous oxide, and ozone. These gases are particularly important for the Earth's energy balance and moisture distribution. For example, water vapor is the most abundant greenhouse gas and plays a key role in regulating the planet's temperature through the water cycle. Similarly, carbon dioxide is a significant contributor to the greenhouse effect, trapping heat and thereby influencing global climate patterns. The Earth's atmosphere is stratified into layers, each with distinct characteristics and properties based on temperature variations with altitude. Within these layers, gases are categorized as either constant or variable. Permanent gases, such as nitrogen, oxygen, argon, neon, and helium, remain relatively stable (Table 6.1). Variable gases, including water vapor, carbon dioxide, methane, nitrous oxide, ozone, and chlorofluorocarbons, can fluctuate in concentration due to natural processes and human activities (Table 6.2). These variable gases are crucial to understanding climate change, as even small changes in their concentrations can lead to significant shifts in the Earth's climate system. Moreover, the atmospheric composition has not remained static throughout Earth's history. Significant changes have occurred over geological time, particularly with the advent of photosynthetic life, which drastically increased oxygen levels and reduced carbon dioxide concentrations. Today, human activities, such as fossil fuel combustion and deforestation, are causing rapid changes in the concentrations of key atmospheric gases, leading to a range of environmental challenges, including global warming, ocean acidification, and loss of biodiversity.

Table 6.1 Permanent gases in the atmosphere and their relative abundances

Chemical element/gas	Symbol	Percentage volume representation in air
Nitrogen	N_2	78.08%
Oxygen	O_2	20.95%
Argon	A_r	0.93%
Neon	N_e	Trace
Helium	H_e	Trace
Krypton	K_r	Trace
Hydrogen	H_2	Trace

Table 6.2 Variable gases in the atmosphere

Chemical element/gas	Symbol	Percentage volume representation in air
Water	H_2O	0–4% (varies)
Carbon dioxide (CO_2)	CO_2	~0.038% (or 380 ppmv)
Methane (CH_4)	CH_4	~0.0002% (or 1.8 ppmv)
Nitrous oxide (N_2O)	N_2O	~0.00003% (or 0.3 ppmv)
Ozone (O_3)	O_3	~0.000004% (or 0.04 ppmv)

Nitrogen (N_2) is the most abundant gas in the atmosphere, comprising 78.08% of its volume. It enters the atmosphere primarily through volcanic eruptions, biomass burning, and nitrification. Nitrogen is returned to the Earth's surface by nitrogen-fixing bacteria and lightning, serving as a vital nutrient for plants.

Oxygen (O_2) making up 20.95% of the atmosphere is a diatomic molecule released as a byproduct of photosynthesis. During this process, carbon dioxide and water react with sunlight to produce oxygen and glucose. Photosynthesis, carried out by plants, algae, and certain bacteria, is the primary source of oxygen in the atmosphere. The general equation for photosynthesis is:

$$6CO_2 + 6H_2O + \text{light energy} \rightarrow C_6H_{12}O_6 + 6O_2 \quad (6.4)$$

The oxygen produced is essential for respiration in living organisms and is also consumed during the decomposition of organic matter.

Water vapor (H_2O) is the most significant variable component of the atmosphere, with concentrations ranging from 0.0% to 4.0% by volume. It plays a crucial role in the greenhouse effect, trapping infrared radiation and warming the Earth. Water vapor also redistributes energy in the atmosphere and cycles through precipitation and evapotranspiration, with higher concentrations found over oceans, seas, and tropical rainforests.

Carbon dioxide (CO_2), although only 0.038% by volume, is the fifth most abundant gas in the atmosphere and a key greenhouse gas. It is released into the atmosphere through volcanic eruptions, biomass and fossil fuel burning, cement production, and deforestation. CO_2 is absorbed during photosynthesis and dissolved into oceans, where it can form carbonic acid. Over the past 300 years, atmospheric CO_2 levels have risen by approximately 25%.

Ozone (O_3) is a naturally occurring but unstable molecule consisting of three oxygen atoms (Fig. 6.8). In the stratosphere, ozone forms a protective layer that

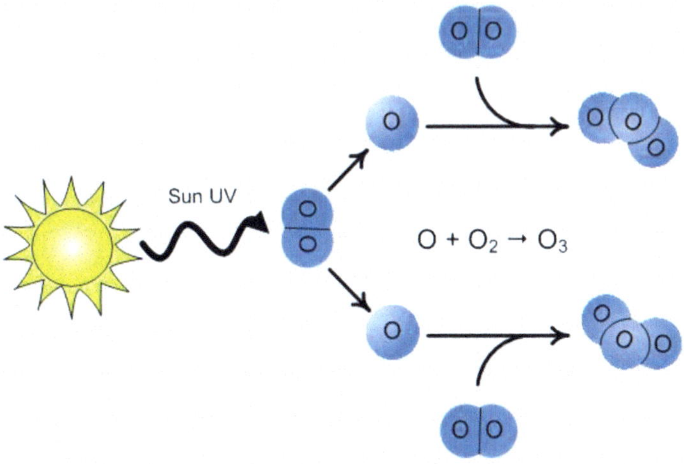

Fig. 6.8 Ozone formation in the atmosphere, showing the production of ozone (O_3) from molecular oxygen (O_2)

shields the Earth from harmful ultraviolet radiation, thereby preventing skin cancer, cataracts, and immune system damage in humans. However, ozone can cause respiratory problems at lower altitudes, contribute to smog, and damage building materials. Recently, ozone has also been recognized as a greenhouse gas.

The atmosphere also contains small amounts of other gases, including argon (0.93%), neon, and helium, each playing a role in the Earth's complex atmospheric system.

The concentration of carbon dioxide (CO_2) in the atmosphere is typically expressed in parts per million by volume (ppmv) or parts per billion by volume (ppbv), which are not SI units. One ppmv corresponds to 0.0001%, while one ppbv equals 0.0000001%. As of the latest update in January 2022, the global atmospheric concentration of CO_2 is approximately 414 ppmv (0.00000414%). These values fluctuate over time due to various factors, including human activities, such as fossil fuel combustion and natural processes like volcanic eruptions.

In contrast, stratospheric CO_2 concentrations, at altitudes between 11 and 50 km, are much lower, ranging from about 5 to 12 ppmv. The concentration of gases in the atmosphere varies with altitude, and the atmospheric layers reflect this variation. Notably, 99% of the total mass of the atmosphere is contained within the first 32 km above the Earth's surface.

Table 6.3 presents the mixing ratios of the most important components of air, how these ratios change with altitude, and the key processes that maintain or break down this mixture.

Table 6.3 Composition of the atmosphere (Salby, 2012)

Element	Tropospheric volume ratio of the mixture	Change of the mixture ratio with altitude	Processes by which the mixture is maintained
Nitrogen, N_2	0.7808	Homogeneous	Mixing with altitude
Oxygen, O_2	0.2095	Homogeneous	Mixing with altitude
Water vapor, H_2O^a	$\leq$0.030	Rapidly declines in the troposphere; increases in the stratosphere; highly variable	Evaporation, condensation, transport; production via CH_4 oxidation
Argon, A_r	0.0093	Homogeneous	Mixing with altitude
Carbon dioxide, CO_2^a	345 ppmv[b]	Homogeneous	Mixing with altitude; production from natural and anthropogenic processes
Ozone, O_3^a	10 ppmv[c]	Increases rapidly in the stratosphere; highly variable	Photochemical production in the stratosphere; destroyed through transport from the Earth's surface
Methane, CH_4^a	1.6 ppbv	Homogeneous in the troposphere; decreases in the middle atmosphere	Produced by terrestrial processes; through H_2O oxidation

(continued)

Table 6.3 (continued)

Element	Tropospheric volume ratio of the mixture	Change of the mixture ratio with altitude	Processes by which the mixture is maintained
Nitrous oxide, N_2O[a]	350 ppbv[d]	Homogeneous in the troposphere; decreases in the middle atmosphere	Produced at ground level and through anthropogenic processes; dissociation in the middle atmosphere; production via NO transport
Carbon monoxide, CO[a]	70 ppbv	Decreases in the troposphere; increases in the stratosphere	Anthropogenic production and transport of oxidized CH_4
Nitrogen oxide, NO	0.1 ppbv[c]	Increasing	Produced through dissociation of N_2O; catalytic destruction of O_3
Chlorofluorocarbons CFC–11[a], CFC–12[a]	0.2 ppb, 0.3 ppb	Homogeneous in the troposphere; decreases in the middle atmosphere	Industrial production; mixing in the troposphere; photodissociation in the stratosphere

[a] Actively interacts with radiation
[b] Millionth of the volume
[c] Values in the stratosphere
[d] Billionth part of the volume

6.4 Vertical Structure of the Atmosphere

The atmosphere is composed of distinct layers that vary in temperature, pressure, and composition. These layers include the troposphere, where weather occurs; the stratosphere, characterized by the ozone layer; the mesosphere, known for its cold temperatures; the thermosphere, where temperatures rise significantly; and the exosphere, the transition zone to outer space (Lutgens & Tarbuck, 2012). Each layer plays a vital role in Earth's climate and weather systems (Wallace & Hobbs, 2006). The division of these layers is determined primarily by changes in temperature as one moves away from the Earth's surface (Ahrens, 2019).

6.4.1 *Vertical Stratification of the Atmosphere*

Before delving into the dynamic and thermodynamic characteristics of the atmosphere, it is useful to first examine some general properties of the atmosphere. One key feature is the distribution of mass with height, which leads to the variation in pressure as altitude increases. The atmosphere is generally in mechanical equilibrium, or hydrostatic balance, expressed by the equation:

$$\frac{dp}{dz} = -g\rho \tag{6.5}$$

where p is the air pressure, and ρ is the air density. By substituting the density using the equation of state and integrating, we get:

$$p = p_0 e^{-\int_0^z \frac{dz}{H(z)}} \quad (6.6)$$

where p_0 is the pressure at ground level (or at z_0), and

$$H(z) = \frac{RT(z)}{g} \quad (6.7)$$

is the scale height. This scale height represents the characteristic thickness of the atmospheric mass, ranging from about 8 km at ground level to 6 km in colder regions of the atmosphere.

Globally, air pressure and density decrease exponentially with height. For example, pressure decreases from 1000 mb (10^5 Pa = 10^3 hPa = 10^2 kPa) at ground level to just 10% of that value at an altitude of 15 km. By hydrostatic balance, 90% of the atmosphere's mass is concentrated within this layer. Beyond this altitude, pressure decreases tenfold every 15 km.

Similarly, air density declines with altitude, from approximately 1.2 kg/m^3 at the surface. This rapid change in pressure with height (much faster than changes in the horizontal direction) implies that isobaric surfaces (where p = const) are generally horizontal.

A deviation from the horizontal suggests a slight pressure gradient in the horizontal direction, which drives atmospheric motion. Even at altitudes above 100 km, pressure and density continue to decrease according to the exponential law, though the rates of change gradually vary with height. A significant shift occurs at around 100 km, indicating a transition in the processes governing the stratification of mass and the composition of the air. The mean free path of a molecule, determined by the frequency of collisions, is inversely proportional to air density. As a result, the mean free path increases exponentially with altitude, from approximately 10^{-7} m at ground level to around 1 m at an altitude of 100 km. Molecular diffusion, which is governed by the mean free path, affects critical atmospheric properties, such as viscosity and thermal conductivity. Diffusion, along with the associated transfer of momentum and heat, plays a significant role in dissipating atmospheric movements. This dissipation disrupts velocity and temperature gradients, leading to air stagnation.

At altitudes below 100 km, the mean free path is short enough that molecular diffusion cannot significantly hinder the movement of turbulent elements. At these heights, turbulent air transport dominates over the diffusion transport of atmospheric components. Turbulent air movement mixes different gases in the same way, resulting in a homogeneous ratio of passive components. For this reason, it is said that the air is well mixed in the atmospheric layer up to 100 km in altitude. In this region, the atmosphere is homogeneous, with a nitrogen mixture ratio of approximately

$rN_2 \approx 0.78$ and an oxygen mixture ratio of $rO_2 \approx 0.21$. This well-mixed layer, extending up to 100 km, is referred to as the *homosphere*.

Above 100 km, the mean free path becomes longer than the scale of turbulent air movement. As a result, turbulent motions are strongly dampened by the diffusion of momentum and heat, making diffusion transport the dominant mechanism by which various properties are transmitted vertically. The transition from a layer dominated by turbulent transport to one governed by diffusive transport occurs at an altitude of about 100 km and is known as *homopause* or *turbopause*. This transition can be easily observed by tracking the vapor trail of a rocket passing through this region. Below 100 km, the trail is visibly turbulent, while above approximately 107 km, the vapor trail remains laminar, demonstrating the absence of turbulence and spreading only due to molecular diffusion.

The region between the homopause and an altitude of 500 km is called the heterosphere. In this region, airflow is largely laminar, and molecular diffusion separates gases by their molar weight. Heavier components, such as O_2, decrease in concentration more rapidly with height than lighter components like N_2. As a result, the composition of the air changes with altitude in the heterosphere. Up to 100 km, the average molar mass $\bar{M}$ remains constant, but around 100 km, it starts to decrease sharply and continues to decline monotonically with altitude.

Diffusive separation is the primary factor responsible for the stratification of atmospheric components in the heterosphere, but it is not the only one. Photodissociation also contributes to this process. Energetic ultraviolet (UV) rays, upon reaching the upper atmosphere, dissociate oxygen molecules, which generate a significant amount of oxygen atoms at these altitudes. By dissociating these molecules, UV radiation is filtered out of the solar spectrum.

Now, let's briefly consider the outermost layer of the atmosphere, the boundary with space, sometimes referred to as the airless space "the vacuum." Previously, this region was thought to be empty and of little scientific interest. However, recent satellite measurements have shown it to be filled with various charged particles, whose numbers fluctuate constantly. This discovery has led to the development of a new field known as *space weather*. While the study of space weather is undoubtedly important, its effects on terrestrial or "tropospheric" weather—where we live—are not yet fully understood. Nevertheless, since the study of space weather falls outside the scope of this book, we will return to our discussion of the atmospheric boundary layer.

In the homosphere and heterosphere, air molecules frequently collide. Above the altitude of 500 km, known as the critical level, molecular collisions become so rare that many molecules pass through this part of the atmosphere without experiencing a single collision. This region is called the exosphere. It consists of molecules that have left the denser layers of the atmosphere and are moving into space. These molecules follow ballistic trajectories determined by their velocity at the critical level and the planet's gravitational pull. Many molecules, drawn back by gravity, return on roughly parabolic paths. However, some molecules with higher velocities escape Earth's gravity, venturing permanently into the vast expanse of space.

The required speed of the molecule, V_c, can be determined from the following expression:

$$\frac{1}{2} m V_c^2 = \int_a^{\infty} m g_0^* \left(\frac{a}{r} \right)^2 \mathrm{d}r \tag{6.8}$$

where m is the mass of the molecule, g_0^* is the gravitational acceleration at the Earth's surface, and a is the radius of the Earth. This equation is derived from the condition that the kinetic energy of the molecule is equal to the work needed to move the molecule against the gravitational force from the Earth's surface to infinity. The integration begins at a, as the critical level corresponds to the Earth's radius. From Eq. (6.7), we obtain:

$$V_c = \sqrt{2 g_2^* a} \tag{6.9}$$

For Earth, this critical speed is approximately 11 km/s. Importantly, this speed is independent of the molecular weight; it is the same for all molecules. However, molecules do not all travel at this speed. Within a mixture of heavy and light molecules, there exists a spectrum of velocities—heavier molecules tend to move more slowly, while lighter ones move faster. As a result, lighter molecules escape the atmosphere more frequently than heavier ones.

The critical level of Earth's atmosphere is around 500 km above the surface. At this altitude, the temperature during normal solar activity is about 1000 K, but it can reach up to 2000 K during periods of solar disturbance.

6.4.2 The Main Atmospheric Layers

The Earth's atmosphere is stratified into distinct vertical layers, each with unique characteristics that define its structure and temperature profile. The primary atmospheric layers, as shown in Fig. 6.9, are:

- Troposphere
- Stratosphere
- Mesosphere
- Thermosphere
- Exosphere

The troposphere is the lowest layer of the atmosphere, containing approximately 80% of its total mass. This layer is characterized by a rapid decrease in temperature and water vapor content with altitude. As shown in Fig. 6.10, the troposphere is where most significant atmospheric processes and phenomena occur. Water vapor plays a crucial role in regulating air temperature by absorbing incoming solar energy and radiating heat from the Earth's surface. The troposphere holds about 99% of the

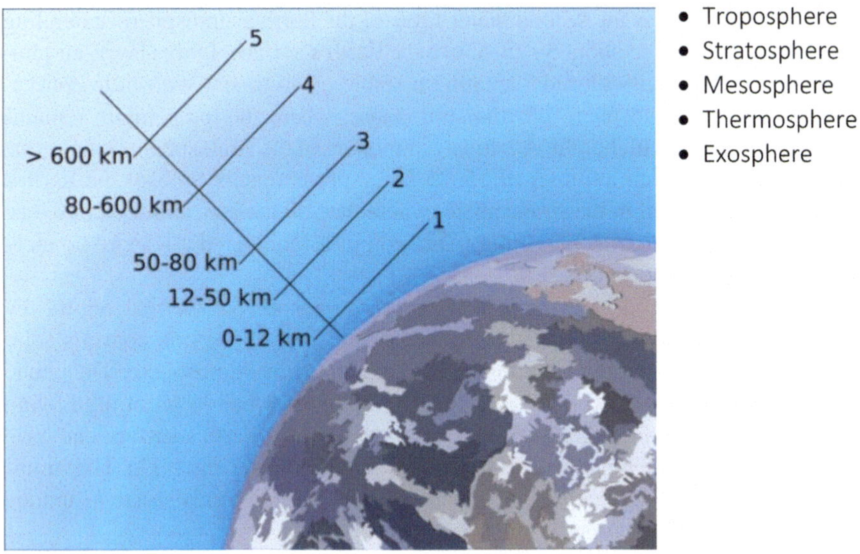

Fig. 6.9 Vertical profile of the Earth's atmosphere, showing the distinct layers: troposphere, stratosphere, mesosphere, thermosphere, and exosphere

Fig. 6.10 The troposphere: the lowest atmospheric layer, where most significant atmospheric processes and phenomena occur

atmosphere's water vapor. Concentrations vary with latitude, being highest in the tropics (up to 3%) and lower in polar regions. Weather phenomena predominantly occur in the troposphere, though disturbances can sometimes extend into the lower stratosphere.

The term "troposphere" derives from the Greek word for "mixing," reflecting the strong convective currents present in this layer. The boundary between the troposphere and the stratosphere is known as the tropopause, which ranges in altitude from 8 km near the poles to 18 km above the equator. The height of the tropopause varies seasonally, being highest in summer and lowest in winter.

The stratosphere is the second major layer of the Earth's atmosphere, extending from the tropopause to about 50 km above the Earth's surface. In this layer, air temperature increases with altitude, creating a stable environment with little convection. Above the stratosphere lies the stratopause, where the temperature remains relatively constant with height. A critical component of the stratosphere is the ozone layer, located between altitudes of 15–25 km. Approximately 90% of the Earth's ozone is concentrated in the stratosphere, where the volume concentration of ozone is about 10 ppmv (parts per million by volume), compared to approximately 0.04 ppmv in the troposphere.

Ozone plays a vital role in regulating the stratosphere's thermal regime by absorbing ultraviolet (UV) radiation from the Sun. This absorption converts solar energy into kinetic energy, resulting in the heating of the stratosphere. The ozone layer effectively absorbs UV-B radiation (290–320 nm), which is harmful to living organisms. Increased levels of UV radiation reaching the Earth's surface can have detrimental effects on plant life and the broader environment (Fig. 6.11). The stratosphere's low water vapor content and the presence of ozone contribute to its unique thermal and chemical properties.

The mesosphere is the third layer of the Earth's atmosphere, positioned above the stratosphere and below the thermosphere (Fig. 6.12). It extends from approximately 50–85 km above the Earth's surface. In contrast to the stratosphere, where temperatures rise with altitude due to the ozone layer, the mesosphere is characterized by a decrease in temperature with altitude, which can plunge to as low as −90 °C. The boundary between the mesosphere and the thermosphere, known as the mesopause, is the coldest point in the Earth's atmosphere.

The mesosphere is also where most meteorites disintegrate upon entering the atmosphere, creating streaks of light called meteors or "shooting stars" due to high-speed collisions with atmospheric particles. During the summer months, noctilucent clouds, composed of ice crystals, can be observed at dusk. The mesosphere features extremely low atmospheric pressure and density, making it a challenging environment for aircraft and weather balloons. Together with the stratosphere, the mesosphere is sometimes referred to as the "middle atmosphere."

The thermosphere is located above the mesosphere (Fig. 6.13). In this layer, temperatures generally increase with height, ranging from 600 K to 2000 K, depending

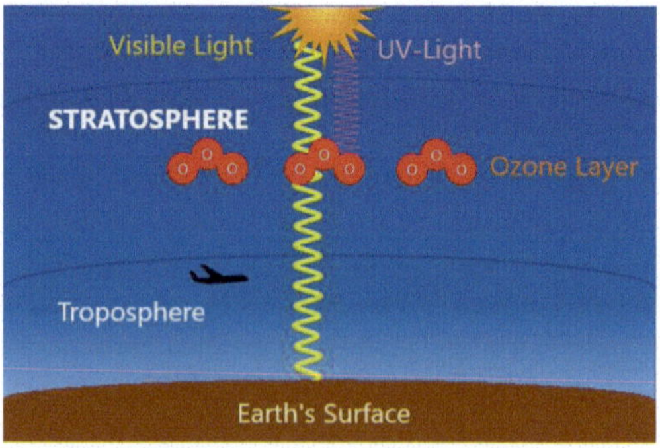

Fig. 6.11 The stratosphere (ozonosphere): the layer containing 90% of the Earth's ozone, located between altitudes of 15 and 25 km

Fig. 6.12 The mesosphere: a key layer in Earth's atmosphere characterized by extreme conditions, including low temperatures and pressure. It acts as a transition zone between the stratosphere below and the thermosphere above

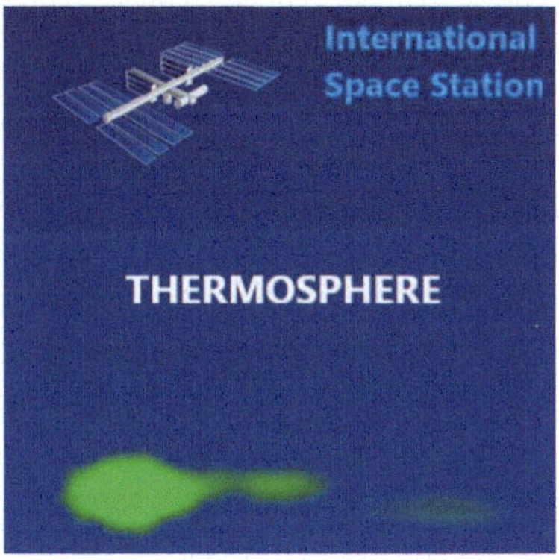

Fig. 6.13 The thermosphere: a key layer for studying space weather and Earth–Sun interactions, characterized by extremely high temperatures and low density

on solar activity. This warming is due to the absorption of intense solar radiation by the sparse molecular oxygen present at these altitudes. The gas molecules in the thermosphere are widely spaced, and the atmosphere becomes increasingly enriched with lighter gases, such as atomic oxygen, helium, and hydrogen, as altitude increases. Above 100 km, the chemical composition of the air is highly altitude-dependent, with atomic oxygen becoming the dominant component above 160 km. At very high altitudes, gases begin to stratify according to molecular weight due to gravitational separation.

The exosphere is the outermost layer of the Earth's atmosphere (Fig. 6.14). In this region, molecules have sufficient velocity to either escape into space or be pulled back to Earth by gravity, with minimal chances of colliding with other molecules. The lower boundary of the exosphere, known as the thermopause or exobase, ranges from approximately 250 to 500 km above the Earth's surface, varying with solar activity. The upper limit of the exosphere is theoretically determined by the distance at which the gravitational pull of Earth is countered by solar radiation pressure on hydrogen atoms, which can extend up to about 205,680 km. From a space-based perspective, the observable boundary of the exosphere, known as the

Fig. 6.14 The exosphere: the outermost layer of the Earth's atmosphere, extending from the upper boundary of the thermosphere to the transition into outer space. The exosphere is characterized by extremely low atmospheric density, where particles move into the near-vacuum conditions of space. Interaction between charged particles in the exosphere and the Earth's magnetic field can contribute to auroras, particularly in polar regions

Earth's corona, is about 102,840 km from the Earth's surface. The exosphere serves as a transition zone between the Earth's atmosphere and interplanetary space.

6.5 Thermal and Dynamic Characteristics of the Atmosphere

Understanding the specific behavior of air is much easier when viewed within the broader context of the atmosphere's global dynamics. To this end, we will delve further into the description of the atmosphere's eight key global thermal and dynamic characteristics. These characteristics have been well-documented through various direct and remote measurement methods. In recent decades, advancements in technology—such as the use of rockets and, for almost 40 years, meteorological satellites—have significantly enhanced our understanding of these atmospheric features (Marshall & Plumb, 2020; Vallis, 2017).

The most straightforward thermal profile of the atmosphere can be depicted by examining how the globally averaged temperature varies with height, with this temperature averaging performed across the horizontal plane (Salby, 2012). This approach provides valuable insights into the vertical distribution of thermal energy and helps elucidate the atmospheric layers' behavior.

The atmosphere's layers exhibit vertical temperature changes, primarily driven by the absorption of solar radiation. At the Earth's surface, visible light warms the ground, while near-ultraviolet radiation affects the middle atmosphere, and ultraviolet radiation influences the upper layers (Thuburn, 2020). The vertical temperature distribution, or stratification curve, of various atmospheric layers is illustrated in Fig. 6.15. From the ground up to about 10 km, the temperature decreases with an

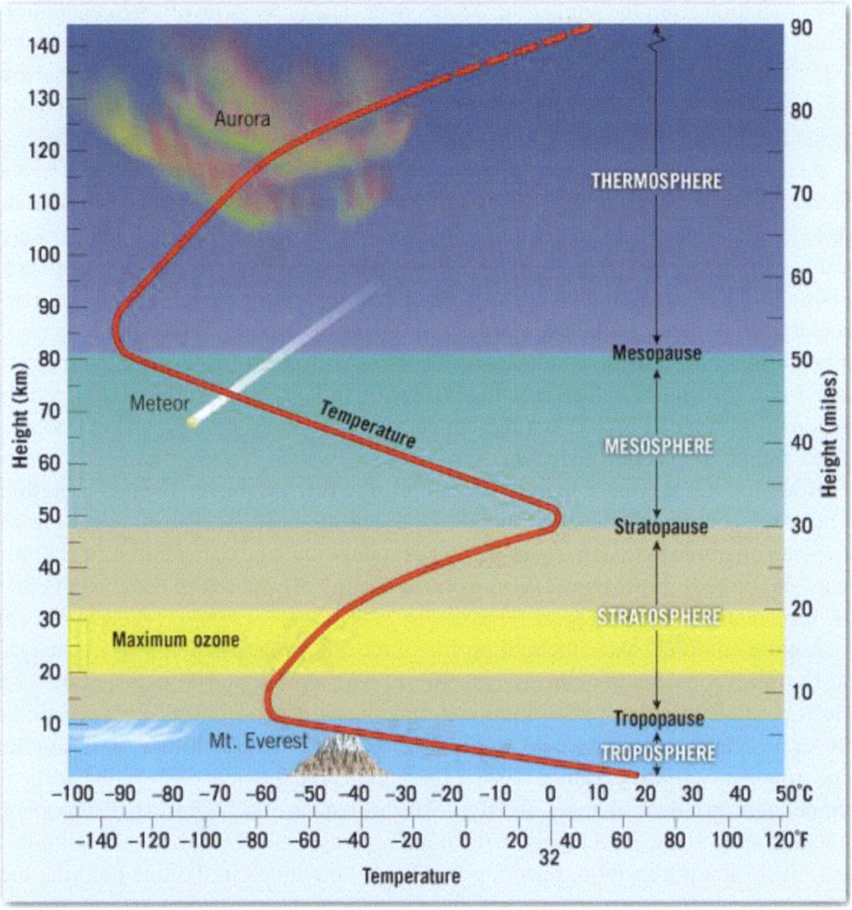

Fig. 6.15 Vertical thermal structure of the atmosphere

approximately constant vertical gradient of around 6.5 °C/km. This region, known as the troposphere, is the lowest layer of the atmosphere and is marked by strong convection, which is its defining characteristic. Most significant weather phenomena occur within the troposphere, driven by the heating of Earth's surface (Lutgens & Tarbuck, 2012). At the upper boundary of this layer, at around 10 km in altitude (or at a pressure of about 100 mb), lies the tropopause, a thin transitional zone where the temperature gradient shifts abruptly.

Above the tropopause, the region extending up to about 85 km is referred to as the middle atmosphere. Within and just above the tropopause, the temperature becomes nearly constant with height, before it starts to rise in the next layer, the stratosphere. This increase in temperature with altitude (negative vertical temperature gradient, $\gamma = -\frac{\partial T}{\partial z}$) in the stratosphere is due to the absorption of ultraviolet

(UV) radiation from the sun by the ozone layer. The process involves oxygen molecules (O_2) being split and recombined into ozone (O_3) under the influence of solar radiation. This absorption of energy results in a warming of the stratosphere, where the temperature reaches a maximum of approximately 0 °C at the stratopause, located at a height of about 50 km (1 mb).

Above the stratopause lies the mesosphere, where the temperature once again decreases with altitude due to the loss of ozone-driven heating. This layer extends up to the mesopause, located at around 85 km (0.01 mb), where the temperature reaches its coldest point, near −90 °C. Beyond the mesopause is the thermosphere, where temperatures rise dramatically due to the absorption of solar radiation, particularly by ionized particles. Unlike the lower atmospheric layers, the thermosphere cannot be considered electrically neutral, as free electrons and ions interact with Earth's magnetic and electric fields, heavily influenced by solar activity. In this region, at an altitude of 200 km, the temperature can reach 1200 K, and at 600 km, it can rise to 2200 K during the day.

However, despite these high temperatures, the thermosphere contains such thin air that these values do not correspond to the sensation of heat. The air density is so low—comparable to that in a near-perfect vacuum—that the transfer of heat is minimal. Therefore, the effects of thermospheric heating are not felt in the layers below the mesopause.

A more comprehensive thermal profile of the atmosphere is obtained by analyzing the mean zonal temperature as a function of both altitude and latitude. Figure 6.16 illustrates this temperature distribution during the Northern Hemisphere's winter. In the troposphere, temperature decreases with both increasing altitude and latitude. The tropopause (indicated by a dashed line) marks the boundary where the vertical temperature gradient changes its sign. The tropopause is highest in the tropics, reaching up to 16 km, and lowest in the polar regions, around 8 km. In mid-latitudes, there is no sharp transition, resulting in a noticeable break in the line marking the tropopause.

In the stratosphere, the temperature begins to increase with altitude. The highest stratospheric temperatures occur above the South Pole, while the lowest are found above the North Pole. The stratopause is also marked by a dashed line in the figure. Moving into the mesosphere, the temperature once again decreases with height, and the meridional temperature gradient is reversed compared to the stratopause. The coldest region is observed over the South Pole, despite the continuous daylight and summer conditions there, which may initially seem counterintuitive. However, this temperature distribution highlights the critical role of atmospheric dynamics in shaping thermal patterns, rather than just the amount of solar radiation received.

The temperature distribution shown in Fig. 6.16 is closely aligned with the mean zonal wind patterns, as illustrated in Fig. 6.17 for the same period. In the troposphere, the subtropical jet streams dominate both hemispheres, intensifying up to the tropopause. These jet streams are westerly in both hemispheres, with air moving from west to east. Above the subtropical jets, the zonal wind first weakens with increasing altitude and then strengthens again, but with opposite directions in the two hemispheres.

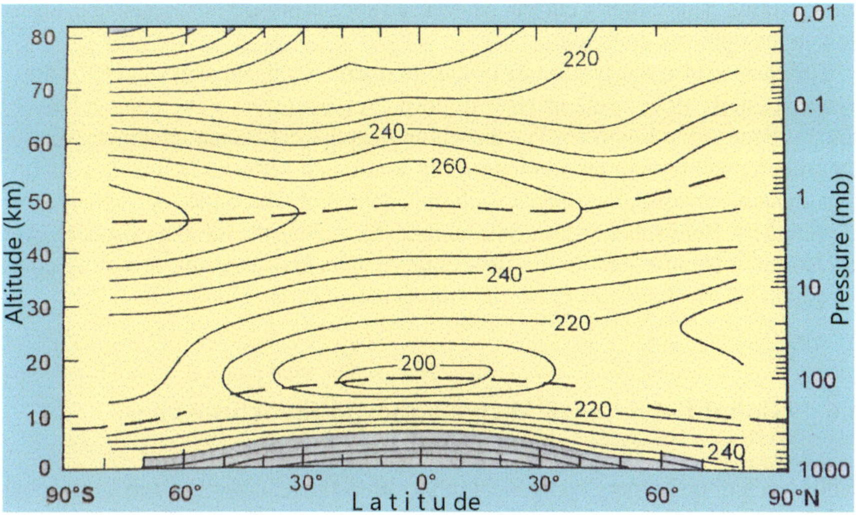

Fig. 6.16 Mean zonal temperature (K) during northern winter as a function of altitude and latitude

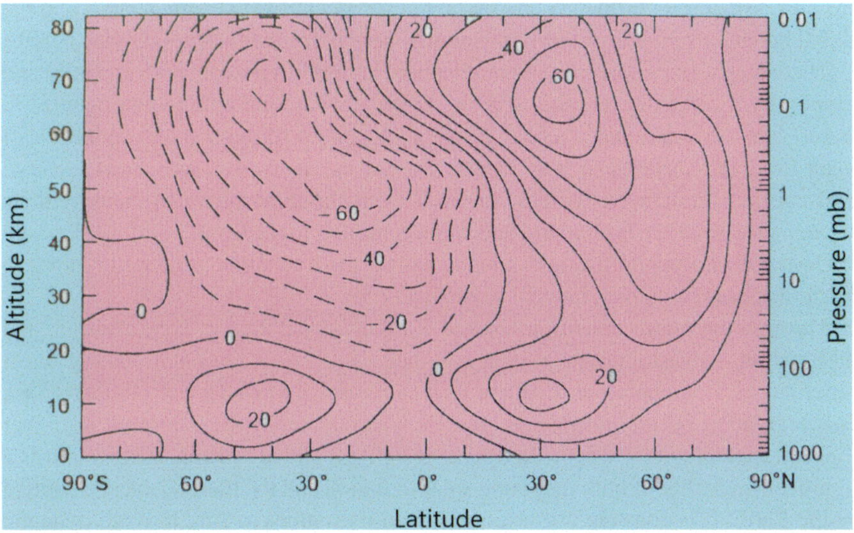

Fig. 6.17 Zonal velocity component, like Fig. 6.16. Isotachs are expressed in m/s, with positive values indicating westward motion and negative values indicating eastward motion

In the Northern Hemisphere during winter, the westerly currents strengthen, reaching a maximum speed of about 60 m/s in the lower mesosphere. This current is known as the polar jet stream or more specifically, the polar night jet stream, as it occurs during the polar night when the region experiences continuous darkness. In contrast, over the Southern Hemisphere at the same altitude, an easterly polar jet stream forms. This easterly current can exceed speeds of −70 m/s, creating a strong flow in the opposite direction.

This powerful easterly current in the Southern Hemisphere is coupled with a weaker easterly flow in the tropical troposphere, where the air remains relatively cold, as depicted in Fig. 6.16. The air within the polar jet stream undergoes significant mixing due to both meridional (north–south) and vertical circulation patterns. This dynamic mixing contributes to the phenomenon where the mesosphere over the Southern Hemisphere (during its summer) is colder than the mesosphere over the Northern Hemisphere during its seemingly paradoxical result, driven by complex atmospheric dynamics rather than direct solar heating alone.

6.6 Global Behavior of Water Vapor in the Atmosphere

While dry air, the primary component of the atmosphere, is well-mixed up to about 100 km due to atmospheric motions, the same cannot be said for water vapor. Despite being mixed with dry air, water vapor's distribution is not homogeneous, unlike certain other atmospheric constituents such as ozone. Given its vital role in life on Earth and its influence on key meteorological processes, understanding the global behavior of water vapor in the atmosphere is crucial (Rogers & Yau, 1989).

The water vapor content in the atmosphere is highly variable, both spatially and temporally. Unlike dry air, atmospheric movements are not able to homogenize it. Water vapor is continually added to the atmosphere through the evaporation of water from vast surfaces such as oceans, seas, lakes, rivers, wetlands, and terrestrial landscapes. Simultaneously, it is removed from the atmosphere in many regions through precipitation, both liquid and solid, returning to the Earth's surface (Peixoto & Oort, 1992). Despite this constant exchange, the total amount of water vapor in the atmosphere remains relatively stable over time.

Through the phase transitions of water vapor—evaporation, condensation, and sublimation—a vast amount of heat is either absorbed or released, making water vapor a critical temperature regulator in the atmosphere (Wallace & Hobbs, 2006). This regulation occurs not only in regions where water vapor is present in higher concentrations (up to 4% by volume) but also has a global impact. Water vapor is a significant absorber of both incoming solar radiation and the thermal energy emitted by the Earth. It is constantly transported from its sources to sink through dynamic atmospheric processes while also playing a key role in generating atmospheric dynamics itself (Hartmann, 2015).

This is particularly evident in the subtropics, where the formation of clouds induces upward motion in the atmosphere. These vertical movements, initiated by water vapor, contribute to further atmospheric circulation, transmitting energy to higher latitudes via horizontal flows.

Figure 6.18 illustrates the variation of mean zonal water vapor content with height and latitude. Much of the water vapor is concentrated in the troposphere, with the average mixing ratio of water vapor, $\overline{r}_{H_2O}$, decreasing uniformly with

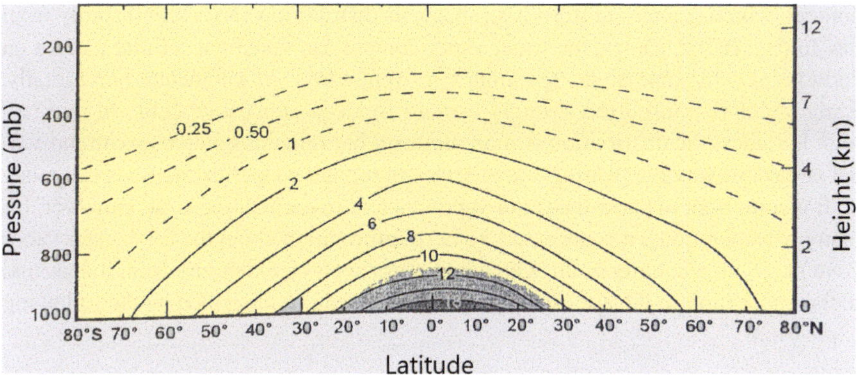

Fig. 6.18 Lines representing the mean zonal ratio of water vapor to absolute humidity, with shaded areas indicating 20%, 40%, and 60% of the maximum value, as a function of altitude and latitude

altitude in the tropics. Starting at a maximum of around 20 g/kg near the Earth's surface, it drops to just a few parts per million at the tropopause. The absolute humidity $\bar{\rho}_{H_2O}$, which is depicted as the shaded area in the figure, decreases even more rapidly with height, as the density of dry air declines exponentially with increasing altitude.

This can be seen from the expression:

$$\bar{\rho}_{H_2O} = r_{H_2O}\rho_s \tag{6.10}$$

where ρ_s is the density of dry air. The mean mixing ratio $\bar{r}_{H_2O}$ also decreases with latitude, having a value of less than 5 g/kg at the sixtieth degree of latitude.

These values illustrate where the greatest production of water vapor occurs (in tropical regions), its redistribution through atmospheric circulation, and its loss through condensation and precipitation at higher altitudes, as well as in mid and high latitudes. Considering the mechanisms of its formation and removal, tropospheric water vapor can be regarded as a short-lived substance. The characteristic residence time of water vapor is on the order of several days. It can be defined as the time in which an individual air element significantly reduces its r_{H_2O} content. Every few days, an air element is influenced by the warm ocean surface, where it is enriched with water vapor through evaporation, or it encounters a cloudy environment, where it loses water vapor through condensation and precipitation.

Most water vapor forms near the equator, above warm ocean surfaces. Through vertical and horizontal movements (referred to as convection and advection), it is transported and redistributed. Introduced into the lower layers of the tropical atmosphere, water vapor is lifted by deep convective cells and transported horizontally by various processes. Some air parcels retain a consistent amount of water vapor during their movement over a long time, making r_{H_2O} a conservative quantity (tracer), a significant feature of that air mass. Unlike the average water vapor content, which changes uniformly in space, its distribution varies significantly from day to day. In the tropics, the water vapor content can take on a cellular pattern on certain days. In some areas, spanning tens to hundreds of kilometers horizontally, water vapor is transported strongly upward by deep convective cells. In mid and high latitudes, the distribution is more uniform but remains complex, with individual streaks of vapor appearing—sometimes in regular shapes, sometimes irregular.

It is important to know the water vapor content in each air element. However, for many considerations, it is necessary to have information about the total water vapor content, Q_v, in a column of air with a unit cross-section, extending from the ground to the upper limit of the atmosphere. This content can be described by the following expression:

$$Q_v = \int_0^\infty \rho_s r_{H_2O} dz = \frac{1}{g} \int_0^{p_0} r_{H_2O} dp \tag{6.11}$$

Here, p_0 represents the surface pressure, and g is the acceleration due to gravity. This total content is often referred to as "precipitable water vapor." It is measured using modern meteorological geostationary satellites. Equipped with highly sensitive receivers, these satellites, positioned at altitudes of around 36,000 km above the Earth, take measurements of the total water vapor content every 15 min.

For example, the Meteosat satellite covers the Atlantic Ocean, Europe, and Africa, from the Equator to 60° north and south latitude. These precise measurements allow for near-continuous monitoring of water vapor behavior in the atmosphere, providing invaluable data for understanding atmospheric conditions in real time.

6.7 Magneto-Electronic Structure of the Atmosphere

The upper atmosphere is divided into two main regions: the ionosphere and the magnetosphere (also known as the plasmasphere), both of which are defined by the behavior and concentration of free electrons and other charged particles (Kelley, 2009). The ionosphere has a significant concentration of ions and free electrons. These ions, which are atoms and gas molecules that have lost or gained one or more electrons, become positively or negatively charged. They become positively charged when they cannot absorb all the energy from collisions with high-energy particles or solar radiation. The ionosphere plays a critical role in atmospheric effects related to radiation propagation due to the changing concentration of free electrons (Rishbeth & Garriott, 2003).

The magnetosphere, however, is shaped by the Earth's magnetic field and serves as a protective shield against solar wind and cosmic radiation. It influences the motion of charged particles, which can lead to phenomena such as auroras when particles from the solar wind interact with the Earth's atmosphere (Gombosi, 2004).

Both regions are essential for understanding the dynamics of space weather and their impact on communication systems, satellite operations, and global positioning systems (GPS).

The plasmasphere, part of the magnetosphere, is a region composed primarily of plasma—the fourth state of matter, consisting of hydrogen ions (protons) and electrons. Despite its name, the plasmasphere is more like a torus than a sphere and is separated by a distinct boundary called the plasmapause. The plasmapause is located about 19,000–32,000 km from the Earth's surface on the sunward side, and it defines the area where the Earth's magnetic field strongly influences the behavior of plasma. Below this boundary, the ionosphere extends into the plasmasphere, where protons replace atomic oxygen as the dominant species in the ionospheric plasma, typically at an altitude of about 1000 km.

The magnetosphere extends beyond the plasmasphere and is shaped by the Earth's magnetic field as it interacts with the solar wind—a stream of charged particles emitted by the Sun (Fig. 6.19). The outer boundary of the magnetosphere, known as the magnetopause, is where the Earth's magnetic field is balanced by the solar wind and interplanetary magnetic field. This boundary is generally located about 56,000 km above the Earth's surface. The magnetosphere acts as a protective shield, deflecting solar wind particles and preventing them from directly impacting the Earth's atmosphere.

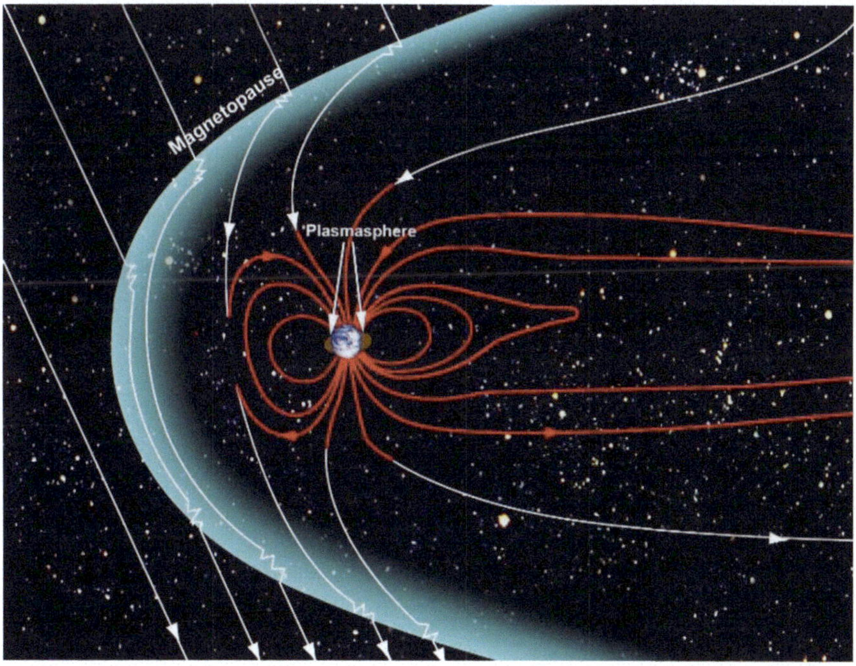

Fig. 6.19 The magnetosphere and plasmasphere. The magnetosphere, shaped by Earth's magnetic field, protects against solar radiation. The plasmasphere contains plasma within the magnetosphere and is bounded by the plasmapause

6.8 Classification of the Atmosphere According to Chemical Composition

The atmosphere can be classified into layers based on its chemical composition (Fig. 6.20). In the absence of external sources and sinks of air components, two processes determine this composition: mixing due to motion and molecular diffusion. The mixing process creates a homogeneous atmosphere, forming the layer known as the homosphere, which extends up to about 85 km in altitude. Above this, where molecular diffusion begins to dominate over turbulent transport lies the diffusive separation layer or turbopause.

Beyond the turbopause, the atmosphere is termed the heterosphere. In the heterosphere, photodissociation of oxygen is significant, with most atomic oxygen found above approximately 120 km. At 500 km altitude, atomic oxygen is prevalent, with minor amounts of nitrogen molecules and lighter gases like helium and hydrogen. Above 1000 km, helium and hydrogen dominate. The temperature in the atmosphere above 200 km varies significantly depending on solar activity. Under a quiet sun, temperatures hover around 600 K, while active solar conditions can push temperatures close to 2000 K, making the thermosphere isothermal at these altitudes. The atmosphere's density can also be described by the mean free path of molecules—the average distance a molecule travels between collisions. At sea level, this distance is about 10^{-7} m and increases steadily with altitude. At 150 km, the mean free path reaches 100 m, and at 500 km, it exceeds 10 km (Fig. 6.21). In these regions, molecules and atoms move along ballistic trajectories, following curved paths under the influence of gravity. This upper layer of the atmosphere is called the exosphere, marking the conceptual upper boundary of the atmosphere. At these altitudes, atoms or molecules may achieve an escape velocity of approximately 11 km/s, allowing

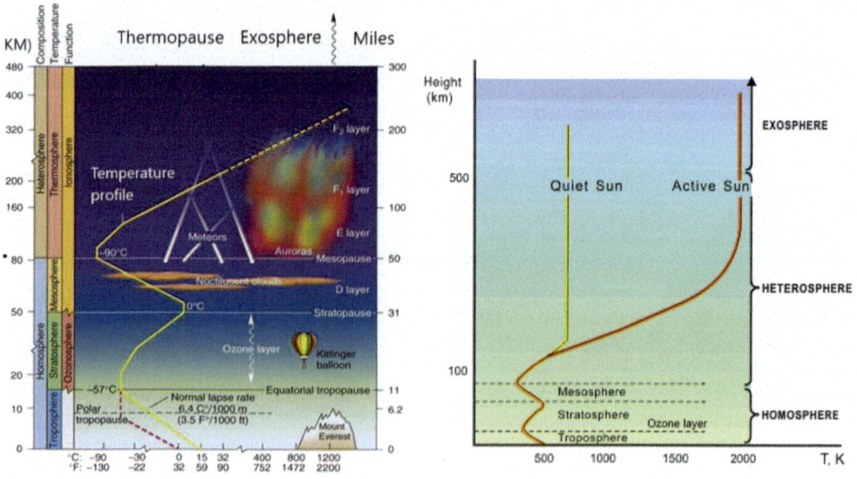

Fig. 6.20 Classification of the atmosphere according to chemical composition

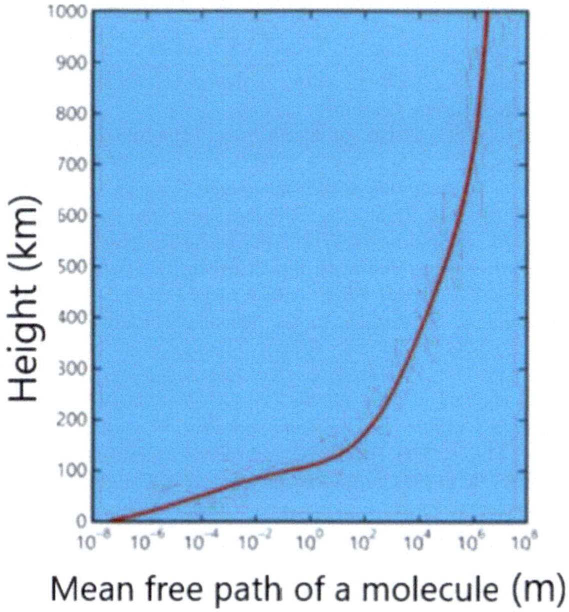

Fig. 6.21 The mean free path length of molecules as a function of height

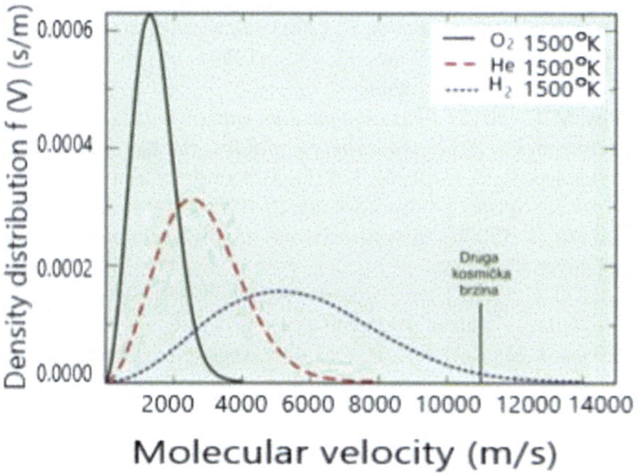

Fig. 6.22 Velocity density distribution of oxygen, helium, and hydrogen atoms as a function of atomic velocity at a temperature of 1500 °K

them to leave Earth's gravitational field. The likelihood of this occurring can be estimated using the Maxwell–Boltzmann distribution of molecular velocities.

Figure 6.22 illustrates this distribution for oxygen, helium, and hydrogen atoms at a temperature of 1500 K, showing that hydrogen atoms, with the highest typical speeds, can exceed 5 km/s. The time required for atmospheric components to escape Earth's gravity varies, with hydrogen likely to leave much faster than the age of Earth, while the loss of atomic oxygen is negligible. Theoretically, the last hydrogen atom could escape at an altitude of 205,680 km, representing the upper limit of Earth's atmosphere.

References

Ahrens, C. D. (2019). *Meteorology today: An introduction to weather, climate, and the environment*. Cengage Learning.

Andrews, D. G. (2010). *An introduction to atmospheric physics*. Cambridge University Press.

Brasseur, G. P., & Solomon, S. (2005). *Aeronomy of the middle atmosphere: Chemistry and physics of the stratosphere and mesosphere*. Springer.

Canup, R. M., & Righter, K. (2000). *Origin of the Earth and Moon*. University of Arizona Press.

Ćurić, M., & Spiridonov, V. (2023). The earliest past of the Earth and the atmosphere. In *History of meteorology*. Springer. https://doi.org/10.1007/978-3-031-45032-7

Gombosi, T. I. (2004). *Physics of the space environment*. Cambridge University Press.

Halliday, A. N. (2001). In the beginning: Earth's early atmosphere and oceans. *Nature, 412*(6848), 24–26.

Hartmann, D. L. (1994). *Global physical climatology*. Academic Press.

Hartmann, D. L. (2015). *Global physical climatology* (2nd ed.). Elsevier.

Jacob, D. J. (1999). *Introduction to atmospheric chemistry*. Princeton University Press.

Kelley, M. C. (2009). *The Earth's ionosphere: Plasma physics and electrodynamics*. Academic Press.

Lambeck, K. (1991). *Earth's variable rotation: Geophysical causes and consequences*. Cambridge University Press.

Lissauer, J. J., & de Pater, I. (2013). *Fundamental planetary science: Physics, chemistry, and habitability*. Cambridge University Press.

Lutgens, F. K., & Tarbuck, E. J. (2012). *The atmosphere: An introduction to meteorology*. Pearson.

Marshall, J., & Plumb, R. A. (2020). *Atmosphere, ocean and climate dynamics: An introductory text*. Academic Press.

Peixoto, J. P., & Oort, A. H. (1992). *Physics of climate*. American Institute of Physics.

Rishbeth, H., & Garriott, O. K. (2003). *Introduction to ionospheric physics*. Academic Press.

Rogers, R. R., & Yau, M. K. (1989). *A short course in cloud physics* (3rd ed.). Butterworth-Heinemann.

Salby, M. L. (2012). *Physics of the atmosphere and climate*. (2nd ed.). Cambridge University Press.

Schumann, U. (2012). *Atmospheric physics: Background, methods, trends*. Springer.

Spiridonov, V., & Ćurić, M. (2021). Meteorology as a natural science. In *Fundamentals of meteorology*. Springer. https://doi.org/10.1007/978-3-030-52655-9

Thuburn, J. (2020). *An introduction to the dynamics of atmosphere and ocean*. Cambridge University Press.

Vallis, G. K. (2017). *Atmospheric and oceanic fluid dynamics: Fundamentals and large-scale circulation*. Cambridge University Press.

Wallace, J. M., & Hobbs, P. V. (2006). *Atmospheric science: An introductory survey*. Elsevier.

Chapter 7
Solar Radiation: Driving Atmospheric Change

Solar radiation is the primary source of energy for Earth's atmosphere, surface, and climate systems. It plays a crucial role in sustaining life by providing the energy required for processes such as photosynthesis, atmospheric circulation, and ocean currents. The Sun's radiant energy influences the Earth's weather, climate, and ecosystems, driving fundamental processes in the atmosphere, hydrosphere, and lithosphere. As this energy reaches the Earth, it is absorbed, reflected, and scattered in

© The Author(s), under exclusive license to Springer Nature Switzerland AG 2025
V. Spiridonov et al., *Atmospheric Perspectives*,
https://doi.org/10.1007/978-3-031-86757-6_7

various ways, contributing to the distribution of heat and the formation of different weather patterns. Understanding solar radiation and how it interacts with the Earth's atmosphere is essential for explaining the planet's energy balance, climate variability, and the dynamics of weather systems. Solar radiation not only governs temperature patterns but also triggers the formation of air masses, weather fronts, and large-scale atmospheric systems that shape global circulation patterns. It is the driving force behind seasonal changes, weather extremes, and shifts in climatic conditions.

This section explores the fundamental principles of solar radiation, including its definition, the mechanisms by which it reaches the Earth, and its varying effects across different latitudes and timescales. We will examine the electromagnetic spectrum, the role of solar irradiance and insolation, and how solar energy influences global temperature distributions and atmospheric processes. Additionally, this section covers key concepts such as blackbody radiation, Planck's law, Stefan–Boltzmann law, and Wien's law, all of which provide important insights into the behavior of radiation emitted by the Sun and Earth. By understanding the characteristics of solar radiation and its impact on the Earth's atmosphere, we gain a deeper appreciation for the forces that govern our planet's climate and weather systems.

7.1 Basic Features of Solar Radiation

The energy emitted by the Sun provides nearly 99.9% of the total energy that heats the Earth's surface. This solar energy warms the Earth and its atmosphere, sustaining life. Some energy that reaches the Earth is converted into heat, which is then transported by wind in the atmosphere and currents in the oceans and seas. Understanding how the Sun heats the Earth and how this warming varies across geographic space and time is crucial. Radiation is the emission or transmission of energy through space or matter through waves. Solar energy, the primary energy source, drives atmospheric circulation and various physical processes on Earth (Liou, 2002; Visconti, 2016). It warms the atmosphere, water, and Earth's surface, contributing to the formation of air masses and atmospheric systems and the generation of winds and ocean currents. Through photosynthesis, solar energy supports all life on our planet and influences the atmosphere, hydrosphere, and lithosphere (Houghton, 2015). Solar radiation is the most important external factor in climate formation (Wild et al., 2019). Estimates suggest that the total thermal energy on Earth is about 300 times less than the solar energy that reaches the upper atmosphere daily (Ramanathan & Vogelmann, 2017). The heat from the Earth's interior reaching the surface is approximately 5000 times less than the heat from the Sun. Significant changes in solar heat drive winds in the atmosphere and currents in the seas and oceans (Wallace & Hobbs, 2006). To understand how the Sun heats the Earth and how this warming varies geographically and temporally, we must study heat and temperature as variables and explore the mechanisms of their transfer (Goody & Yung, 1989). The exosphere, located between 700 and 10,000 km above the Earth's surface, is the uppermost layer of the atmosphere, merging with the solar wind. The solar wind, formed by the outward spread of plasma (a collection of charged particles) from the Sun's corona (its outermost layer), is heated to the point where the Sun's gravity can no longer contain it. This plasma travels along the Sun's magnetic field lines, which extend radially outward (Parker, 1958).

7.2 Electromagnetic Radiation

Radiant energy travels through space as a series of electromagnetic waves. Electromagnetic radiation (EM radiation) is a form of energy produced by moving electrically charged particles through matter or a vacuum or by oscillating magnetic and electrical disturbances. The magnetic and electric fields (Fig. 7.1) are oriented at 90° to each other, and the combined waves travel perpendicular to both fields, creating a disturbance. All these waves make up the electromagnetic radiation spectrum (Fig. 7.2), which is characterized by wavelength, ranging from radio waves (10^3 m) to gamma rays (10^{-14} m) (Hawking, 2015). The Sun emits all forms of radiation but in varying amounts.

The main part of the visible spectrum includes ultraviolet to infrared radiation. Electromagnetic radiation represents the distribution of solar energy across different wavelengths. High-energy waves have short wavelengths (measured in billionths of a meter or less), while low-energy waves have long wavelengths (measured in meters or kilometers) (Serway & Jewett, 2018). The shortest wavelengths are known as gamma rays, while slightly longer wavelengths are called X-rays. Within the visible spectrum, light is perceived by the human eye and can be further divided into colors: violet, blue, green, yellow, and red. In this sequence, violet has the shortest wavelength, while red has the longest within the visible range. These colors are seen

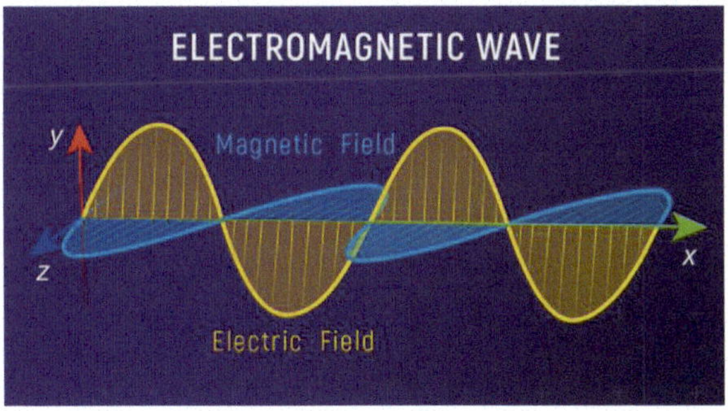

Fig. 7.1 An electromagnetic wave showing perpendicular electric and magnetic fields. (Credit: webbtelescope.org)

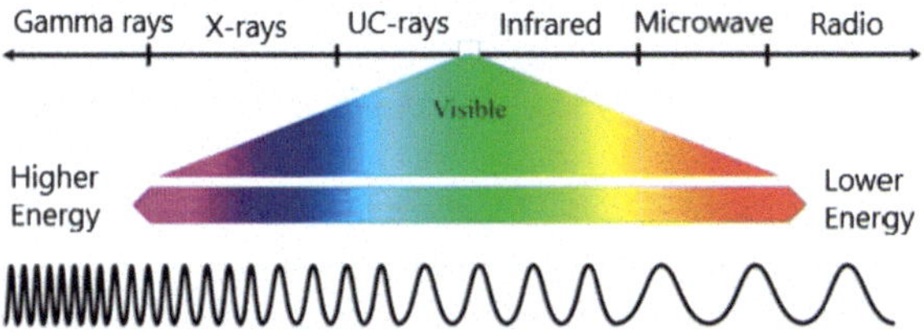

Fig. 7.2 The electromagnetic spectrum illustrates the range of wavelengths from radio waves to gamma rays

in a rainbow. All wavelengths shorter than infrared are considered short waves (Tipler & Mosca, 2021).

Terms and Definitions

- ***Electromagnetic radiation*** *is a form of energy produced by oscillating electrical and magnetic disturbances.*
- ***Thermal radiation*** *is electromagnetic radiation generated by the thermal movement of particles in matter. All objects with a temperature above absolute zero emit thermal radiation across a spectrum of wavelengths.*
- ***Sunlight****, commonly called light, is the portion of electromagnetic radiation visible on the Earth's surface.*
- ***Solar irradiance*** *is the output of solar energy from the entire disk of the Sun, measured on Earth. It is a measure of solar power (intensity).*
- ***Solar insolation*** *is the amount of solar radiation received over a specific period—typically 1 day.*
- ***The solar constant*** *is the total radiation energy received from the Sun per unit of time and unit area perpendicular to the Sun's rays at an average distance from the Earth. The value of the solar constant is approximately 1.366 kW/m*^{2}.

Solar Radiation and the Earth's Energy Balance: Useful Insights and Ideas

1. ***Solar energy and gravitational energy*** *are the primary energy sources for the Earth's climate system. In an idealized scenario (referred to as a "black body"), matter absorbs all incident energy in the form of electromagnetic waves, resulting in heating and becoming a source of radiation. This energy exchange leads to a state of equilibrium, where outgoing radiation balances incoming energy.*
2. *The energy radiated by a black body is distributed across all wavelengths, forming a "bell-shaped" curve. The maximum energy is radiated at a wavelength inversely proportional to the absolute temperature.*
3. *The total energy radiated by a black body (integrated over all wavelengths) is proportional to the fourth power of its absolute temperature.*
4. *The energy flux radiating from a point source decreases proportionally to the square of the distance from it, which explains why light dims rapidly as it moves away from its source.*
5. *By applying basic radiation laws and knowing the temperature of the Sun, we can calculate the "effective" temperature of any surrounding planet.*
6. *The Earth and other planets are not perfect black bodies, as they do not absorb all incoming solar radiation but instead reflect some of it into space. The ratio of reflected to incoming energy is called* ***planetary albedo****.*
7. *Due to its spherical shape, incoming solar radiation is not equally distributed across the planet. At any given moment, the Sun illuminates only half of the Earth's surface, with maximum radiation occurring at local noon and less at other times of the day.*
8. *Total daily radiation decreases from the equator to the poles, resulting in a warmer surface at the equator compared to the poles.*
9. *The Earth's axis is tilted at 23.5° relative to its orbit, causing seasonal changes. When the axis points toward the Sun, it is summer in that hemisphere; when it*

tilts away, it is winter. During spring and autumn, day and night are approximately equal in length. In summer, the North Pole is closer to the Sun, resulting in longer days, while in winter, it is farther, leading to shorter days.

7.2.1 Blackbody Radiation

Every substance, whether solid, liquid, or gas, emits radiation according to its thermodynamic temperature (T), a unit in SI Kelvin (symbol K). The temperature in the Celsius scale (t) is obtained as $t = \frac{T}{\mathrm{K}} - 273.15°\mathrm{C}$ (a temperature difference of 1 °C is equal to a temperature difference of 1 K). The efficiency with which radiation is emitted varies by substance; however, at any given temperature, there is an upper limit to how much radiation can be emitted. A black body is a hypothetical object that:

- Emits radiation with the maximum possible intensity for each wavelength
- Completely absorbs all incident radiation, hence the term "black"

7.2.2 Planck's Law

Planck's law, named after the German physicist Max Planck, describes the spectral distribution of electromagnetic radiation emitted by a black body at a given temperature (Fig. 7.3).

A black body is an idealized physical entity that absorbs all incidents of electromagnetic radiation, regardless of frequency or angle of incidence.

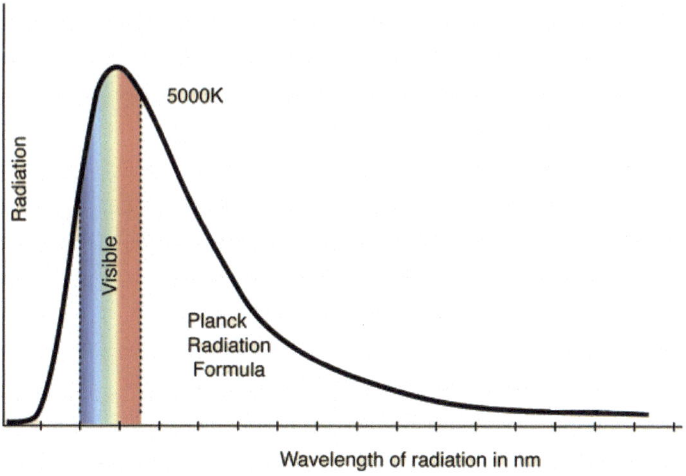

Fig. 7.3 Planck's law. This figure illustrates Planck's law, which describes the spectral distribution of electromagnetic radiation emitted by a black body at a specific temperature. It highlights the relationship between wavelength, temperature, and the intensity of radiation

$$E_\lambda^* = \frac{C_1}{\lambda^5\left[\exp\left(C_2/\lambda T\right)-1\right]} \tag{7.1}$$

In Eq. (7.1), the asterisk (*) indicates blackbody radiation. Here, λ is the wavelength, T is the absolute temperature, $c_1 = 3.741771852 \bullet 10^{-16}$ W • m^{-2} is the first radiation constant, and $c_2 = 1.438776877 \bullet 10^{-2}$ m • K is the first radiation constant. Planck's law illustrates the spectral dependence of the energy emitted by a black body at different absolute temperatures. While the Sun has an effective radiation temperature of about 6000 K, terrestrial objects have temperatures closer to 300 K. As the radiation temperature changes, the area under the Planck curve shifts, and the spectral quality of the radiation moves along the wavelength axis.

7.2.3 *Stefan–Boltzmann Law*

All objects at a certain temperature emit radiant energy. Warmer objects emit more energy per unit area compared to cooler objects. The total amount of radiant energy emitted by a black body is proportional to the fourth power of its absolute temperature, expressed by:

$$E = \sigma T^4 \tag{7.2}$$

where $\sigma = 5.670374419 \cdot 10^{-8} \frac{\mathrm{W}}{\mathrm{m^2K^4}}$ (exactly) is the Stefan–Boltzmann constant, and T is the absolute temperature. The maximum radiation emitted by warmer bodies occurs at shorter wavelengths (Fig. 7.4). As a result, solar radiation is referred to as shortwave radiation, while Earth's radiation is known as longwave or terrestrial radiation.

7.2.4 *Wien's Law*

Wien's law states that the most probable wavelength in the spectrum of thermal radiation emitted by a black body is inversely proportional to the body's absolute temperature. This relationship is given by:

$$\lambda_{max} = \frac{b}{T} \tag{7.3}$$

where T is the absolute temperature in Kelvin (K), and $b = 2.897771955 \bullet 10^{-3}$ K • m is Wien's constant. According to Wien's law, as the temperature of the body

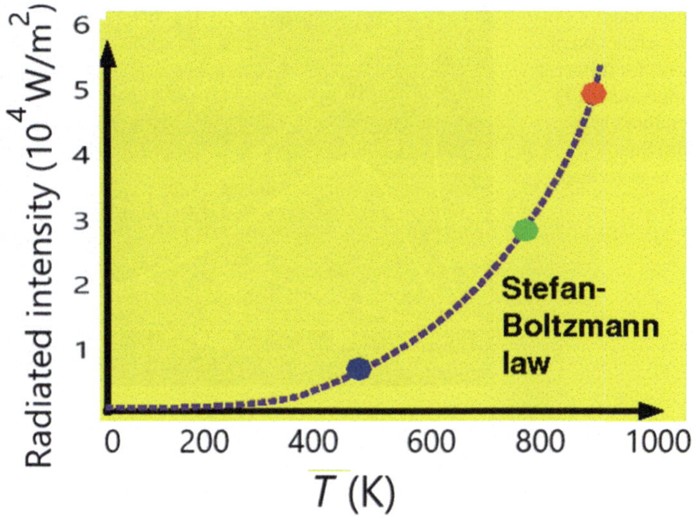

Fig. 7.4 Illustration of the Stefan–Boltzmann law: total emitted energy as a function of temperature and emissivity

increases, the peak of the spectral energy density curve (amount of energy per unit frequency or wavelength) shifts to shorter wavelengths (see Fig. 7.5). This shift demonstrates the significant difference between solar (shortwave) radiation and terrestrial (longwave) radiation. For example:

For example:

- **Sun:** T~6000 K ⇒ E~74 MW/cm^2
- **Earth:** T~300 K ⇒ E~460 W/cm^2

The maximum radiation emitted by warmer bodies is at shorter wavelengths.

$$\lambda_{max} = b / T \left(\text{Wien's law}\right) \tag{7.4}$$

$$\text{Sun}: \lambda_{max} = 0.44 \cdot 10^{-6}\ \text{m} = 0.44\,\mu m,\ \ \text{Earth}: \lambda_{max} = 9.66 \cdot 10^{-6}\ \text{m} = 9.66\,\mu m$$

Objects that are good absorbers of radiation are also good emitters. A perfect absorber/emitter is known as an absolute black body. Both the Sun and Earth are approximate absolute black bodies, absorbing and emitting nearly 100% of radiation. However, gases are selective absorbers and emitters, meaning they absorb or emit radiation only at specific wavelengths. As a result, the atmosphere is transparent to some wavelengths (visible light) but opaque to others. The presence of atmospheric gases and suspended particles affects solar radiation reaching Earth's surface, leading to phenomena such as direct, diffuse, and global solar radiation.

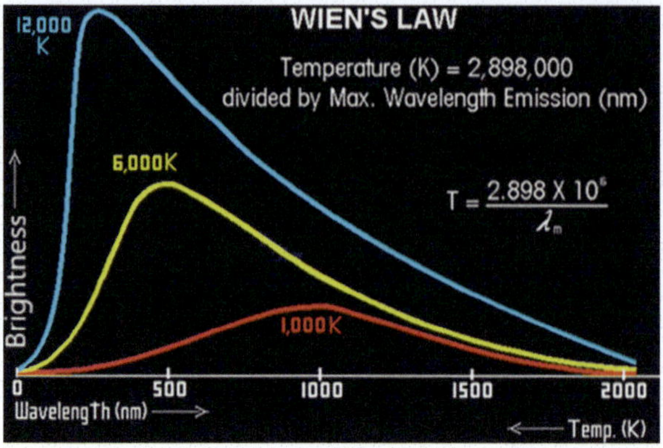

Fig. 7.5 Blackbody emissivity as a function of wavelength (λ) for different absolute temperatures (T). Each temperature curve peaks at a different wavelength, as described by Wien's law

7.3 Solar Radiation

Solar energy is the radiant energy produced by the Sun. Solar radiation refers to the energy emitted by the Sun in the form of electromagnetic waves (Fig. 7.6). Like other stars, the Sun is a massive sphere of gases and plasma—primarily hydrogen and helium—held together by strong gravitational forces (Foukal, 2013).

Solar energy is generated in the Sun's core through nuclear fusion, where hydrogen atoms fuse to form helium. The core extends from the center of the Sun to about a quarter of its radius. In this region, gravity compresses the Sun's mass, creating intense pressure sufficient to fuse atomic nuclei (Stix, 2004). During nuclear fusion, some matter is converted into energy, which is radiated into space as electromagnetic radiation. This energy is carried by photons. It takes over 100,000 years for photons produced in the Sun's core to reach its surface and just over 8 min to travel from the Sun to Earth at the speed of light (Phillips, 2012).

The Sun emits a broad spectrum of electromagnetic radiation, including visible light, infrared radiation, ultraviolet radiation, and other types of waves. This radiation is crucial for life on Earth and drives various natural processes. Approximately half of the Sun's radiation falls within the visible spectrum, while the other half is primarily in the near-infrared, with some in the ultraviolet range. Of the total energy emitted by the Sun, a small fraction reaches the Earth's surface, providing sufficient heat and energy necessary for sustaining life. The Sun is the primary driver of atmospheric processes and phenomena on Earth.

7.3.1 Solar Constant

Due to the vast distance between the Sun and Earth and the relative size of the Earth compared to the Sun, only about 1 part in a billion of the Sun's total energy reaches Earth. To quantify this, the solar constant is defined as the amount of radiant

Fig. 7.6 Solar radiation is radiant energy or electromagnetic energy emitted by the Sun that spreads in all directions into space

electromagnetic energy emitted by the Sun across all wavelengths that fall on one square meter (m^2) outside the Earth's atmosphere each second.

Due to the vast distance between the Sun and Earth and the relative size of the Earth compared to the Sun, only about 1 part in a billion of the Sun's total energy reaches Earth. To quantify this, the solar constant is defined as the amount of radiant electromagnetic energy emitted by the Sun across all wavelengths that fall on one square meter (m^2) outside the Earth's atmosphere each second. The Earth receives only a small fraction of the Sun's total energy. This energy is typically measured at the upper edge of the atmosphere to avoid the effects of atmospheric interference. To assess the solar radiation reaching Earth's surface, we need to account for atmospheric effects. Thus, the solar constant is measured in the upper atmosphere or at the Earth's surface under theoretical conditions, assuming an average distance from the Sun and excluding atmospheric influences.

The solar constant is the total radiation energy received per unit time per unit area on a theoretical surface perpendicular to the Sun's rays at the average distance of Earth from the Sun. Its value is approximately 1366.1 W/m^2 (watts per square meter). However, the solar constant is not truly constant, as it can vary by 2–5% due to factors such as the Earth's elliptical orbit, seasonal changes, sunspots, and other solar perturbations.

Another way to define the solar constant is the amount of energy potentially available to the Earth. Insolation refers to the amount of solar radiation received on a given surface over a specific period. For example, daily insolation is the total solar radiation incident on a horizontal surface per square meter, integrated over 1 day. Insolation can vary seasonally and with diurnal changes in cloud cover. Typically, there is a significant difference between the solar constant and actual insolation at Earth's surface due to atmospheric attenuation, with insolation being less than 50% of the solar constant, meaning about 50% of the solar energy is lost to the atmosphere.

7.3.2 Direct Solar Radiation

The solar radiation covering our sky can be direct, diffuse, or reflected radiation (Fig. 7.7). The Sun always emits light when direct sunlight falls on the Earth's surface.

Sunlight that reaches the Earth's surface unmodified by some of the atmospheric processes is called direct or beam solar radiation. The intensity of direct solar radiation depends on the height of the Sun above the horizon, the height above sea level, cloudiness, transparency of the sky, the amount of water vapor in the air, and other factors. The sum of direct solar radiation (S) and diffuse radiation (D) gives the global radiation (G):

$$G = S + D \tag{7.5}$$

The intensity of global radiation depends on the same factors as the intensity of direct radiation and is diffused. Sunlight falls on the Earth's surface from a different angle and a different direction. Thus, it causes a different heat effect. For example, open solar radiation lasts all day, from sunrise to sunset, and the Sun reaches its maximum value in the afternoon. During the summer, solar radiation has the lowest value in the afternoon. Also in winter, the Sun has the lowest position about the horizon. The value of insolation changes with the change in the angle of inclination of the Earth's surface and the change in latitude. This means that when the Sun's rays strike the equator, they are concentrated over a smaller surface area, resulting in greater heating. In contrast, at higher latitudes, the rays spread over a larger surface area, leading to less heating.

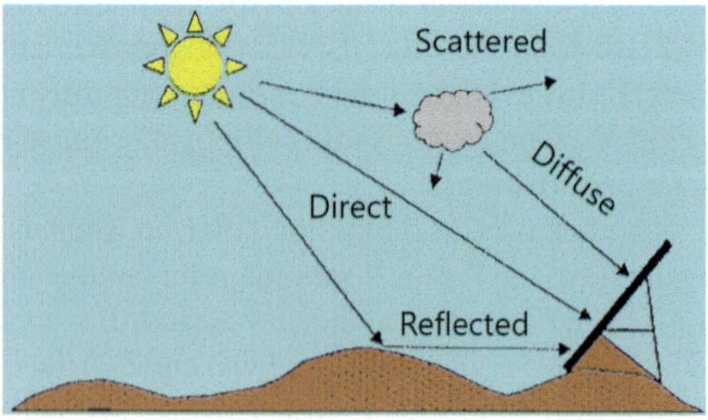

Fig. 7.7 Direct, reflected, and diffuse radiation

7.3.3 *Diffuse Solar Radiation* (D)

Beyond direct solar radiation, the Earth also receives diffuse solar radiation. This type of radiation occurs when direct sunlight is scattered as it passes through the atmosphere. As solar rays collide with molecules of various gases, water vapor, and tiny particles like minerals or organic compounds in the atmosphere, they are redirected in multiple directions. This scattering of sunlight, known as diffuse reflectivity, results in a soft illumination that is particularly noticeable under clear skies, especially with short-wavelength rays like blue, violet, and ultraviolet light. Diffuse radiation is present throughout the day, from sunrise to sunset, and its intensity is influenced by several factors: altitude above sea level, cloud cover, atmospheric transparency, and the Sun's position in the sky.

7.4 Solar Radiation and Earth's Energy Balance: Sources of Thermal Energy

The primary source of energy for the Earth is the Sun. Solar radiation drives processes such as photosynthesis, warming the Earth's system, including land, ocean masses, and the atmosphere—and all associated phenomena. The total energy flux from the Sun to the Earth is approximately 174 PW (Trenberth et al., 2009). Another energy source is the Earth's internal heat. The current total heat flow at the Earth's surface is estimated to be 46 ± 3 TW. This includes contributions from heat entering the mantle from the core, mantle cooling, radiogenic heating from the decay of radioactive elements, and other less significant processes (Jaupart et al., 2007).

However, this internal energy flow is about 4000 times smaller than the solar energy input and can be considered negligible in the Earth's overall heat balance (Pollack et al., 1993).

Losses of Thermal Energy from the Earth

The Earth radiates energy back into space, consistent with a body at a temperature above absolute zero. The Sun's temperature, as observed from Earth, is about 5500 °C, while the Earth's average temperature, as seen from space, is approximately −20 °C.

Solar radiation does not heat the Earth uniformly. This uneven heat occurs because the Sun's rays strike different parts of the Earth at varying angles, depending on geographic latitude and time. The amount of solar energy received by a given surface influences climate patterns, with much of the energy, aside from what is converted into chemical energy via photosynthesis, being redistributed by the oceans and atmosphere. This redistribution, known as the climate heat engine, moves energy from the equator to the poles, from the Earth's surface to the atmosphere, and eventually back into space. When the incoming solar energy is roughly balanced by the energy radiated back into space, the Earth achieves a state of equilibrium, maintaining an approximately constant temperature (Hartmann, 2016).

7.4.1 Characteristics of the Radiation Reaching the Earth

To accurately explain Earth's climate, reliable radiation measurements are essential. These measurements are largely obtained from satellite systems that monitor the Earth's energy balance, known as the Earth's radiation budget (ERB). According to the latest data, the solar radiation flux entering Earth's atmosphere is approximately 1361.1 W/m^2, as indicated by the AM0 spectrum. This corresponds to the radiation emitted by an ideal black body at a temperature of about 5572 K.

The AM0 spectrum represents the distribution of spectral irradiance, which is depicted in Fig. 7.8. Although this spectrum closely resembles that of an ideal black body with a temperature of 5572 K, it is not identical. The slight differences are due to the unique characteristics of solar radiation as it passes through space before reaching Earth's atmosphere.

The AM0 spectrum is particularly significant for extraterrestrial applications, such as the use of solar cells on orbital stations outside Earth's atmosphere, where accurate knowledge of solar irradiance is crucial.

The atmosphere, or air mass (AM), plays a crucial role in absorbing and scattering radiation in two directions: the incoming solar radiation and the outgoing radiation from Earth into space.

As solar radiation passes through the atmosphere from space to Earth's surface, part of it is absorbed and scattered, altering the spectrum of the radiation that ultimately reaches the surface. The AM1.5 spectrum is considered typical for solar radiation that has traveled through the atmosphere at an angle of 48.19° relative to the zenith (where sec (48.19°) = 1.5). This angle indicates that the solar radiation has passed through a path 1.5 times longer than the atmospheric thickness.

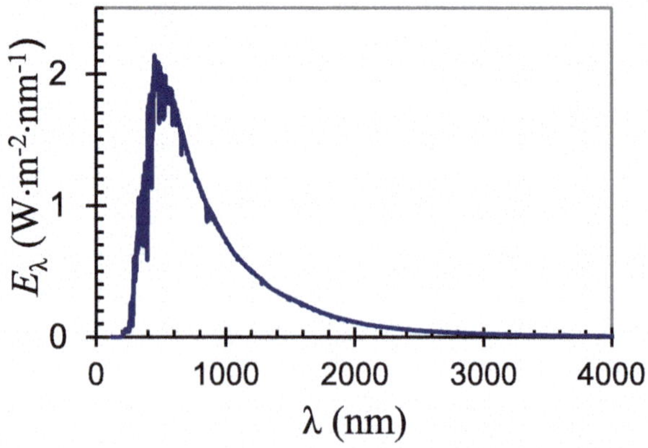

Fig. 7.8 Spectrum of radiation at the entrance to the atmosphere (AM0)

Two types of AM1.5 spectra are important: Direct Spectrum: This spectrum represents the solar radiation that reaches the Earth's surface directly after being absorbed and scattered by the atmosphere. It is measured along the path of solar radiation as it passes through the atmosphere and reaches the surface (illustrated in Fig. 7.9).

Global Spectrum: The AM1.5 global radiation spectrum encompasses the total radiation reaching the Earth's surface from all directions. This includes both the direct solar radiation that has been attenuated by the atmosphere and the diffuse radiation scattered from all directions (depicted in Fig. 7.10).

The AM1.5 global radiation spectrum is critical for calculations involving the conversion of solar energy into electricity for terrestrial installations. It is also relevant to processes such as photosynthesis and the melting of ice in polar regions.

Fig. 7.9 Spectrum of radiation falling on the Earth's surface (AM1.5)—direct

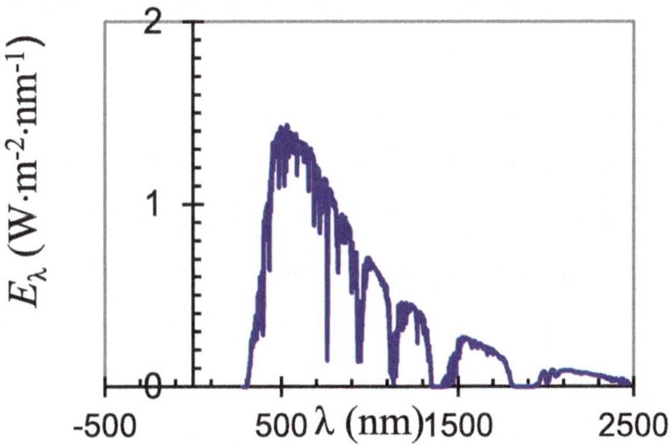

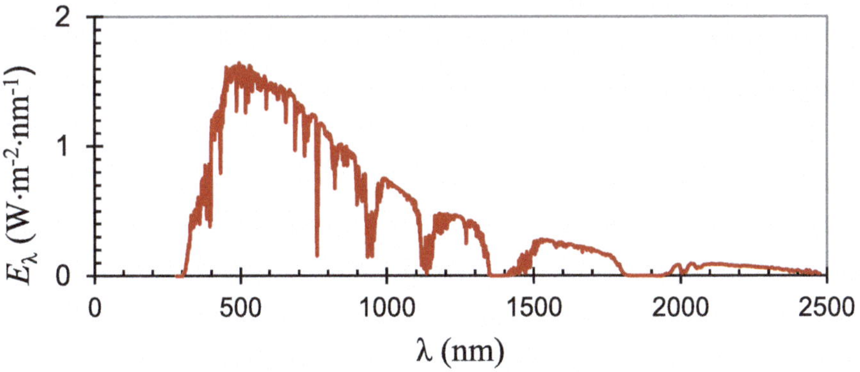

Fig. 7.10 A spectrum of radiation falling on the surface of the Earth (AM1.5)—global

7.5 Solar Radiation Factors

Absorption: The atmosphere plays a critical role in the interaction with both incoming solar radiation and outgoing terrestrial radiation. The extent to which the atmosphere interacts with radiation varies by wavelength and is crucial in heating the lower atmosphere. Absorption is the process by which atmospheric gases and particles reduce the amount of solar or terrestrial radiation by transferring energy, which increases molecular movement. Due to the chemical composition of the Earth's atmosphere, much of the infrared radiation emitted by the warm surface is not directly lost to space. Instead, it is reflected or absorbed by greenhouse gases, such as carbon dioxide, water vapor, and methane. When these gases absorb infrared radiation from the Earth's surface, the atmosphere is warming and the energy reflected to the surface further increases its temperature. This cycle of absorption and re-emission creates a feedback loop that maintains the warmth of both the atmosphere and the Earth's surface. The lower atmosphere is primarily composed of gases with varying absorption capabilities. For example, nitrogen (N_2) is a poor absorber, while oxygen (O_2), ozone (O_3), and water vapor (H_2O) are effective absorbers, contributing significantly to the overall absorption in the atmosphere.

Reflection is the process by which solar radiation is redirected toward space without being absorbed by the Earth's surface. The amount of sunlight reflected depends on the albedo, which is the reflectivity of the Earth's surface and the intensity of the incoming sunlight. Surfaces with high albedo, such as ice or snow, reflect more sunlight, whereas darker surfaces, like oceans or forests, absorb more radiation and reflect less.

Albedo is a measure of a surface's ability to reflect solar radiation. An albedo value of 1 represents an ideal reflector, meaning that all incident solar radiation is reflected. Table 7.1 provides the albedo values for various surfaces on Earth.

The highest albedo, indicating the greatest reflection of solar radiation, is observed in freshly fallen snow, with values ranging from 75% to 95%. In contrast, the lowest albedo values, between 5% and 10%, are found in dark surfaces such as asphalt streets and dark land areas. Albedo also varies with the angle of the Sun: water surfaces can have an albedo ranging from 5% at high Sun angles to 80% at low Sun angles. On a global scale, the overall albedo of Earth and its atmosphere is approximately 30%. This value is influenced predominantly by the extensive

Table 7.1 Reflection of solar radiation—albedo

Subject or surface	Albedo (%)
Fresh snow	75–95
Old snow	40–60
Sand	20–30
The land (grass/vegetation)	15–25
Black street	5–10
Thick cloud	70–80
Thin cloud	25–30

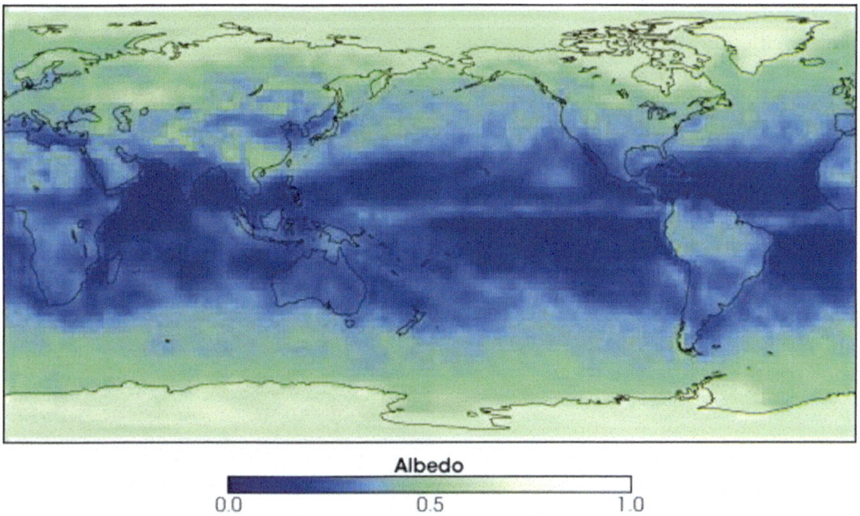

Fig. 7.11 Earth's average albedo for March 2005: This figure shows the Earth's average albedo for March 2005, as measured by the Clouds and the Earth's Radiant Energy System (CERES) instrument on NASA's Terra satellite. (Source: NASA Earth Observatory)

coverage of clouds and is lower than the albedo of oceanic areas but higher than that of terrestrial surfaces (Fig. 7.11).

Scattering refers to the redirection of incoming solar radiation in all directions, not just back into space. Solar radiation travels in a straight line but can be redirected or scattered by gases (Rayleigh scattering) or aerosols (Mie scattering). Scattering changes the direction of light without altering its wavelength. Three primary types of scattering mechanisms are observed:

Rayleigh Scattering: This occurs when light interacts with particles much smaller than the wavelength of the incident light, typically less than one-tenth of the wavelength. Rayleigh scattering is wavelength-dependent, with shorter wavelengths (blue and violet) being scattered more effectively than longer wavelengths. This process explains the blue color of the sky.

Non-Selective Scattering: This type of scattering is independent of the wavelength of the incident light. It occurs when the particles responsible for scattering are much larger than the wavelength of the light. Non-selective scattering affects all wavelengths equally, which contributes to the whiteness of clouds and fog.

Mie Scattering: This occurs when light interacts with particles whose size is comparable to or larger than the wavelength of the incident light. Unlike Rayleigh scattering, Mie scattering affects a broad range of wavelengths and is responsible for phenomena like the white appearance of clouds and haze. It is caused by particles such as pollen, dust, smoke, and water droplets in the lower atmosphere.

Figure 7.12 visually represents these three scattering mechanisms, highlighting how different types of particles interact with light.

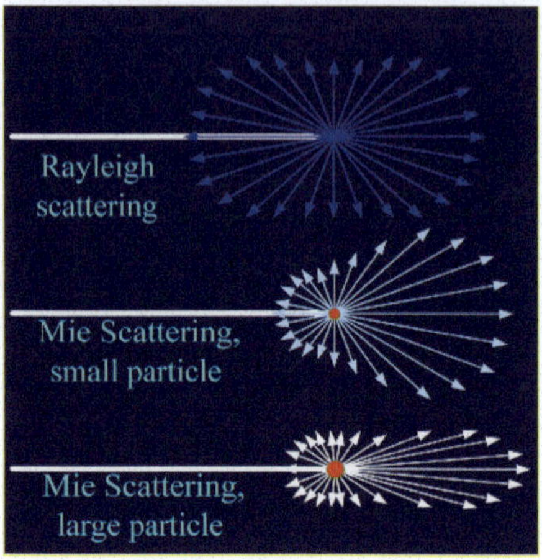

Fig. 7.12 Rayleigh vs. Mie scattering of light

7.6 Temporal and Spatial Changes in Insolation

The Earth's energy balance can be conceptualized as an arbitrary 100 units of incoming solar radiation. Insolation, or the amount of solar radiation received at the Earth's surface, varies over time and space. These variations occur due to seasonal changes and differences in latitude, leading to geographic and seasonal fluctuations in temperature (*T*). To understand these variations, it is essential to consider key physical principles and relationships within the Earth–Sun system:

Rotation of the Earth: The Earth's rotation on its axis causes daily changes in insolation, creating a cycle of day and night. This rotation influences the distribution of solar radiation across the surface throughout the day.

Revolution of the Earth: The Earth's orbit around the Sun, known as its revolution, affects seasonal changes. Although the distance between the Earth and the Sun does vary slightly over a year, this variation has a minimal impact on seasonal temperature changes. For instance, during the Northern Hemisphere's winter, the Earth is closer to the Sun, yet this proximity has little effect compared to the tilt of the Earth's axis, which plays a more significant role in seasonal temperature variations.

Seasonal and Latitudinal Changes in Insolation: Seasonal variations in geographic insolation are significantly influenced by the altitude of the Sun, which refers to the angle at which sunlight strikes the Earth's surface. During summer, the midday Sun is high in the sky, providing more direct solar radiation, whereas in winter, the Sun is lower on the horizon. This variation in the Sun's altitude affects the amount of solar energy reaching the Earth's surface. When the Sun is at a lower angle, its rays pass through a greater thickness of the atmosphere, increasing the likelihood of absorption or reflection of solar radiation. The Earth's spherical shape further impacts insolation. Vertical rays of the Sun reach only specific latitudes. As one moves north or south, the angle of the Sun and the

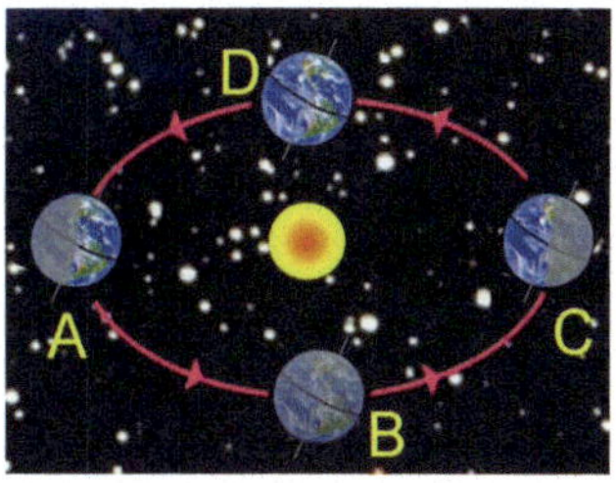

Fig. 7.13 The Earth's revolution: equinox, solstice, seasons, perihelion, and aphelion

length of daylight change, altering the amount of solar energy received at the surface. This variation is due to the tilt of the Earth's axis, which is inclined at 23.5° relative to its orbital plane around the Sun. For example, during the June solstice, the Northern Hemisphere is tilted toward the Sun, resulting in longer days and more direct sunlight. Conversely, during the December solstice, the Northern Hemisphere is tilted away from the Sun, leading to shorter days and less direct sunlight. These changes create the opposite conditions in the Southern Hemisphere. The Tropics of Cancer (23.5°N) and Capricorn (23.5°S) mark the latitudinal boundaries where the Sun can be directly overhead at noon during the solstices. At the Arctic Circle (66.5°N) during the summer solstice, regions experience continuous daylight, while the Antarctic Circle (66.5°S) undergoes continuous darkness. The opposite occurs during the winter solstice. These variations in daylight and solar energy contribute to the seasonal differences in temperature, with higher temperatures typically observed in summer due to the more direct solar radiation. Figure 7.13 depicts the Earth's revolution around the Sun, highlighting the equinoxes, solstices, seasons, perihelion, and aphelion, and illustrates how these factors influence seasonal and latitudinal changes in insolation.

7.7 Optical Radiation

Optical radiation is a segment of the electromagnetic spectrum that spans wavelengths from 100 to 10,000 nm. This range of wavelengths can be divided into three primary areas: ultraviolet (UV), visible light, and infrared (IR) radiation.

- Ultraviolet (UV) belt: 100–400 nm
- Visible (VIS) band: 400–770 nm
- Infrared (IR) band: 770–10,000 nm

Each of these regions has distinct characteristics and effects on both the environment and biological systems. UV radiation occupies the shortest wavelengths within this range, followed by visible light, which is the only portion perceptible by the human eye, and finally, infrared radiation, which carries heat and extends into the longest wavelengths of the optical spectrum. Optical radiation plays a critical role in various natural processes, including photosynthesis and the regulation of Earth's

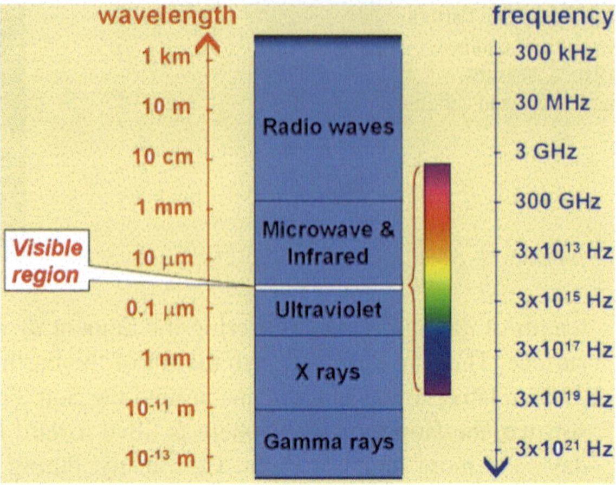

Fig. 7.14 Electromagnetic spectrum: wavelength and frequency

temperature, and has numerous applications in science, technology, and medicine (Kollias & Stamatas, 2002). Figure 7.14 illustrates the division of optical radiation into these three areas, showing how each segment of the spectrum is utilized in various scientific and practical applications.

7.7.1 Visible Light

Visible light extends from approximately 400 to 700 nm in the electromagnetic spectrum and is the range detectable by the human eye. This segment of the spectrum encompasses the colors visible to us, ordered from shortest to longest wavelength: violet, blue, green, yellow, orange, and red. The perception of these colors allows us to experience the diverse hues of our environment, with each color corresponding to a specific wavelength within this range. Visible light plays a crucial role in human vision and various practical applications, such as photography, colorimetry, and the study of light properties.

7.7.2 Infrared (IR) Radiation

Infrared (IR) radiation spans from approximately 700 to 10,000 nm in wavelength and is emitted by objects based on their temperature. It plays a crucial role in thermal imaging, heat transfer, and various other applications. Infrared radiation is categorized into three main regions based on wavelength:

1. **Near-Infrared (NIR):** Ranges from about 700 to 1400 nm. This portion is closest to visible light and is commonly used in applications such as night-vision technology, remote sensing, and telecommunications.

2. **Mid-Infrared (MIR):** Spans from approximately 1400 to 3000 nm. MIR radiation is widely used in thermal imaging, environmental monitoring, and chemical analysis due to its interaction with molecular vibrations.
3. **Far-Infrared (FIR):** Extends from around 3000 to 10,000 nm. FIR radiation is associated with heat emitted by objects and is utilized in thermal cameras, heating systems, and certain industrial processes.

Compared to other forms of electromagnetic radiation, infrared radiation is generally less harmful. As the wavelength increases, the potential for harmful effects decreases, making IR radiation relatively safe for most applications.

7.7.3 Ultraviolet (UV) Radiation

Ranging from 100 to 400 nm, UV radiation is responsible for phenomena such as sunburn and has effects on biological systems and materials. UV radiation plays a crucial role in both biological and environmental processes, highlighting the importance of the ozone layer in filtering out harmful UV rays (Diffey, 2002; Rivas & Fernández, 2007). It is further subdivided into UV-A, UV-B, and UV-C based on wavelength. UV radiation is known for its effects on biological systems and materials, such as causing sunburn and influencing the health of living organisms. It is further subdivided into three regions based on wavelength:

UV-A (320–400 nm): The longest wavelengths in the UV spectrum, UV-A radiation penetrates deeply into the skin and is associated with skin aging and long-term skin damage. It is commonly used in black lights and certain types of fluorescent lamps.

UV-B (280–320 nm): UV-B radiation is responsible for sunburn and can cause DNA damage in living cells, which is linked to skin cancer and other health issues. It is partially absorbed by the ozone layer and has significant effects on the environment.

UV-C (100–280 nm): The shortest wavelengths, UV-C radiation is highly energetic and effectively absorbed by the ozone layer and atmosphere. It is used in germicidal applications, such as disinfection and sterilization, due to its ability to kill microorganisms.

Figure 7.15 illustrates the range of UV radiation that reaches the Earth's surface, highlighting the variation in wavelengths compared to the broader UV spectrum. The diagram shows the attenuation of UV radiation as it passes through the atmosphere and the different UV regions that impact the Earth's surface. UV radiation is a natural component of sunlight but can also be artificially generated by certain light sources, such as UV lamps used in various industrial and medical applications.

UV-C radiation is completely absorbed in the upper atmosphere by oxygen and ozone molecules. Most UV-B radiation is absorbed by ozone molecules in the stratosphere, with only a small fraction reaching the Earth's surface. Since ozone is the primary absorber of UV-B radiation, its intensity at the Earth's surface is closely linked

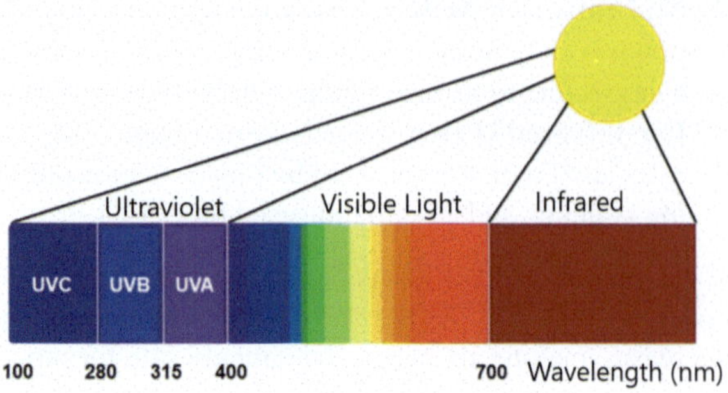

Fig. 7.15 Ultraviolet (UV) spectrum at the Earth's surface

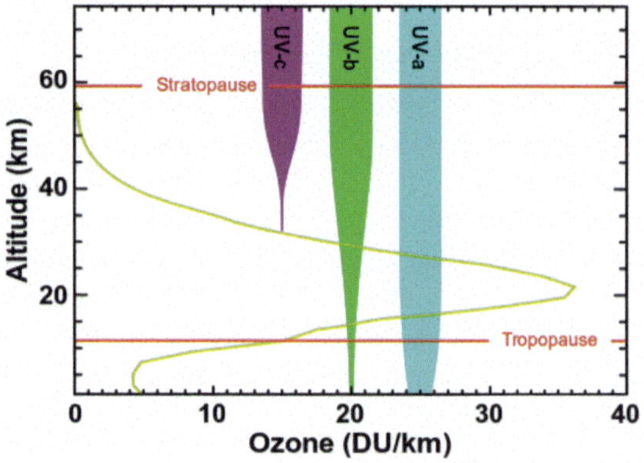

Fig. 7.16 Ozone levels at different altitudes and UV radiation blocking. (Credit: NASA, Public domain, via Wikimedia Commons)

to the total amount of ozone in the atmosphere and its distribution within the ozone layer (Fig. 7.16). UV-B radiation is biologically harmful and can cause significant damage to living organisms. On the Earth's surface, UV-A radiation is more prevalent.

Although UV-A is less harmful than UV-B, it can still cause skin darkening in humans. High-intensity UV radiation can overwhelm the natural protective mechanisms of biological systems, leading to potential damage to the skin and eyes.

7.7.4 UV Index

To mitigate the risks associated with UV exposure, the UV index (UVI) is provided to inform the public about UV radiation levels and appropriate precautions. The UV index measures the irradiance or radiation energy received per unit area, typically expressed in watts per square meter (W/m^2) and indicates the potential for harm from UV radiation.

Figure 7.17 depicts the global distribution of the UV index, illustrating the variation in solar UV radiation intensity across different regions. The UV index values show the potential risk of UV-related skin and eye damage, highlighting areas with higher and lower UV exposure. As the index value increases, the risk of UV-related harm rises, and the time required for damage to occur decreases.

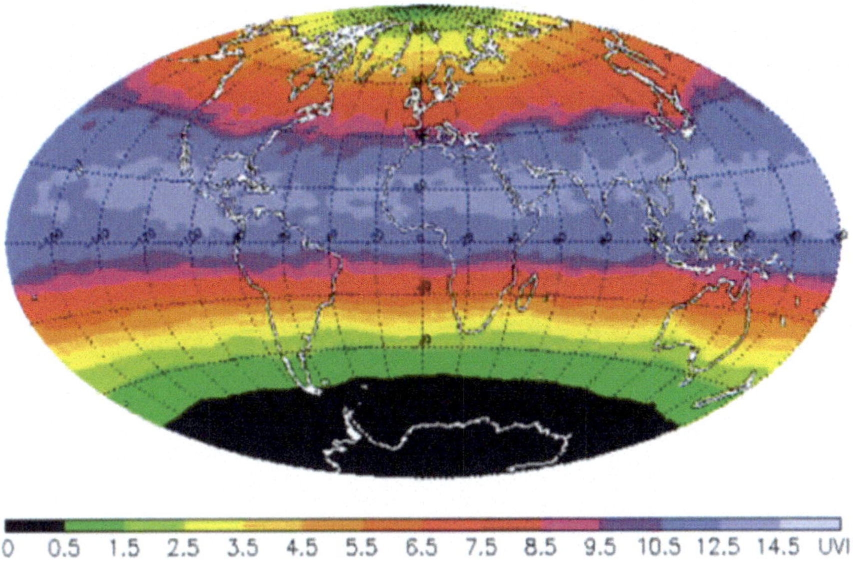

Fig. 7.17 Global distribution of the UV index

7.7.5 *Factors Affecting the UV Index*

Several key factors influence the intensity of UV radiation:

Sun's Elevation Angle: The altitude of the Sun refers to its angle above the horizon. A higher Sun angle means the Sun's rays travel a shorter path through the atmosphere, resulting in more direct UV radiation reaching the Earth's surface. The Sun is highest at midday and during summer, particularly near the equator, leading to greater UV intensity. In contrast, at higher latitudes, the angle of sunlight is lower, and UV intensity is reduced.

Altitude: UV radiation increases with elevation above sea level due to the thinner atmosphere, which absorbs fewer UV photons. For every 1000 m of altitude, UV intensity can increase by 6–8%.

Atmospheric Scattering: UV radiation is scattered by gases and particles in the atmosphere. Direct radiation passes through the atmosphere without scattering, while diffuse radiation has been scattered before reaching the Earth's surface. Scattering is wavelength-dependent; UV radiation is more diffusely scattered than visible light, contributing to its presence even on cloudy days. UV-B radiation at the surface typically comprises nearly equal parts of direct and diffuse radiation.

Cloud Cover and Atmospheric Humidity: UV intensity is generally higher on clear days, as clouds reduce UV radiation. However, thin or broken clouds may have a minimal impact on UV levels. Low clouds can sometimes enhance UV radiation, while atmospheric fog and haze absorb and scatter UV radiation, leading to reduced UV levels.

Surface Reflection: The reflection of UV radiation from the Earth's surface varies by surface type. Fresh snow can reflect up to 80% of UV radiation, significantly increasing UV exposure, particularly at high altitudes and latitudes. Sand reflects about 25% of UV radiation, and clear water allows about 95% of UV radiation to penetrate, with up to 50% penetrating to depths of 3 m.

In summary, the UV index is affected by a combination of the Sun's angle, altitude, atmospheric scattering, cloud cover, and surface reflectivity, each playing a role in determining UV exposure levels.

7.7.6 Former and Present Status of Ozone in the Troposphere

Ozone, although present in trace amounts in the atmosphere, is essential in protecting life on Earth. For every ten million molecules of air, there are about three molecules of ozone. Despite its rarity, ozone is concentrated in two main regions of the atmosphere: the stratosphere and the troposphere. Approximately 90% of atmospheric ozone resides in the stratosphere, between 10 and 50 km above the Earth's surface, in a region known as the ozone layer or ozonosphere. The remaining ozone layer is found in the troposphere, the lower atmosphere closest to the Earth's surface. While stratospheric and tropospheric ozone molecules are chemically identical (O_3), their roles and effects differ significantly. Stratospheric ozone absorbs biologically harmful ultraviolet (UV-B) radiation from the Sun, preventing excessive UV-B from reaching the surface (Molina & Rowland, 1974). This absorption also contributes to the stratosphere's temperature structure, where temperatures increase with altitude. In contrast, tropospheric ozone, especially near the Earth's surface, is a major component of photochemical smog, a significant air quality issue in urban and rural areas alike.

Ozone depletion has been a major environmental concern, particularly with the discovery of the "ozone hole" over Antarctica in the mid-1980s (Farman et al., 1985). For example, on September 3, 2000, a NASA instrument detected an Antarctic ozone hole spanning 28 million km^2 (Fig. 7.18a). The lowest recorded ozone level, about 85 Dobson units, was observed on October 8, 2006 (Solomon, 1999). Stratospheric ozone depletion is primarily caused by human-made chemicals, including chlorofluorocarbons (CFCs), carbon tetrachloride, chloroform, and methyl bromide. These compounds contain chlorine, fluorine, bromine, carbon, and hydrogen and are widely used in refrigeration, air conditioning, electronics cleaning, and as solvents.

Recent reports by NASA and NOAA highlight significant progress in the recovery of the ozone layer. In 2024, the Antarctic ozone hole was relatively small compared to earlier decades (Fig. 7.18b). During the peak depletion season (September 7 through October 13), the average area of the hole ranked as the seventh smallest since recovery efforts began in 1992, following the implementation of the Montreal Protocol. This landmark international agreement, signed in 1987, mandated the phase-out of ozone-depleting substances. At its largest in 2024, the ozone hole reached 22.4 million km^2 (September 28), a size nearly three times that of the contiguous United States (NASA, 2024). This improvement reflects continuing declines in CFC emissions and an unexpected influx of the ozone layer carried by air

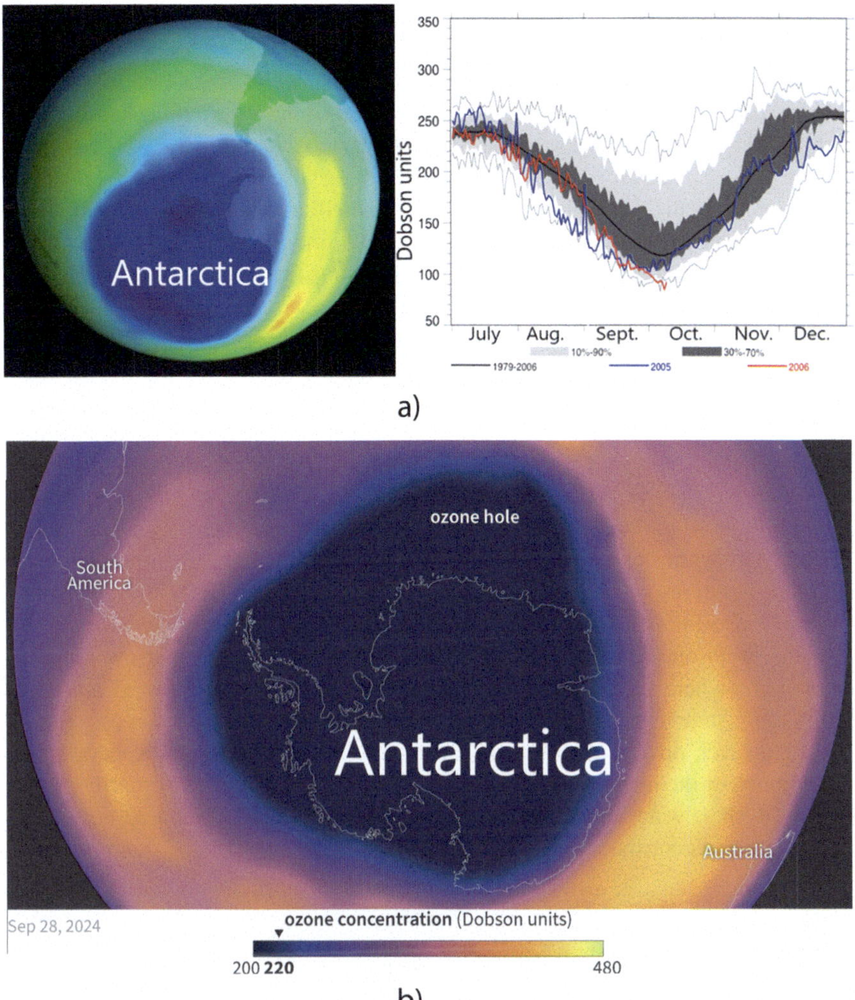

Fig. 7.18 (**a**) The largest area of ozone depletion over Antarctica captured in September 2000. (Data sourced from the Total Ozone Mapping Spectrometer (TOMS) on NASA's Earth Probe satellite); (**b**) The Antarctic ozone hole on September 28, 2024. The dark blue region shows where the stratospheric ozone concentration is less than 220 Dobson Units. (Image credit: NOAA Climate.gov)

currents from north of Antarctica. However, the 2024 ozone hole still ranked as the 20th smallest since satellite monitoring began in 1979, underscoring the long recovery timeline. As existing CFC levels decline over the coming decades, ozone in the upper atmosphere is expected to rebound, with full recovery projected by 2066 (NOAA, 2024). This recovery is critical, as the ozone layer serves as a planetary sunscreen, shielding life from harmful UV radiation. Depleted ozone levels increase UV exposure, which can lead to higher rates of skin cancer, cataracts, and ecosystem damage, including reduced agricultural yields and harm to aquatic organisms.

The gradual recovery demonstrates the success of international cooperation and the efficacy of the Montreal Protocol. As Paul Newman, NASA's ozone research team leader, remarked, "The gradual improvement we've seen in the past two decades shows that international efforts that curbed ozone-destroying chemicals are working." However, NOAA scientist Stephen Montzka noted, "The ozone layer is still far from being fully healed." To track progress, scientists rely on advanced monitoring systems, including NASA's Aura satellite and NOAA's Suomi NPP, NOAA-20, and NOAA-21 satellites. These tools provide essential data to guide future environmental policies and ensure the continued recovery of the ozone layer.

7.8 Earth's Long-Wave Radiation

The Earth emits energy in the long-wave spectrum, distinct from the Sun's short-wave energy emission. This long-wave radiation, often referred to as terrestrial radiation (TR), is emitted by the Earth's surface and atmosphere. The Earth's surface typically warms up to around 20 °C on average through the absorption of solar radiation, and in response, it emits thermal infrared radiation (IR) with wavelengths ranging from 5 to 50 μm.

In the atmosphere, there exists an "atmospheric window," a narrow range of wavelengths where the atmosphere is relatively transparent to infrared radiation. Within this window, some of the Earth's thermal infrared radiation can escape into space. However, outside this window, absorption bands due to atmospheric gases absorb much of the infrared radiation. These absorption bands are primarily caused by gases such as water vapor, carbon dioxide, and methane, which trap heat in the atmosphere and contribute to the greenhouse effect.

7.8.1 Distribution of Longwave Radiation

The distribution of the mean global outgoing longwave radiation from 1979 to 1995 is shown in Fig. 7.19. The amount of longwave radiation escaping into space is influenced by various factors including latitude, season, time of day, cloudiness, aerosol backscattering, and surface characteristics. Water vapor (H_2O) and carbon dioxide (CO_2) are the primary absorbers of longwave radiation, with water vapor being five times more effective in absorbing than other gases. Ozone (O_3), methane (CH_4), and other trace gases also contribute to absorption.

The greenhouse effect occurs because the atmosphere is transparent to solar radiation but absorbs and re-emits terrestrial radiation. Gases that absorb Earth's long-wave radiation and re-radiate it in all directions are known as greenhouse gases. Some of this re-emitted energy returns to the Earth's surface, further warming it. All forms of electromagnetic radiation, including longwave radiation, form part of the electromagnetic spectrum. Bodies at typical atmospheric temperatures emit radiation in the infrared part of the spectrum, while bodies at higher temperatures emit

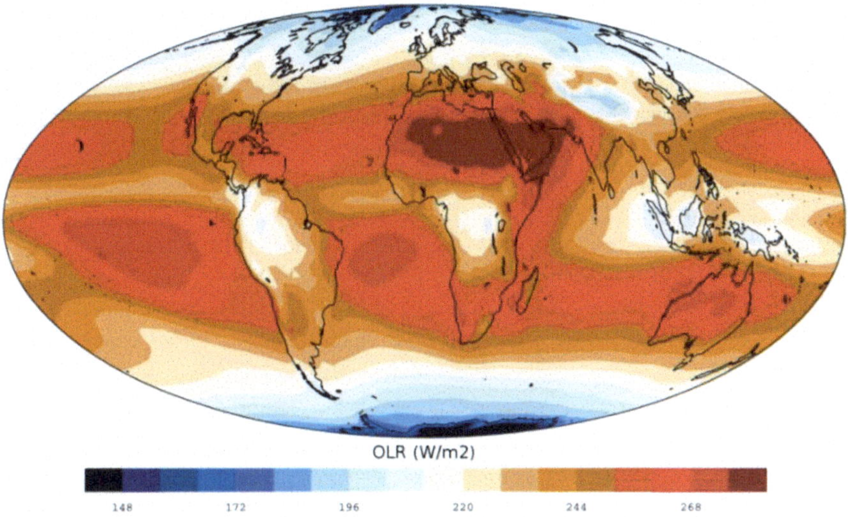

Fig. 7.19 Outgoing longwave radiation

visible light. The solar spectrum, which represents the distribution of electromagnetic radiation from the Sun as a function of wavelength, interacts with the upper atmosphere and influences the Earth's energy balance.

7.9 Earth's Global Energy Budget

Earth's global energy budget represents the balance between incoming solar energy and outgoing terrestrial radiation. This balance is crucial for maintaining the Earth's climate system and is visualized through a schematic diagram of the Earth's energy budget shown in Fig. 7.20. Understanding this energy balance is essential for studying climate dynamics, as it dictates the Earth's temperature and weather patterns. NASA (2018) has conducted extensive research on the Earth's energy budget, providing valuable insights into how solar energy is absorbed, reflected, and emitted back into space. In addition, comprehensive assessments, such as those compiled by Houghton et al. (2001), further elucidate the interactions within the climate system and emphasize the importance of maintaining this energy equilibrium.

Components of the Energy Budget

1. **Incoming Solar Radiation:** The Sun provides the Earth with energy at an average rate of 342 W/m^2. This energy reaches the Earth as shortwave radiation.
2. **Reflection and Absorption:** Of the incoming solar radiation, approximately 77 W/m^2 is reflected into space by the Earth's atmosphere, including clouds and aerosols. Aerosols are fine particles or droplets suspended in the atmosphere, including substances like sulfuric acid formed from atmospheric oxidation of volatile sulfur compounds. Clouds, a specific type of aerosol, also play a significant role in reflecting solar radiation.

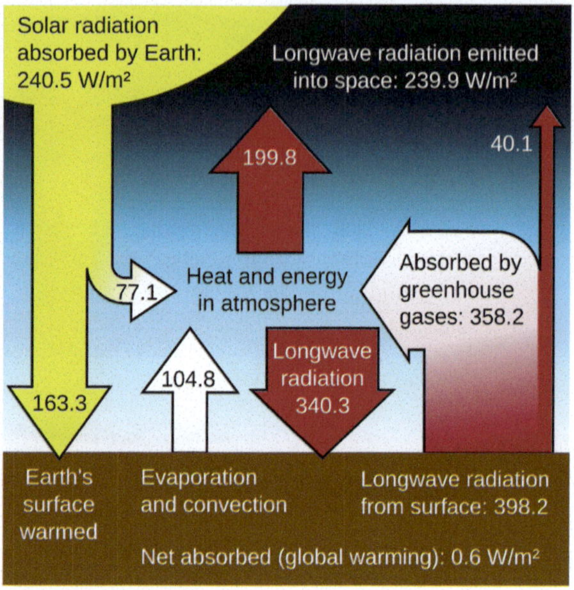

Fig. 7.20 Diagram showing Earth's energy budget, including the greenhouse effect. Energy flows between space, the atmosphere, and Earth's surface, with greenhouse gases in the atmosphere absorbing and emitting radiant heat, affecting Earth's energy balance. (Credit: https://commons.m.wikimedia.org/)

3. **Absorption by Greenhouse Gases:** The atmosphere absorbs about 67 W/m^2 of solar energy. This absorption primarily occurs due to greenhouse gases, which include water vapor, carbon dioxide, and methane. In the stratosphere, approximately 10 W/m^2 of this energy is absorbed by oxygen molecules, which dissociate into two oxygen atoms under the influence of high-energy ultraviolet photons.
4. **Outgoing Longwave Radiation:** The energy that is not reflected or absorbed is eventually emitted back into space as longwave (infrared) radiation. This outgoing radiation is essential in regulating the Earth's surface temperature and is influenced by the greenhouse effect, where certain gases trap heat in the atmosphere.

The Earth's energy budget is a dynamic system where the energy inputs from the Sun are balanced by the energy outputs from the Earth. This balance is essential for maintaining the climate and supporting life on Earth.

Reactive oxygen atoms combine with oxygen molecules (O_2) to form ozone (O_3) through the reaction: (O) + (O_2) → (O_3). This process is crucial for life on Earth because ozone absorbs the majority of incoming ultraviolet (UV) radiation from the Sun. The remaining solar radiation that is not absorbed or reflected by the atmosphere reaches the Earth's surface, where 30 W/m^2 is reflected into space, and 168 W/m^2 is absorbed by the surface. This absorbed energy heats the Earth's surface, which then emits infrared radiation back into the atmosphere. The Earth's energy budget illustrates how incoming solar radiation is distributed and how outgoing longwave radiation is emitted across the planet. There is a noticeable imbalance in energy distribution across latitudes: higher latitudes experience a deficit of energy, leading to cooling, while the equatorial region has an excess of energy, resulting in warming. This imbalance drives the transfer of heat from the equatorial regions to the polar areas, helping to balance the global energy distribution.

7.9.1 *Latitudinal Variations in Solar Radiation*

The seasonal variations in solar radiation received at different latitudes are a fundamental driver of global energy, air, and moisture circulation. The equatorial regions receive the highest levels of solar radiation throughout the year, leading to a net surplus of heat between the equator and the mid-latitudes (approximately 30°–35° north and south). In contrast, the higher latitudes, from the mid-latitudes to the poles, experience a net radiation deficit, where outgoing longwave radiation exceeds the incoming shortwave solar radiation (Fig. 7.21). Despite the stark differences in radiation received at different latitudes, the tropics do not continuously become hotter, nor do the poles become increasingly colder. This is due to the dynamic heat transport mechanisms that redistribute energy from the equatorial regions to the poles, maintaining a balance in Earth's climate system. As a function of latitude, the variation in mean annual solar radiation shows that equatorial regions receive more incoming shortwave radiation than they lose in outgoing longwave radiation. This radiation surplus extends roughly between 30° and 35° north and south latitudes. Beyond these latitudes, the balance shifts, and regions from 35° to the poles experience a radiation deficit. This imbalance between the tropics and the poles creates a radiation gradient, which drives the large-scale heat transport processes within the atmosphere and oceans.

The movement of energy from the equator toward the poles is not uniform throughout the year. Seasonal variations, influenced by Earth's axial tilt, lead to changes in solar radiation distribution, causing shifts in the strength and patterns of atmospheric and oceanic circulation. These variations influence global weather patterns, such as the formation of monsoons, trade winds, and jet streams, which are essential for redistributing heat and moisture across the globe.

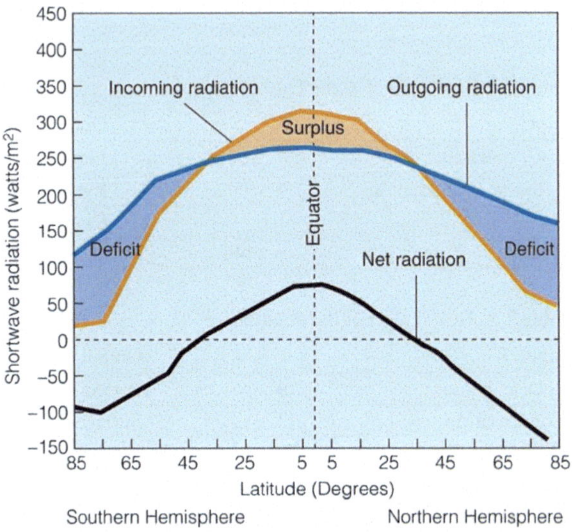

Fig. 7.21 Latitudinal distribution of net radiation, depicting the surplus of incoming solar radiation between the equator and approximately 30–35° north and south, contrasted with the deficit toward the poles. This radiation gradient facilitates heat transfer from the tropics to the poles, shaping atmospheric and oceanic circulation patterns

7.10 Thermal Balance of the Earth

The Earth's thermal balance, or heat balance, is the equilibrium between incoming and outgoing radiation that helps regulate the planet's mean temperature. Despite seasonal variations and temperature fluctuations caused by cold and heat waves, this balance remains relatively constant. However, this equilibrium is not uniform across latitudes. Between 36°N and 36°S, known as the tropics, the Earth receives more solar radiation than it loses to space. In contrast, higher latitudes receive less solar energy compared to the longwave radiation emitted by the Earth's surface. This imbalance leads to the transfer of excess heat from the tropics to the polar regions through global winds and ocean currents. Tables 7.2, 7.3, and 7.4 illustrate the Earth's energy budget across its surface, atmosphere, and the entire planet. These tables describe a simple model of the radiation energy budget, with values expressed in watts per square meter W/m^2. The Sun emits an average of 342 W/m^2 to the Earth's surface, which, in this model, is represented as 100 units. Table 7.2 shows that 51 units of solar radiation reach the Earth's surface directly, while the atmosphere radiates 95 units, totaling 146 units. The Earth's longwave radiation amounts to 116 units. Through processes like evaporation and convection, 30 units are released from the surface, maintaining a balance of 146 units between incoming and outgoing heat at the Earth's surface. Table 7.3 details the heat budget of the atmosphere. The majority of the incoming radiation in the atmosphere comes from the Earth's longwave radiation, totaling 110 units, while only 19 units are received directly from solar radiation.

Most of the outgoing radiation from the atmosphere is directed back toward the Earth's surface. This creates a balance between the incoming and outgoing radiation within the atmosphere, where 159 units of radiation are received, and 159 units are emitted. This balance is crucial in maintaining the Earth's thermal equilibrium,

Table 7.2 Heat budget of the Earth's surface

Input		Output	
Solar radiation	51	Earth's radiation	116
Atmospheric radiation	95	Evaporation	23
		Convection	7
	146		**146**

Table 7.3 Heat budget of the atmosphere

Input		Output	
Solar radiation	19	Radiation in space	64
Condensation	23	Surface radiation	95
Convection	7		
Earth's radiation	110		
	159		159

Table 7.4 Heat budget of the entire planet

Input		Output	
Solar radiation	100	Reflected/scattered radiation from clouds or gases	30
		Atmospheric radiation in space	64
		Longwave radiation from Earth to space	6
	100		**100**

ensuring that the energy absorbed from the Sun is effectively redistributed and radiated back into space.

The heat budget of the entire planet (Table 7.4) represents a balance between the incoming shortwave radiation from the Sun, in total 100 units, and the outgoing radiation. The outgoing radiation includes 6 units of longwave radiation emitted by the Earth into space, 64 units of atmospheric radiation directed into space, and 30 units of reflected or scattered radiation from clouds and gases, summing to a total of 100 units.

There is a noticeable imbalance in radiation across different latitudes. The northern latitudes of both hemispheres experience an energy deficit, leading to atmospheric cooling, while the equatorial regions receive excess energy, resulting in atmospheric heating. This disparity in energy distribution causes heat to be transferred from the equatorial regions to the polar areas.

7.10.1 Earth's Heat Balance and Planetary Temperature

As previously discussed, the flux of solar energy that reaches the Earth is referred to as the solar constant.

$$S_0 = \frac{Q}{4\pi r^2} = 1366.1 \frac{\mathrm{W}}{\mathrm{m}^2} \tag{7.6}$$

This value represents the average amount of solar energy received per unit area at the top of the Earth's atmosphere. The solar constant is crucial in understanding the Earth's heat balance and overall planetary temperature. The solar constant provides a baseline for calculating how much energy is absorbed by the Earth and how it is distributed between various components of the climate system. This incoming solar energy is absorbed by the Earth's surface, atmosphere, and clouds and is subsequently re-radiated as longwave infrared radiation. The balance between incoming solar radiation and outgoing longwave radiation is fundamental in determining the Earth's mean temperature and climate patterns. Figure 7.22 illustrates the distribution of solar energy as it reaches the Earth's surface, emphasizing the relationship between solar energy input and the Earth's thermal output. This balance is critical for maintaining the Earth's climate system and ensuring that the average global temperature remains within a range that supports life.

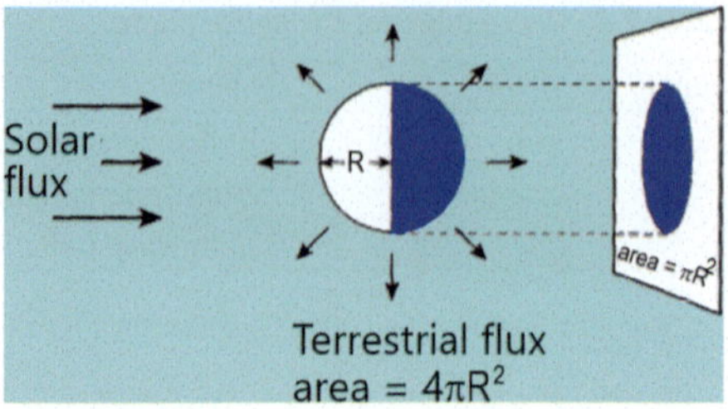

Fig. 7.22 Solar and terrestrial flux

The amount of radiant energy emitted by a body and the wavelength at which it is emitted depends on temperature, according to Wien's law and the Stefan–Boltzmann law. According to the Stefan–Boltzmann law, the total emitted radiation energy per unit area is given by σT^4, where T is the absolute temperature and σ is the Stefan–Boltzmann constant, valued at $5.670374419 \cdot 10^{-8} \frac{\mathrm{W}}{\mathrm{m}^2\mathrm{K}^4}$. This relationship helps determine the solar shortwave radiation, which can be expressed as:

$$E_{\mathrm{Solar}} = S_0 \pi a^2 \tag{7.7}$$

If the fraction of incident solar energy that is reflected is denoted by (α), which represents the albedo (planetary albedo = 0.3), the absorbed shortwave solar radiation is

$$E_{\mathrm{Solar}_{\mathrm{abs.}}} = (1-\alpha) S_0 \pi a^2 \tag{7.8}$$

If the Earth emits radiation as a black body at a uniform temperature, then T_e represents the effective temperature corresponding to the emitted radiation per unit area

$$E_{\mathrm{Earth}} = \sigma T_e^4 \tag{7.9}$$

or the total emitted radiation from the Earth is

$$E_{\mathrm{Earth}_{\mathrm{tot.}}} = 4\pi a^2 \sigma T_e^4 \tag{7.10}$$

Since the absorbed solar radiation is equal to the emitted radiation, it follows that the temperature of the planetary emission

$$T_e = \left[\frac{S_0 (1-\alpha)}{4\sigma} \right]^{\frac{1}{4}} \approx 255\,\mathrm{K} \tag{7.11}$$

7.10.2 Earth Emission Spectrum and Atmospheric Absorption

The atmosphere is transparent to visible solar radiation but becomes opaque to ultraviolet radiation and exhibits variable transparency in the infrared spectrum. This transparency varies across different wavelengths, with some regions being completely opaque and others fully transparent. Figure 7.23 illustrates blackbody curves, absorption bands in the Earth's atmosphere (middle panel), and their effects on both incoming solar radiation and outgoing thermal radiation (top panel). The bottom panel displays absorption spectra for major greenhouse gases and Rayleigh scattering.

Both the Earth and the Sun emit electromagnetic radiation that approximates a blackbody spectrum. For the Sun, emissions peak in the visible region and correspond to a temperature of approximately 5500 K. In contrast, Earth's emissions peak in the infrared, reflecting temperature variations at different locations and altitudes. The position and extent of absorption bands are determined by the chemical properties of atmospheric gases. Currently, water vapor is the most significant greenhouse gas, followed by carbon dioxide and other minor greenhouse gases. Additionally, Rayleigh scattering, which causes the sky to appear blue, scatters some incoming sunlight. Collectively, these processes capture and redistribute 25–30% of the direct sunlight energy passing through the atmosphere. In contrast,

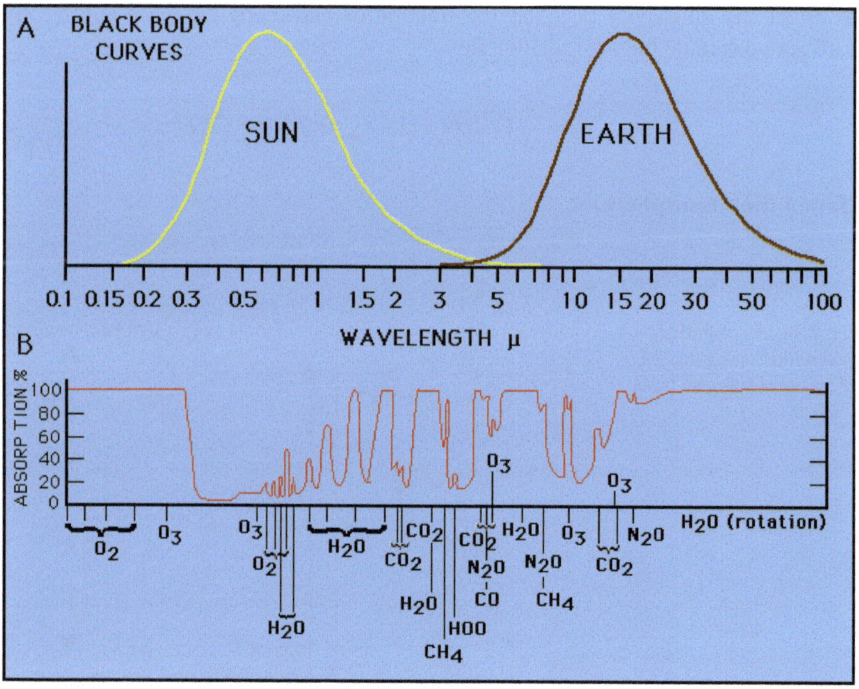

Fig. 7.23 (a) Black body curves for the Sun and Earth, and (b) Absorption spectra in the Earth's atmosphere

greenhouse gases capture 70–85% of the energy in the outgoing heat radiation emitted from the Earth's surface. Dominant atmospheric components, such as nitrogen and oxygen, are remarkably transparent in the important spectral ranges, with nitrogen not absorbing at all and oxygen absorbing only in the far UV, where solar flux is minimal. The absorption of terrestrial radiation is primarily due to triatomic molecules (O_3, H_2O, CO_2) that have rotational and vibrational modes easily excited by infrared radiation. These atmospheric constituents, known as "greenhouse gases," play a crucial role in the Earth's climate system.

7.11 A Simple Model of a Greenhouse

Figure 7.24 illustrates a basic model of greenhouse gases. This model begins with the estimation of the average incoming solar energy E_{Solar} per unit area. The figure demonstrates how greenhouse gases affect the energy balance by trapping heat, like how a greenhouse traps warmth.

$$E_{\text{solar}_{\text{aver.}}} = \frac{E_{\text{solar}}}{4\pi a^2} = \frac{S_0 \pi a^2}{4\pi a^2} = \frac{S_0}{4} \tag{7.12}$$

If we consider absorptivity ($\varepsilon_{\text{атмосфера}}$) as the fraction of longwave radiation absorbed and emissivity ($\varepsilon_{\text{earth}}$) as the fraction of blackbody radiation emitted by Earth, we calculate the net radiative flux

$$E_{\text{Total-Solar(top)}} = \frac{S_0}{4}(1-\alpha) = (1-\varepsilon_{\text{atm}})\sigma T_s^4 + \varepsilon_{atM}\sigma T_a^4 \tag{7.13}$$

(Top of the Atmosphere)

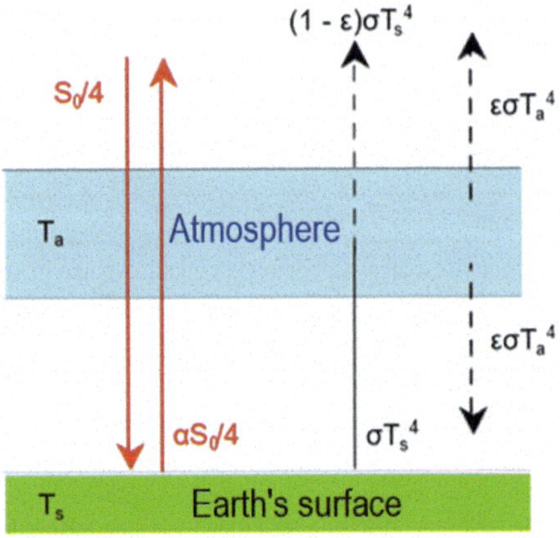

Fig. 7.24 A simplified model illustrating the effect of greenhouse gases

$$E_{\mathrm{Total\_Solar\_Surface}} = \frac{S_0}{4}(1-\alpha) + \varepsilon_{0tm}\sigma T_{\mathrm{a}}^4 = \varepsilon_{\mathrm{Earth}}\sigma T_{\mathrm{s}}^4 \tag{7.14}$$

which must disappear.

Hence the final expression for the mean temperature of the Earth's surface

$$T_{\mathrm{s}} = \left[\frac{S_0(1-\alpha)}{2\sigma(2-\varepsilon_{\mathrm{Earth}})}\right]^{1/4} = \left(\frac{2}{2-\varepsilon_{\mathrm{Earth}}}\right)^{1/4} T_{\mathrm{e}} \tag{7.15}$$

The average temperature of the atmosphere is

$$T_{\mathrm{s}} = \left(\frac{1}{2}\right)^{1/4} T_{\mathrm{s}} \tag{7.16}$$

If the average emissivity $\varepsilon = 0.8$, then the surface temperature $T_{\mathrm{s}} = 290$ K, and the atmospheric temperature $T_{\mathrm{a}} = 244$ K. This T_{s} value is close to the global average temperature of the Earth's surface, which is approximately 288 K.

References

Diffey, B. L. (2002). Ultraviolet radiation physics and the skin. *Physics in Medicine & Biology, 47*(14), R41–R57.

Farman, J. C., Gardiner, B. G., & Shanklin, J. D. (1985). Large losses of total ozone in Antarctica reveal seasonal ClOx/NOx interaction. *Nature, 315*(6016), 207–210.

Foukal, P. (2013). *Solar astrophysics* (3rd ed.). Wiley-VCH.

Goody, R. M., & Yung, Y. L. (1989). *Atmospheric radiation: Theoretical basis* (2nd ed.). Oxford University Press.

Hartmann, D. L. (2016). *Global physical climatology* (2nd ed.). Elsevier.

Hawking, S. W. (2015). *The universe in a nutshell*. Bantam Books.

Houghton, J. T. (2015). *Global warming: The complete briefing* (5th ed.). Cambridge University Press.

Houghton, J. T., Ding, Y., Griggs, D. J., Noguer, M., Van Der Linden, P. J., & Dai, X. (2001). *Climate change 2001: The scientific basis*. Cambridge University Press.

Jaupart, C., Labrosse, S., & Mareschal, J.-C. (2007). Heat flow and thermal structure of the Earth. *Treatise on Geophysics, 7*, 217–253.

Kollias, N., & Stamatas, G. N. (2002). Ultraviolet and infrared radiation: The effects on skin. *Clinics in Dermatology, 20*(1), 21–28.

Liou, K. N. (2002). *An introduction to atmospheric radiation* (2nd ed.). Academic Press.

Molina, M. J., & Rowland, F. S. (1974). Stratospheric sink for chlorofluoromethanes: Chlorine atom-catalyzed destruction of ozone. *Nature, 249*(5460), 810–812.

NASA. (2018). *Earth's energy budget*. NASA Earth Observatory.

NASA. (2024). *Ozone hole continues healing in 2024*. NASA Earth Observatory. https://earthobservatory.nasa.gov/images/153523/ozone-hole-continues-healing-in-2024.

NOAA. (2024). *Antarctic ozone hole ranks 7th-smallest since recovery began*. National Oceanic and Atmospheric Administration. https://www.noaa.gov/news-release/2024-antarctic-ozone-hole-ranks-7th-smallest-since-recovery-began.

Parker, E. N. (1958). Dynamics of the interplanetary gas and magnetic fields. *The Astrophysical Journal, 128*, 664–676.
Phillips, K. J. H. (2012). *Guide to the sun*. Cambridge University Press.
Pollack, H. N., Hurter, S. J., & Johnson, J. R. (1993). Heat flow from the Earth's interior: Analysis of the global data set. *Reviews of Geophysics, 31*(3), 267–280.
Ramanathan, V., & Vogelmann, A. M. (2017). Greenhouse effect, atmospheric solar absorption and the Earth's radiation budget: From the Arrhenius-Langley era to the 1990s. *Philosophical Transactions of the Royal Society A: Mathematical, Physical and Engineering Sciences, 372*(2019), 20140326.
Rivas, J. M., & Fernández, M. (2007). Effects of ultraviolet radiation on health and ecosystems. *Photochemistry and Photobiology, 83*(1), 1–14.
Serway, R. A., & Jewett, J. W. (2018). *Physics for scientists and engineers* (10th ed.). Cengage Learning.
Solomon, S. (1999). Stratospheric ozone depletion: A review of concepts and history. *Reviews of Geophysics, 37*(3), 275–316.
Stix, M. (2004). *The Sun: An introduction* (2nd ed.). Springer.
Tipler, P. A., & Mosca, G. (2021). *Physics for scientists and engineers* (7th ed.). W. H. Freeman.
Trenberth, K. E., Fasullo, J. T., & Kiehl, J. (2009). Earth's global energy budget. *Bulletin of the American Meteorological Society, 90*(3), 311–323.
Visconti, G. (2016). *Fundamentals of physics and chemistry of the atmosphere*. Springer.
Wallace, J. M., & Hobbs, P. V. (2006). *Atmospheric science: An introductory survey* (2nd ed.). Elsevier.
Wild, M., Folini, D., Hakuba, M. Z., & Schar, C. (2019). The global energy balance as represented in CMIP6 climate models. *Geophysical Research Letters, 46*(17–18), 10251–10259.

Chapter 8
Earth's Temperature Distribution

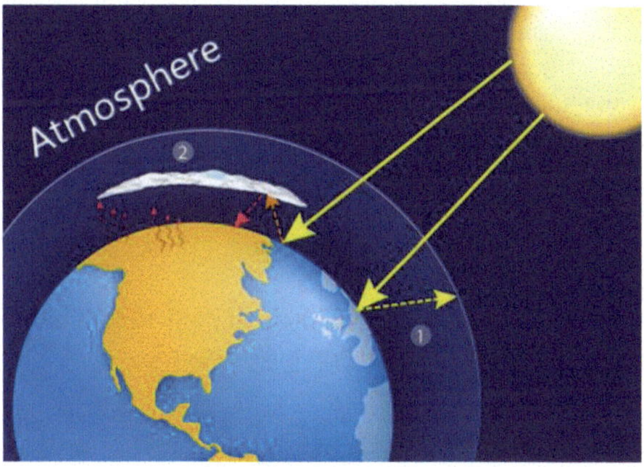

8.1 Thermal Characteristics of the Earth's Surface and Atmosphere

The key properties influencing various environments' heating and cooling processes include albedo, permeability, thermal conductivity, and specific heat. Albedo determines the reflectivity of a surface, affecting how much solar energy is absorbed or reflected. Higher albedo values indicate a greater reflection of sunlight, while lower values suggest increased absorption, impacting local temperature and climate patterns (Baker et al., 2020).

Thermal conductivity refers to a material's ability to conduct heat, quantified by the thermal conductivity coefficient. This coefficient is the amount of heat that passes through a unit area per unit of time at a temperature gradient of 1 °C/m. A

© The Author(s), under exclusive license to Springer Nature Switzerland AG 2025
V. Spiridonov et al., *Atmospheric Perspectives*,
https://doi.org/10.1007/978-3-031-86757-6_8

higher thermal conductivity coefficient indicates more rapid heat transfer by molecular means, allowing the distribution of the same heat over a larger volume (Gonzalez et al., 2021).

Additionally, specific heat reflects how much energy is required to change a material's temperature. This property is crucial for understanding the thermal dynamics of land and water surfaces, as materials with high specific heat can store and release energy more slowly, moderating temperature fluctuations (Liu et al., 2022). Permeability influences how water moves through soil and rock, affecting heat transfer processes in the subsurface environment, particularly groundwater flow and thermal regulation (Mason & Harvey, 2023).

This efficiency in heat transfer causes temperature changes to penetrate deeper layers more quickly. Thermal diffusivity, defined as the ratio of thermal conductivity to volumetric heat capacity (heat capacity per unit volume), further describes how quickly heat spreads within a material. Specific (volumetric) heat capacity of a substance is the amount of heat that must be added to one unit of mass (volume) of the substance to cause an increase of one unit in temperature. Materials with higher specific (volumetric) heat capacities heat up more slowly since they require more energy for a given temperature increase.

The Earth's surface is composed of land, water, and ice. The heating and cooling of these surfaces, along with the heat exchange between them and the atmosphere, are highly dynamic processes. These processes vary significantly depending on the type of substrate and its current properties. Table 8.1 presents the thermal characteristics of various soil, water, and air types.

There are substantial differences in the thermal properties of water and land. For instance, water can transmit solar radiation into deeper layers, whereas soil cannot.

Table 8.1 Thermal characteristics of different types of earth substrates and air

	Land									
Matter	Sand (porosity) 40%		Clay (porosity) 40%		Peat (porosity) 40%		Snow		Water	Air
	Dry	Wet	Dry	Wet	Dry	Wet	Fresh	Old	($T = 4$ °C)	($T = 10$ °C)
Density (kg/m^3) • 10^3	1.60	2.00	1.60	2.00	0.30	1.10	0.10	0.48	1.00	0.0012
Specific heat capacity (Jkg^{-1}K^{-1}) • 10^3	0.80	1.48	0.89	1.55	1.92	3.65	2.09	2.09	4.18	1.01
Volumetric heat capacity (Jm^{-3}K^{-1}) • 10^3	1.28	2.96	1.42	3.10	0.58	4.02	0.21	0.84	4.18	0.0012
Heat conductivity coefficient (Wm^{-1}K^{-1})	0.30	2.20	0.25	1.58	0.06	0.50	0.08	0.42	0.57	0.025
Thermal diffusivity coefficient (m^{2}s^{-1})	0.24	0.74	0.18	0.51	0.10	0.12	0.10	0.40	0.14	20.50

Water also has a significantly higher specific heat than soil—more than four times higher—meaning that a much larger amount of heat is required to raise the temperature of water by unit temperature per unit mass. The thermal conductivity of wetlands is higher than that of water. Additionally, water can exist in all three states of matter, and in water, heat can be transferred through mixing, unlike in solid land where particles are immobile.

Being permeable to solar radiation, air has a much lower volumetric-specific heat and thermal conductivity than soil and water. However, air exhibits a much higher thermal diffusivity, enabling heat to be transferred efficiently through mixing, which is far more effective than molecular heat transfer.

8.2 Heating and Cooling of the Earth's Surface: Land

The heating and cooling of the Earth's surface are significantly influenced by factors such as albedo, mechanical composition, soil moisture, and the presence of vegetation or snow cover. Land surfaces generally experience rapid temperature changes due to their low specific heat, which varies with soil composition and moisture content (see Table 8.1).

Soil moisture plays a critical role in determining the specific heat of soil. As soil moisture increases, the specific heat approaches that of water, which is notably high (Hillel, 2019). Conversely, dry soil, containing air in its pores, exhibits a lower specific heat because the specific heat of air is significantly lower than that of water. When soil becomes saturated with water, it displaces the air in its pores, leading to a notable increase in the soil's specific heat (Chen & Dudhia, 2001).

Given the variability in land surfaces, volumetric-specific heat is often utilized in analyses to account for these differences. Water, being a superior conductor of heat compared to air, allows wet soil to conduct heat more effectively than dry soil (Oleson et al., 2010; refer to Table 8.1). Research indicates that a 10% increase in soil moisture can lead to up to a tenfold increase in thermal conductivity (Ma & Wang, 2020).

The coefficient of thermal conductivity of soil is influenced by the same factors affecting specific heat and thermal conductivity. Initially, as soil moisture increases, the coefficient of thermal conductivity rises sharply due to the enhanced heat conduction capabilities of water. However, at higher moisture levels, this coefficient may begin to decrease as the increase in specific heat outweighs the enhancement in thermal conductivity.

Among different types of soil, wet sandy and clayey soils exhibit the highest thermal conductivity, meaning that temperature changes are felt more deeply within these soils. Peat, even when moist, has the lowest thermal conductivity compared to other soil types. Peat also experiences the greatest temperature fluctuations, warming up significantly during the day and cooling down markedly at night due to its low albedo.

Vegetation cover affects soil temperature by reducing the extremes of heating and cooling compared to bare soil. Vegetation reflects and absorbs solar radiation, and it also diminishes soil radiation by absorbing and reflecting long-wave radiation from the soil.

Snow cover significantly influences both land and air temperatures, as well as the temperature of the underlying soil. Fresh, dry, loose snow has a high albedo, reflecting much of the solar radiation, which prevents it from melting significantly during the day. The shortwave radiation that penetrates the snow cover reaches greater depths than long-wave radiation. Snow's high absorption and emissivity of infrared radiation mean that radiative cooling occurs only in a thin surface layer. Consequently, the highest temperature is found just below the snow's surface rather than on the surface itself. When melting occurs, it is most intense at the depth where the temperature is highest. Snow melts more rapidly near dark objects like tree branches or roof edges, which have a lower albedo and greater absorption of solar radiation. These objects then emit infrared radiation, which is absorbed by the snow and accelerates melting.

Under a thin snow cover (up to approximately 15 cm thick), the soil can absorb transmitted solar radiation, which may lead to the melting of the snow from below. The insulating properties of snow, due to its low thermal conductivity and the presence of air pockets, protect plants from frost. Consequently, there is always a flow of heat from the soil's depth to the surface beneath the snow cover. In regions with relatively short cold spells, the temperature at the bottom of the snow cover can be around 0 °C, leading to melting from the ground up, while the temperature at the top of the snow cover can be as low as −10 °C or lower.

8.3 Diurnal Trend of Earth's Surface Temperature

The temperature of the Earth's surface is influenced by various factors, but its diurnal variation is primarily governed by the radiation balance, which refers to the ratio of incoming solar radiation (insolation) to outgoing terrestrial radiation (Pinker & Laszlo, 1992). Figure 8.1 illustrates the relationship between this radiation balance and the diurnal temperature pattern of the soil.

Soil temperature increases when insolation exceeds outgoing radiation, resulting in a positive radiation balance. Conversely, when insolation is less than outgoing radiation, a negative radiation balance leads to a decrease in soil temperature. Typically, soil temperature rises from sunrise until around 13:00 local time, reaching its daily maximum when insolation equals outgoing radiation (Kirkby, 1985). At this point, soil radiation is highest, as the soil surface temperature peaks due to the accumulation of heat.

The diurnal temperature cycle is influenced not only by the amount of insolation but also by factors such as surface characteristics, including albedo, vegetation cover, and moisture content (Bristow & Campbell, 1984). For example, areas with high albedo reflect more solar energy, resulting in lower surface temperatures compared to darker surfaces that absorb more heat. Additionally, vegetation can

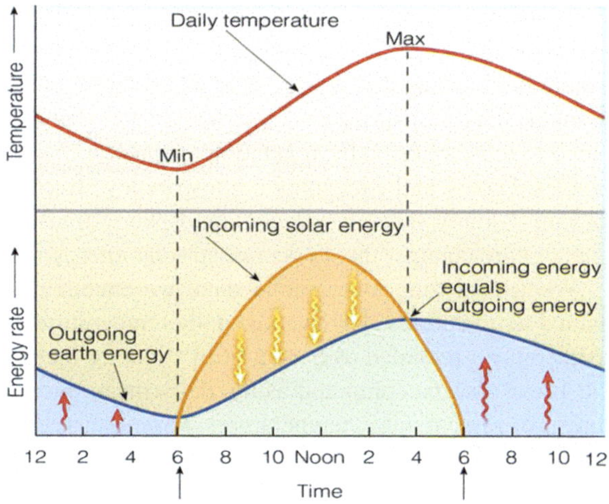

Fig. 8.1 Diurnal and annual variation in soil surface temperature and radiation balance

moderate temperature fluctuations by providing shade and releasing moisture through transpiration, which helps cool the surrounding environment (Granier et al., 2000).

Understanding the diurnal trends in surface temperature is crucial for predicting local weather patterns and assessing the impacts of climate change on terrestrial ecosystems.

The daily minimum temperature occurs in the early morning hours, shortly after sunrise, when the soil surface has cooled overnight.

The daily maximum and minimum soil surface temperatures, along with the daily temperature fluctuation (amplitude), which is the difference between the highest and lowest temperatures of the day, are influenced by several factors. These include latitude, season, altitude, terrain orientation, soil composition, and condition (dry, wet, vegetated, or snow-covered), as well as atmospheric turbulence and transparency.

Daily temperature fluctuations decrease with increasing latitude because, at lower latitudes, there is more intense solar heating during the day due to the Sun's higher position and cooler temperatures at night due to higher soil temperatures. Additionally, daily temperature variations are greater in summer than in winter due to the Sun's higher position and longer daylight hours. Increased altitude results in greater daily temperature fluctuations because mountain tops heat up more intensely during the day due to stronger solar radiation and cool down more at night due to lower temperatures and greater air permeability. South-facing slopes experience more significant diurnal temperature changes than north-facing slopes due to higher solar radiation. Western and eastern slopes show larger daily temperature fluctuations than northern slopes, with western slopes experiencing greater fluctuations because they receive more solar energy and heat up more during the day.

Soil temperature fluctuations are also influenced by soil color and composition. Dark soils, with high shortwave absorption and long-wave emissivity, heat and cool more than lighter soils. Soils with higher specific heat and thermal conductivity warm up and cool down more slowly. Consequently, wet soils exhibit lower diurnal temperature fluctuations compared to dry soils, due to the energy required for water evaporation during the day and heat release from condensation or vapor deposition at night. Bare soil surfaces experience greater daily temperature fluctuations compared to those covered by vegetation or snow.

Cloud cover and atmospheric transparency affect daily temperature fluctuations as well. Cloudy skies or reduced atmospheric transparency result in smaller daily temperature changes due to decreased solar energy input and reduced radiation.

When analyzing temperature data, we encounter average daily, monthly, and annual temperatures. The average daily temperature is calculated by averaging the temperatures recorded over 24 h or by dividing the sum of hourly temperatures by 24. It can also be computed as the daily mean maximum and minimum temperatures. Average monthly temperatures are determined by summing the daily average temperatures for each day of the month and dividing by the number of days in the month. The average annual temperature is obtained by averaging the 12-monthly mean temperatures for the year.

8.4 Annual Trend of the Earth's Surface Temperature

The annual trend of land surface temperature, like the daily trend, is primarily influenced by the radiation balance (see Fig. 8.1). In the temperate latitudes of the Northern Hemisphere, insolation typically exceeds outgoing radiation from January (approximately 3 weeks after the winter solstice) until late July (about 5 weeks after the summer solstice). This period spans from the annual minimum to the maximum temperature of the soil surface.

The annual fluctuation in soil surface temperature—the difference between the mean monthly temperatures of the warmest and coldest months—is influenced by factors such as latitude, altitude, terrain orientation, substrate type and condition, and atmospheric turbidity and transparency. For instance, areas at higher altitudes generally experience lower annual mean temperatures due to decreased air pressure and density, leading to reduced thermal retention (López-Bustins et al., 2008).

In general, the annual fluctuation in land surface temperature increases with latitude due to greater variations in solar radiation energy between warmer and cooler seasons. Higher latitudes receive less solar energy during the winter months, leading to more significant temperature drops compared to the summer months. This seasonal variation is exacerbated in continental areas, where land heats and cools more rapidly than bodies of water (Huang et al., 2018). Moreover, the urban heat island effect can significantly alter local temperature trends, particularly in metropolitan areas where human activity, infrastructure, and reduced vegetation contribute to elevated temperatures (Oke, 1982). Urban areas may experience annual temperature increases that differ from surrounding rural regions, highlighting the influence of anthropogenic factors on the thermal dynamics of the Earth's surface. Additionally, climate change is impacting annual temperature trends globally, leading to a rise in average temperatures and altering seasonal patterns. For example, prolonged heat waves and milder winters have been observed in many regions, causing shifts in ecosystems and affecting agricultural practices (IPCC, 2021). Just as with daily temperature fluctuations, annual temperature variations are affected by similar factors, including geographic and environmental conditions. Understanding these trends is crucial for predicting the impacts of climate change on natural and human systems, as well as for developing effective mitigation strategies.

8.5 Variation in Soil Temperature with Depth

Solar radiation absorbed by the Earth's surface is transferred into the soil primarily through heat conduction, a relatively inefficient process compared to heat transfer in water. Consequently, temperature variations with depth in the soil are considerably smaller than those observed in water. Figures 8.2 and 8.3 illustrate the daily and annual variations in soil temperature with depth. The daily and annual variations in soil temperature are most pronounced at the surface and decrease with increasing depth. The extent of daily temperature variation and the depth at which these variations are felt depend on factors such as soil type, moisture content, solar radiation intensity, and geographic latitude.

In soils with high thermal conductivity, such as moist sandy and clay soils, temperature variations are more significant. Daily temperature variations typically extend to depths of 25–100 cm, with maximum and minimum temperatures occurring with an average delay of 2–3 h for every 10 cm of depth (Fig. 8.2a). Annual temperature variations penetrate to depths of 8–30 m. In tropical regions, this depth is approximately 10 m, while in polar regions, it reaches about 20 m, with moderate latitudes averaging around 15 m. Annual temperature extremes occur about 20–30 d later for each meter of depth. For example, at a depth of 10 cm, the maximum temperature is usually reached in July, whereas at a depth of 10 m, it is reached in December (Fig. 8.2b).

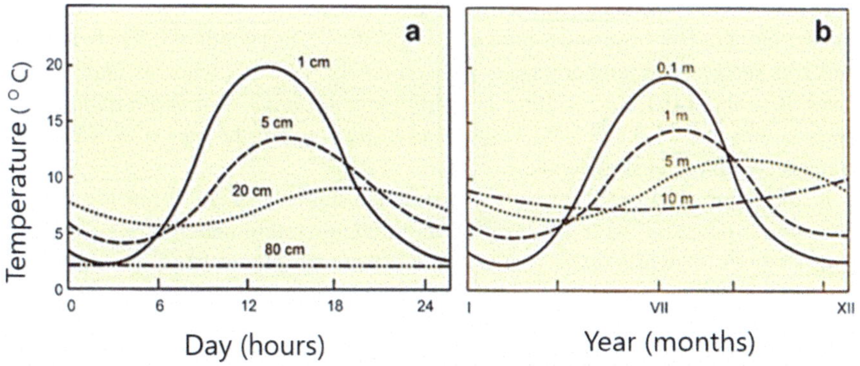

Fig. 8.2 Changes in soil temperature with depth. (**a**) During the day and (**b**) during the year

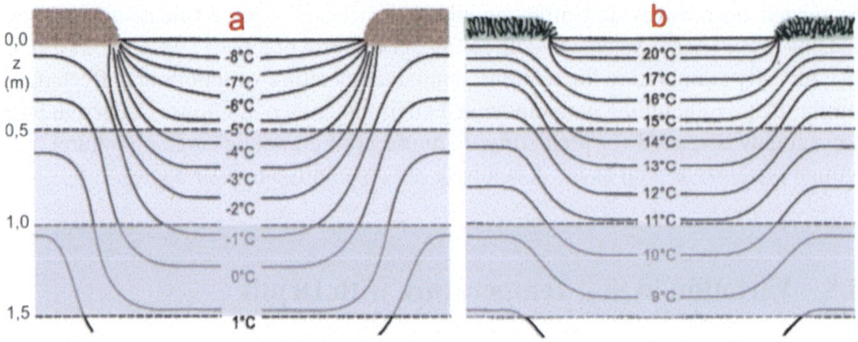

Fig. 8.3 Variation in soil temperature with depth below bare soil surface and under. (**a**) Snow cover in February and (**b**) plant cover in June

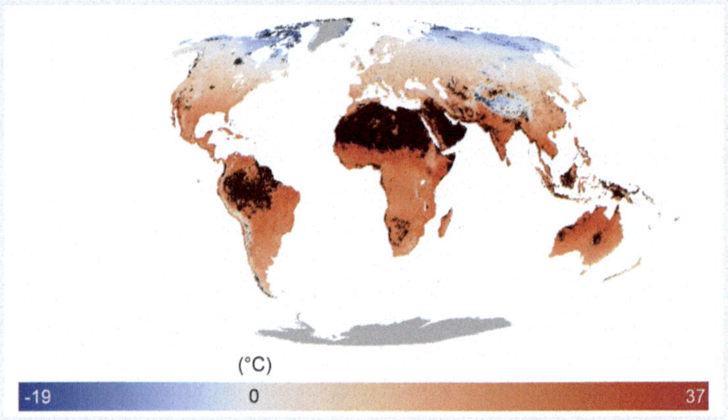

Fig. 8.4 Mean global soil temperature (°C)

The daily and annual fluctuations in soil temperature with depth are significantly reduced when the soil is covered with snow (Fig. 8.3a) or vegetation (Fig. 8.3b). During the winter months, soil temperatures at the same depths under snow are higher compared to those under bare soil. For instance, at a depth of 0.5 m below the bare soil surface, the temperature is approximately −4.5 °C, whereas under snow cover, it is about 0.5 °C, resulting in a difference of 5 °C. At a depth of 1 m, the temperatures are −1.0 and 2 °C, respectively, showing a difference of 3 °C, which decreases with increasing depth.

In the summer (Fig. 8.3b), soil temperatures under vegetation are lower than those under bare soil. At a depth of 0.5 m, the temperature under bare soil ranges from 14 to 15 °C, while under vegetation, it is between 11 and 12 °C, resulting in a difference of about 3 °C. At a depth of 1 m, the temperatures are 11 and 9 °C, respectively, with a difference of 2 °C.

Figure 8.4 shows a global soil temperature map (0–15 cm depth) at a 1 km^2 resolution, based on the global SoilTemp database, which includes over 8500 time series of in situ soil temperature measurements. It illustrates that, over the year, soil in cold and/or dry biomes is on average 3.6 °C warmer, while soils in hot and humid environments are 0.7 °C cooler than the air temperature measured at a standard height (1.25–2 m) at meteorological stations.

8.6 Heating and Cooling Processes in Ocean Waters

Water surfaces on Earth heat up slowly and weakly during the day and in the warmer part of the year when the heat balance is positive. This slow warming occurs due to water's distinct thermal and physical characteristics compared to adjacent land surfaces (Houghton & Goodall, 2017). Water has a high heat capacity, which allows it to absorb and store large amounts of heat energy. Consequently, ocean and lake temperatures exhibit smaller daily and annual fluctuations compared to land surfaces (Levitus, et al. 2012). This slow rate of heat loss at night and during colder periods results in less pronounced temperature changes in aquatic environments. The transition between water's states—solid, liquid, and gaseous—involves substantial amounts of heat being absorbed or released. For example, during freezing, water's density decreases by approximately 9%, and pure water reaches its maximum density at 4 °C, a phenomenon known as the water anomaly. The water anomaly arises from the asymmetric structure of the water molecule, where the two positively charged hydrogen atoms are not aligned with the central doubly charged oxygen atom but form an angle of 104.05°. Each water molecule acts as an electric dipole, creating a strong, uneven electric field. This leads to significant intermolecular forces and the formation of molecular clusters. Consequently, a large amount of heat is required to increase the speed of these clusters, resulting in water's high specific heat, heat of vaporization, and unique freezing and boiling points. These thermal properties play a crucial role in supporting life on Earth (Hwang and Hsu, 2019).

As temperature changes, the arrangement of water molecules in clusters also changes. With decreasing temperature, the number of clusters with eight molecules increases, while those with four molecules decrease. This rearrangement affects the density of water in a specific manner. If water molecules were completely disordered, water's density would increase as the temperature decreases, as with most substances. However, due to the formation of larger clusters of eight molecules, water's density decreases below 4 °C. This is because these clusters occupy more space, leading to a decrease in density. Water exhibits an anomalous property in that it becomes less dense in its solid state (ice) compared to its liquid state, with ice having a density of approximately 0.92 kg/dm^3.

Seawater differs slightly from pure water in both physical and chemical properties, which are significant for meteorology and oceanography. The density of seawater is influenced by the concentration of dissolved salts, primarily sodium chloride (NaCl). The amount of dissolved salts in seawater, expressed in parts per million or grams per kilogram, is known as salinity. The average seawater salinity is about

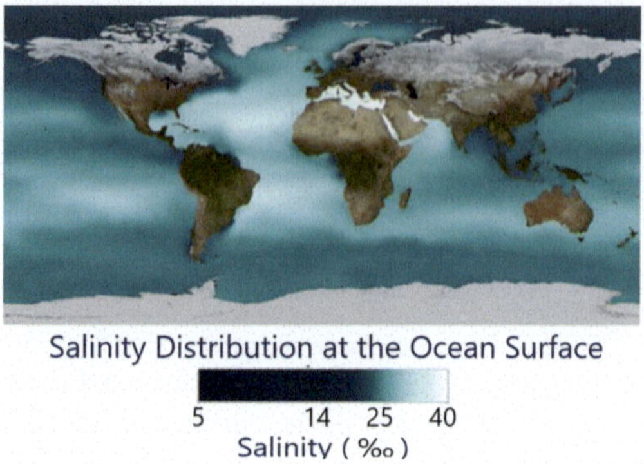

Fig. 8.5 Distribution of surface seawater salinity, expressed in (‰)

35.0‰. While the density of pure water decreases with temperature, seawater's density increases as the temperature drops, reaching its maximum at the freezing point. Therefore, seawater is densest at temperatures lower than 4 °C, depending on its salinity. For seawater with a salinity of 35.0‰, the maximum density occurs at approximately −3.5 °C, and for a salinity of 24.7‰, it is around −1.3 °C. Seawater freezes at around −1.9 or −1.3 °C, depending on its salinity, and it is about 2.4% denser than pure water at the same temperature due to its salt content.

An increase in seawater salinity by 1.0‰ at 10.0 °C results in a density increase equivalent to a 5.0 °C decrease in temperature. At 0.0 °C, this would correspond to a temperature decrease of up to 17.0 °C. Figure 8.5 shows the distribution of mean annual seawater surface salinity. Water is permeable to solar radiation, allowing it to penetrate up to 10 m deep on average and even several hundred meters in clear tropical seas. The permeability of water to solar radiation depends on its chemical composition, the number of plankton and impurities, and the wavelength of the radiation. Water's absorptivity increases with the wavelength of the radiation. As excessive heat and energy warm the ocean, it triggers a series of cascading effects, including ice-melting, sea-level rise, marine heatwaves, and ocean acidification. This phenomenon is closely linked to ocean warming, which plays a critical role in shaping the future of global climate. The resulting temperature changes from increased ocean heat contribute to these unprecedented environmental impacts, further exacerbating global climate challenges (McCarthy et al., 2015).

The water is heated directly by the penetration of solar radiation, while the deeper layers are also warmed by vertical movements. The thermal conductivity of water is relatively low, as it is a poor conductor of heat. For example, the surface of chemically pure water absorbs some of the Sun's radiation, converting it into heat that warms the water particles at the surface. These heated particles become lighter and remain near the surface, becoming more stable as they heat up, and transfer their heat to the deeper, cooler layers primarily through conduction. This process is so slow that, during the day, the effects are felt only to a depth of about 40 cm, and over a year, to approximately 7 m. Heating and cooling of water are not only due to the warming and cooling of the upper layer but also influenced by waves caused by wind, tides, and currents, as well as by the movement of sea and river currents.

Nevertheless, vertical water movement is the most significant mechanism for transferring heat to the deeper layers.

8.7 Diurnal Variations in Water Temperature

Due to the different characteristics of seawater and freshwater, there is a difference in the heating of the surface layer of water. During the day, when the surface layer of seawater is heated, evaporation occurs. Therefore, the concentration of salt and other impurities increases in the given layer. Due to the higher salinity, this layer becomes denser and heavier, so it sinks to a depth where the water has the same density but is colder and has less salinity. The previous, surface layer is replaced by colder, but lighter water. Through this vertical mixing, heat is transferred to deeper layers for the duration of the insolation.

In most of the world's oceans, the density of water in the mixing layer, which is in direct contact with the atmosphere, is significantly lower than the density of the water below it. The largest gradients of seawater density (as well as temperature) are found in a layer called the pycnocline, which extends from a few tens of meters to several hundreds of meters below the ocean surface.

This layer, akin to a temperature inversion, prevents the exchange of heat and salt between the mixing layer and deeper ocean waters. In lower latitudes, pycnocline is often referred to as thermocline, while in polar latitudes, it is termed halocline. Pycnocline encompasses both the halocline (salinity gradients) and thermocline (temperature gradients) and refers to the rapid change in density with depth. Since density is influenced by both temperature and salinity, pycnocline reflects the combined effects of the thermocline and halocline. The strength and depth of the thermocline vary with latitude and time of year, as shown in Fig. 8.6.

In contrast to seawater, daily and annual temperature fluctuations in freshwater are felt at much shallower depths due to its lower impurity levels. Consequently, the density of the upper layer of freshwater decreases with warming, causing it to remain afloat. While the upper layer of freshwater also cools due to evaporation, this only leads to weak mixing in the upper layer without causing it to sink.

When cooled to temperatures above 4 °C, both salt and freshwater behave similarly. During the night, due to the negative heat balance, the surface layer of both types of water cools by radiation, becomes denser, and sinks, with warmer, lighter water rising from below. However, vertical movement in freshwater due to cooling stops when the temperature falls below 4 °C, as the density decreases with further cooling. Consequently, freshwater remains on the surface, cools more intensely, and freezes at temperatures below 0 °C. The water beneath the ice sheet remains relatively warm (around 4 °C), allowing aquatic life to persist during winter. Seawater, due to its lower freezing point, can cool to depths as low as −1.9 °C at average salinity levels without freezing. This characteristic allows the oceans and seas to serve as significant heat reservoirs during winter months. The salt in seawater alters its density and lowers its freezing point, preventing it from freezing at temperatures where

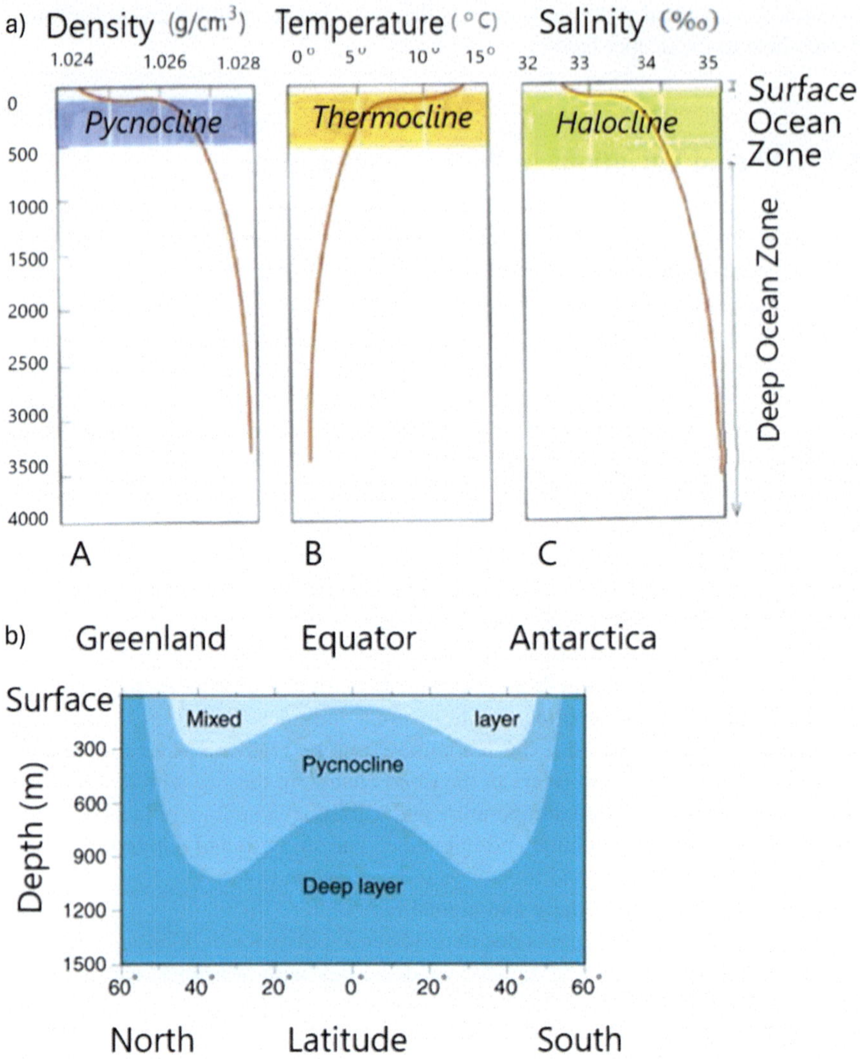

Fig. 8.6 (**a**) Pycnocline, thermocline, and halocline. (**b**) Typical variation of temperature with depth in the oceans for different latitudes and seasons. The area with the strongest temperature gradients corresponds to the thermocline

lakes and rivers in temperate latitudes would typically form ice. Due to its high specific heat, water heats and cools more slowly than adjacent land surfaces. As a result, the maximum surface water temperature is typically reached around 3–4 pm local time, while the minimum occurs 2–3 h after sunrise. The daily temperature fluctuation of the water surface is relatively small, significantly less than that of land surfaces. On the sea and ocean surfaces, it can be up to 1 °C in tropical regions and even lower, about 0.3 °C, in temperate latitudes. The daily temperature fluctuation of lakes and rivers is greater, averaging around 5 °C, due to the influence of adjacent land surfaces. Temperature fluctuations decrease with depth, with daily variations in seas and oceans falling to several dozen meters and in lakes to about 10 m.

8.8 Annual Trends in Water Temperature

The annual maximum and minimum temperatures of the water surface are reached later than those of the land surface. In temperate latitudes, the water temperature typically reaches its minimum in February or March and its maximum in August or sometimes September, occurring about 2–3 months after the summer and winter solstices.

The annual fluctuation of ocean surface temperature increases with latitude due to greater variations in the energy flux from solar radiation between the warmer and cooler parts of the year. In tropical regions, this fluctuation ranges from 2 to 3 °C, while in temperate latitudes it ranges from 5 to 8 °C. Enclosed seas exhibit more significant annual temperature fluctuations; for instance, the Black Sea can experience variations of more than 20 °C, which is like the annual temperature fluctuation of lake surfaces. In lakes and enclosed seas, the annual water temperature fluctuates to depths of up to 60–70 m, whereas in the oceans, it can extend to several hundred meters.

8.9 Air Heating and Cooling

Air is warmed primarily through the absorption of solar radiation and long-wave radiation from the Earth, as well as by the release of latent heat from condensation or freezing, and through the exchange of sensible heat with the environment via thermal conduction, molecular diffusion, turbulent mixing, and convection. Air, serving as a medium for solar radiation, is heated indirectly by the Earth's surface.

The influence of the surface is most pronounced in the planetary boundary layer, which extends from 100 to 3000 m above the ground. This layer has a complex structure:

- **Molecular (Laminar) Boundary Layer**: The layer closest to the Earth's surface, only a few centimeters thick, where heat transfer occurs mainly through thermal conduction and diffusion.
- **Turbulent Surface Layer**: Above the molecular layer, characterized by mechanical turbulence due to the roughness of the substrate. The thickness of this layer varies with wind strength and soil heating, typically around 1 m with weak wind and strong soil heating, and up to tens of meters with strong wind and weak soil heating.

- **Mixed Layer**: The upper part of the planetary boundary layer, where thermal convection becomes the dominant mechanism of vertical heat transport. In this layer, warm air bubbles prevail under weak wind conditions, while stronger winds cause horizontal circulation with a rolling motion.

Above the planetary boundary layer, heat transfer is influenced by large atmospheric disturbances and convective clouds, where deep convection occurs.

The air temperature tends to be lower during the day and in summer and higher at night and in winter compared to the ground temperature. This difference is due to the thermal and radiative properties of the air.

Over freshwater surfaces, the air cools and warms similarly to the underlying water. However, over oceans and seas, air heating and cooling are influenced by the large surface area and the significant presence of water vapor and salt particles. These factors lead to higher daytime temperatures and lower nighttime temperatures of the air above the sea surface compared to the air above land.

8.10 Daily Flow of Air Temperature

Air temperature is typically measured in a meteorological shelter at a height of 2 m. Any reference to air temperature at different heights will be explicitly noted. Due to differences in how land and water surfaces heat and cool, diurnal temperature fluctuations differ between these surfaces.

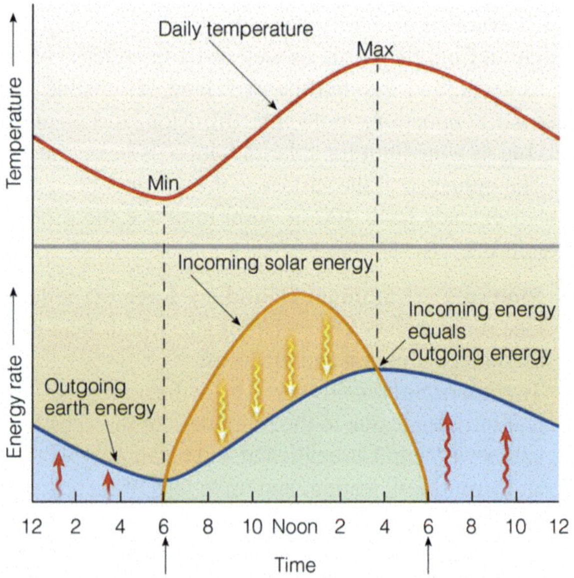

Fig. 8.7 Daily flux of ground air temperature and radiation balance

8.10.1 Daily Flow of Air Temperature Over Land

The air temperature over land reflects the daily heating and cooling of the soil. The maximum air temperature usually occurs 1–2 h later than the maximum soil temperature (Fig. 8.7). The timing of the peak air temperature is influenced by the intensity of soil heating. Increased soil heating leads to a delay in reaching maximum air temperature, as enhanced thermal convection transports heated air to higher layers more effectively. Consequently, maximum air temperatures typically occur around 3 or 4 pm, and in winter, around 1 or 2 pm local time, when thermal convection is less intense.

The minimum daily air temperature is usually observed just before sunrise. The range of daily air temperature fluctuation is affected by factors like those influencing land surface temperature fluctuations, including latitude, season, altitude, topography, orientation, substrate type and condition, and atmospheric turbidity and transparency.

As latitude increases, the daily temperature fluctuation diminishes, particularly starting from subtropical regions. This reduction is due to the high zenith position of the Sun, generally clear weather conditions, and the presence of deserts, where surfaces heat up and cool down rapidly.

In the equatorial regions, pronounced cloudiness reduces the heat flux to the Earth's surface from solar radiation, as well as the radiation itself. The daily fluctuation of air temperature is more significant in summer than in winter due to stronger daytime heating and nighttime cooling of the soil. This fluctuation decreases with altitude, as higher elevations have a smaller contact surface with the air and experience stronger winds compared to lower altitudes. Consequently, the air warms less and cools less at higher altitudes. Conversely, valleys and basins, with larger contact surfaces between the air and the ground and mostly calm air, experience greater daily temperature fluctuations. During the night, colder and denser air descends into these valleys, forming cold water pools. In arid terrains, daily temperature fluctuations are lower than in dry, non-arid areas. Narrow valleys sheltered from the sun by steep mountainsides also exhibit smaller temperature fluctuations compared to more spacious valleys with gentle slopes. Daily air temperature fluctuations are greater than dark soil compared to light soil, dry soil compared to moist soil, and bare ground compared to vegetated soil. Cloudy weather and reduced atmospheric transparency result in smaller daily temperature fluctuations due to decreased solar radiation during the day and reduced soil radiation at night.

8.10.2 Daily Flow of Air Temperature Over Water

In contrast to land surfaces, the maximum air temperature over water occurs between 3 and 5 pm, several hours later than over land. The minimum daily air temperature is reached about 2 h earlier over water compared to land. Due to the thermal characteristics of water, the daily fluctuation of air temperature over water is significantly less than over land. However, this fluctuation is still greater than the daily variation in ocean surface temperature, which averages about 0.5 °C, but can reach approximately 2 °C. This difference is due to the presence of water vapor and salt particles in the Earth's air. The daily fluctuation of air temperature over lakes can be greater due to the influence of adjacent land surfaces, especially if the lakes are small, shallow, and have a lower heating capacity. Despite differences in daily temperature fluctuations between water and air surface temperatures, the influence of large water bodies, particularly oceans, results in minimal differences in surface temperatures of water and air, rarely exceeding 1 °C.

8.11 Annual Changes in Air Temperature

The annual course of air temperature generally follows the annual course of substrate temperature. The timing of extreme values and the annual fluctuation in air temperature are influenced by several factors, including latitude, continentality, altitude, substrate type and condition, cloudiness, and atmospheric transparency. The primary determinants of the annual temperature cycle are latitude, continentality, and altitude. Based on geographical latitude, there are four basic types of annual air temperature variations: equatorial, tropical, temperate, and polar (Fig. 8.8). In the equatorial zone, the Sun reaches the zenith twice a year, at the spring and autumn equinoxes, directly above the equator. Day and night lengths at the equator remain approximately 12 h throughout the year. Consequently, there is minimal annual fluctuation in air temperature in the equatorial region, with variations of less than 1 °C over the ocean and between 5 and 10 °C over land surfaces. This region typically experiences two maxima and two minima in temperature annually.

The maximum occurs after the spring and autumn equinoxes, mostly in May and October. The minimum occurs after the summer and winter solstice, usually in January and July. In tropical regions (from the equator to the northern and southern solstice), there is one maximum and one minimum air temperature after the summer

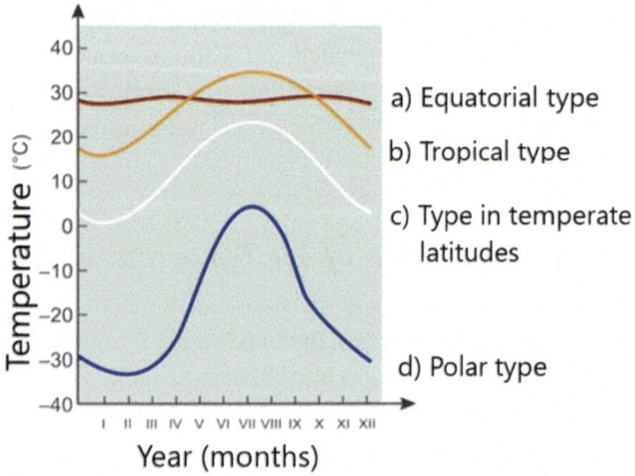

Fig. 8.8 Annual trend of air temperature as a function of latitude

and winter solstice. The annual fluctuation of air temperature is relatively small but greater than in the equatorial zone. It is about 5 °C over the sea surface and up to 20 °C over the land surfaces. However, in areas affected by monsoon circulation, the annual flow of air temperature is altered. There, the minimum occurs in winter and summer and the maximum at the end of the transitional periods.

The type of temperate latitude is characterized by one maximum and one minimum air temperature, which occurs after the summer and winter solstice. They appear in July and January on land and in August and February over the ocean. The annual fluctuation of air temperature is large and increases with increasing latitude. Thus, over the ocean, it is from 10 to 15 °C, and overland from 20 to 40 °C, and even more. In the temperate zone, four seasons are distinguished, which are most pronounced in the central part of the zone. The polar type is characterized by a long winter (starting earlier in autumn and extending into spring) and a short summer. The maximum air temperature is in August, and the minimum is in March in the Northern Hemisphere. The annual fluctuation of air temperature is very large. Water bodies greatly influence the annual temperature flow. In summer, the air temperature above the ocean surface is higher, and in winter, lower than the water temperature. The minimum air temperature occurs 1 month earlier than at the surface of the ocean. The maximum air temperature occurs approximately at the same time. The mean annual temperatures of the ocean surface and the upper atmosphere do not differ. The annual fluctuation of air temperature at a particular location increases with its distance from the sea and ocean. Coastal places have much milder winters than places located deep inland, and temperate latitudes are characterized by warmer autumn than spring. With increasing altitude, the annual fluctuation of air temperature decreases due to the reduced contact surface with the ground. Thus, summers are cooler at higher altitudes, while winters can be milder due to less frequent temperature inversions. The annual fluctuation of air temperature depends on the type of surface and cloudiness in the same way as the daily fluctuation of air. The daily and annual movements of the air temperature are primarily determined by the daily and annual movement of the Sun, which has a periodic character. Changes in air temperature during the day or year, as well as the daily and annual maximum and minimum temperatures, can also occur due to the occurrence of large atmospheric disturbances over a certain place. Such temperature changes are the most frequent and greatest in the temperate zone due to the influx of cold air from the polar regions or warm air from the subtropical regions. At lower latitudes, intrusions of air masses from other areas are much rarer.

8.12 Change in Air Temperature with Height in the Troposphere

The air temperature on average decreases with height in the troposphere. The main reason is that the air there is heated by the substrate. With increasing altitude, the emissivity increases, and the absorptivity of the air decreases, primarily due to the rapid decrease in water vapor content with height in the atmosphere. Also, temperature decreases with height due to dynamic heating and cooling of air due to ascending and descending movements. The average value of the vertical gradient of the air temperature is 0.6 °C/100 m in the free atmosphere and mountainous areas, and it is somewhat lower and amounts to 0.5 °C/100 m due to the reduced influence of the substrate. The vertical gradient of air temperature is characterized by great variability in space, both in the vertical and horizontal directions and in time. The value of the vertical air temperature gradient depends on latitude, relief, type of substrate, time of day, and year. Its value is greatest next to the Earth's substrate. There it can be several degrees per 100 m with intensive heating of the substrate. Due to the large influence of the substrate, the ground layer of the air has the largest fluctuations in the vertical temperature gradient over time. The air temperature drops the fastest in the hottest part of the day and summer and the slowest at night and in winter. A typical diurnal variation of temperature with height and time in clear weather is given in Fig. 8.9.

At night, the ground cools more rapidly than the air due to radiation cooling. Consequently, heat is transferred from the atmosphere to the cooler ground, causing the air layers closest to the surface (Fig. 8.9a) to cool first, followed by the layers above (Fig. 8.9b). Shortly after sunrise, the Earth's surface warms as it absorbs solar radiation, which in turn heats the overlying air layers. The lower layers of air (Fig. 8.9c) warm first, followed by the layers higher up (Fig. 8.9d).

As discussed, temperature inversions—where air temperature increases with height—represent a deviation from the typical vertical temperature gradient in the troposphere. These inversions are relatively common and can occur at any time of day or year and at various altitudes. They are influenced by specific meteorological conditions and other factors, which can increase their frequency, intensity, and thickness.

Temperature inversions are categorized based on altitude and formation mechanism. They can be medium or high, depending on their altitude. Inversions are classified into radiative inversions, air mass descent inversions, and frontal inversions based on their causes. Typical vertical temperature profiles for these inversion types are illustrated in Fig. 8.10.

Radiation inversions form when the ground and the air near the surface cool rapidly due to radiation. Their thickness can range from several meters to hundreds of meters. Ground radiation inversions (Fig. 8.10a) occur during clear nights with little or no wind, caused by radiation from both the ground and the air above it.

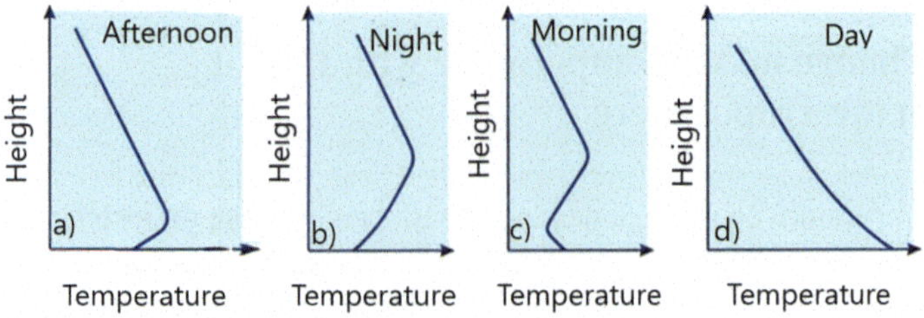

Fig. 8.9 Change in the vertical temperature gradient throughout the day

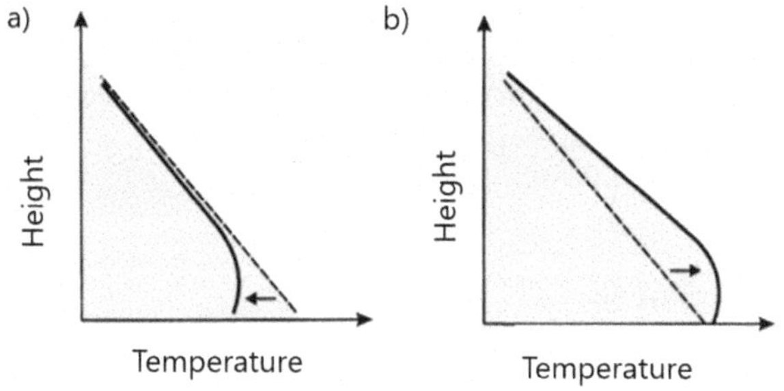

Fig. 8.10 Types of temperature inversions: (**a**) inversion caused by cooling and (**b**) inversion caused by heating

High-altitude radiative inversions typically form above stratus clouds with high solar albedo. As these clouds cool, the air directly above them also cools, creating a temperature inversion.

Radiation inversions are common, especially at night. By afternoon, the ground and the overlying air lose more energy through radiation than they gain, leading to cooling from the negative heat balance (Fig. 8.10b). Since the soil has higher emissivity in the infrared spectrum, it cools faster than the air, causing temperatures to drop shortly after sunset. The ground layer of air transfers heat to the ground via thermal conduction, while the ground loses heat rapidly through radiation. Overnight, the ground and the air layer in contact with it cool further, causing the warmer upper layers of air to lose heat slowly to the cooler lower layers. Minimum daily temperatures are typically reached around sunrise. Cooling can continue after sunrise due to the low angle of solar rays and the remaining higher radiation loss compared to incoming solar energy. Moist soil can further delay warming due to the latent heat of evaporation. The intensity and thickness of radiation inversions depend on the length of the night, humidity, cloud cover, wind speed, relief, and substrate characteristics.

Long-lasting temperature inversions can form at higher latitudes above snow cover, creating thicker layers of air with significant temperature increases. In such cases, the temperature can rise by 20 °C at 20 m above ground. In summer, ground inversions generally occur at night and in vegetated valleys (e.g., clover or corn fields).

These inversions are usually shallow and dissipate quickly after sunrise. Topography plays a crucial role in the formation of radiation inversions. They are most frequently pronounced in basins and valleys where cold air descends from slopes, forming "cold air lakes." Such inversions are especially notable in high-altitude basins and plateaus, where intense radiation and ventilation create ice regions with the lowest temperatures.

Descending inversions result from the dynamic heating of descending air. They are common at higher altitudes, particularly in subtropical anticyclones where air descends (as discussed in "Adiabatic Processes in the Atmosphere"). These inversions are usually more extensive and longer last than ground or radiative inversions. They are intensified by the cooling of ground layers in contact with the cold surface of the Earth. Downslope inversions also occur on windward mountain slopes, especially in winter, where warmer, dynamically heated air spreads over radiatively cooled air in contact with ice or snow.

Frontal inversions occur in frontal zones separating air masses of differing temperatures. In these zones, warmer, lighter air rises above a colder air wedge, resulting in a sharp horizontal temperature gradient and sometimes an increase in temperature with height.

References

Baker, I. T., Luthcke, S. B., & Pritchard, H. D. (2020). Albedo feedback in the Earth system: Observational evidence and future projections. Geophysical Research Letters, 47(2), e2019GL086980.

Bristow, K. L., & Campbell, G. S. (1984). On the relationship between incoming solar radiation and daily maximum and minimum temperature. *Agricultural and Forest Meteorology, 31*(2), 159–166.

Chen, F., & Dudhia, J. (2001). Coupling an advanced land surface–hydrology model with the Penn State–NCAR MM5 modeling system. *Meteorology and Atmospheric Physics, 76*(1–2), 175–193.

Gonzalez, M., Soriano, A., & Cañete, H. (2021). Thermal conductivity of soils and its implications for energy transfer in the environment. *Journal of Environmental Management, 286*, 112238.

Granier, A., et al. (2000). The role of forests in the carbon balance. *Global Change Biology, 6*(6), 687–697.

Hillel, D. (2019). *Soil physics*. Academic Press.

Houghton, R. A., & Goodall, J. (2017). Ocean heat storage and the climate system. *Journal of Climate, 30*(21), 8601–8616.

Huang, J., et al. (2018). The influence of land cover change on local climate in eastern China. *Environmental Research Letters, 13*(5), 054028.

Hwang, A., & Hsu, H. (2019). The role of ocean heat content in climate variability and change. *Nature Climate Change, 9*(2), 95–101.

IPCC. (2021). *Climate change 2021: The physical science basis*. Contribution of Working Group I to the Sixth Assessment Report of the Intergovernmental Panel on Climate Change. Cambridge University Press.

Kirkby, M. J. (1985). A review of the relationships between rainfall and soil moisture in the tropics. *Journal of Hydrology, 76*(1–2), 1–18.

Levitus, S., et al. (2012). World ocean heat content and thermosteric sea level change (0–2000 m), 1955–2010. *Geophysical Research Letters, 39*(10), L10603.

Liu, Y., Wang, W., & Zhang, Q. (2022). Influence of specific heat on temperature regulation in terrestrial ecosystems. *Ecological Modelling, 463*, 109855.

López-Bustins, J. A., et al. (2008). Climate change and elevation: Effects on temperature. *Theoretical and Applied Climatology, 93*(3–4), 227–236.

Ma, X., & Wang, Y. (2020). Impact of soil moisture on thermal conductivity and heat transfer in soils. *Soil Science Society of America Journal, 84*(5), 1234–1245.

Mason, L. A., & Harvey, C. F. (2023). The role of soil permeability in thermal regulation and heat transfer in groundwater systems. *Water Resources Research, 59*(3), e2022WR032649.

McCarthy, G. D., et al. (2015). Ocean warming and the future of global climate. *Nature Climate Change, 5*(5), 368–371.

Oke, T. R. (1982). The energetic basis of the urban heat Island. *Quarterly Journal of the Royal Meteorological Society, 108*(455), 1–24.

Oleson, K. W., et al. (2010). Improvements to the Community Land Model and their effects on the hydrological cycle. *Journal of Geophysical Research, 115*, G04021.

Pinker, R. T., & Laszlo, I. (1992). Modeling the surface radiation budget of the Earth. *Journal of Applied Meteorology, 31*(2), 228–243.

Chapter 9
Integrated Mechanisms of Global Ocean–Atmosphere Circulation

9.1 General Atmospheric Circulation: Definition

The permanent exchange of air between lower and higher latitudes, as well as between the lower and upper layers of the atmosphere, is driven by large-scale atmospheric circulation patterns. This movement facilitates the redistribution of heat, moisture, and momentum across the globe. The term "general circulation of the atmosphere" refers to the complex system of horizontal and vertical air movements that arise from this large-scale exchange, which can be understood by averaging these patterns over long periods (Hartmann, 2016; Holton & Hakim, 2013). A

© The Author(s), under exclusive license to Springer Nature Switzerland AG 2025

V. Spiridonov et al., *Atmospheric Perspectives*,
https://doi.org/10.1007/978-3-031-86757-6_9

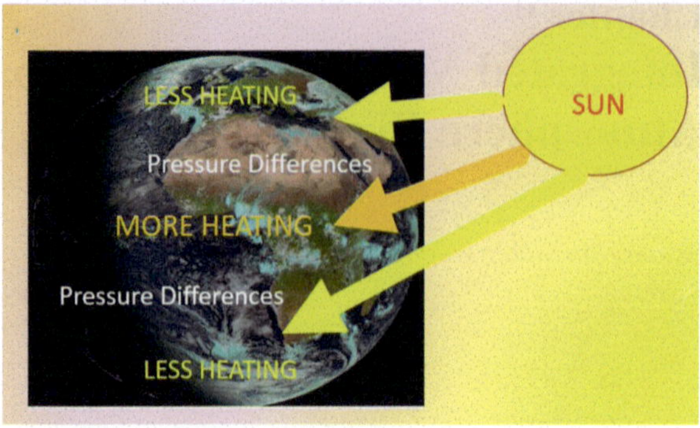

Fig. 9.1 The Sun as the primary driver of atmospheric and oceanic circulation

particularly important feature of the general circulation is the presence of jet streams—narrow, fast-flowing air currents found in the upper levels of the atmosphere. These jet streams, typically located near the tropopause, play a crucial role in shaping various atmospheric processes and phenomena, such as weather patterns, storm development, and the transport of air masses (Trenberth et al., 2018). Polar and subtropical jet streams, in particular, influence temperature gradients and are key drivers in the dynamics of both mid-latitude and tropical weather systems (Vallis, 2017). Global atmospheric and oceanic circulation systems are fundamental in shaping Earth's climate and distributing heat across the globe. These systems are primarily driven by the uneven distribution of solar radiation that the Earth receives, the planet's rotation, and the distinct physical properties of air and water (Peixoto & Oort, 1992). The general circulation of the Earth's atmosphere is governed by two critical factors: the geographic variation in solar energy input to the Earth–atmosphere system and the global distribution of the atmosphere's angular momentum (Held & Hou, 1980). The Sun serves as the principal force behind Earth's global atmospheric and oceanic circulation. The uneven heating of the Earth's surface by solar radiation generates temperature gradients, which in turn drive the movement of air and water within the atmosphere and oceans (Fig. 9.1) (Wallace & Hobbs, 2006). Global atmospheric circulation refers to the large-scale movement of air around the Earth, dictated primarily by solar radiation and the planet's rotation. This circulation pattern is crucial in shaping regional and global weather, distributing thermal energy, and maintaining the dynamic equilibrium within Earth's climate system (Ramanathan et al., 1987).

9.1.1 The Global Ocean Circulation and Its Role in Climate Regulation

Global ocean circulation is a complex system of currents transporting vast amounts of seawater throughout the world's oceans. These currents regulate climate, distribute heat, and influence marine ecosystems. The primary drivers of global ocean

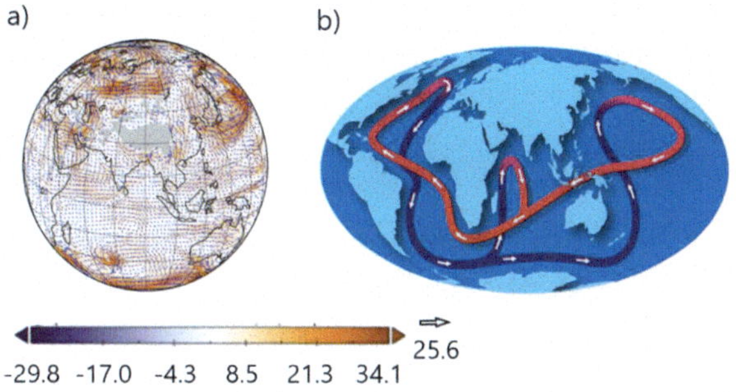

Fig. 9.2 Complex model of (**a**) global atmospheric and (**b**) oceanic circulation

circulation include temperature and salinity gradients, the Coriolis effect, and wind patterns. Ocean circulation significantly impacts Earth's energy balance, general atmospheric circulation, and the interactive processes occurring among the atmosphere, land, and surface water. Approximately half of the heat transport around the globe occurs through the oceans, underscoring their vital role in controlling Earth's climate system.

In Fig. 9.2, the global atmospheric and oceanic circulation systems are depicted, highlighting the interconnected nature of the horizontal flow of air and water. This figure shows the major atmospheric circulation cells and ocean currents that are instrumental in shaping Earth's climate and redistributing heat across the planet. The circulation of air over the Earth is primarily driven by the uneven heating of the Earth's surface.

9.1.2 The Earth's Energy Balance and Its Role in Global Circulation

The Earth's energy balance is a critical component of the climate system, encompassing both incoming solar radiation (S) and outgoing thermal infrared radiation (I). These components together form the net radiation (R_n), which is expressed as:

$$R_n = S(1-\alpha_p)+(I\downarrow - I\uparrow) \tag{9.1}$$

where (α_p) represents the planetary albedo, approximately 0.3; ($I\downarrow$) is the downward infrared radiation; and ($I\uparrow$) is the upward infrared radiation. The maximum transport of both incoming and outgoing radiation occurs around 40° latitude in each hemisphere, with the total energy transport peaking at approximately 6×10^{15} W. Notably, ocean currents contribute significantly to this energy transport, accounting for roughly 6×10^{15} W.

Global atmospheric circulation, pressure distribution, and wind patterns are integral to maintaining the Earth's heat balance and generating global ocean currents (Randall, 2015). Within global atmospheric circulation, warm air is transported from lower latitudes to higher latitudes, while cold air moves from high latitudes toward the equator. This heat exchange is vital for maintaining thermal equilibrium, keeping low-latitude regions warmer due to the continuous influx of energy, and high-latitude regions cooler due to continuous energy loss.

9.2 A Conceptual Model of the General Atmospheric Circulation

The constant exchange of air between lower and higher latitudes, as well as between the lower and higher layers of the atmosphere, is facilitated by large-scale atmospheric movements. The term "general circulation of the atmosphere" refers to a system of horizontal and vertical air movements, averaged over extended periods (Holton & Hakim, 2013). A key feature of the general circulation is the presence of jet streams, which are closely associated with various atmospheric processes and phenomena, such as storm systems, weather patterns, and the transport of air masses (Trenberth et al., 2018).

The first comprehensive model of general atmospheric circulation is often attributed to George Hadley, an English meteorologist, in the eighteenth century. In 1735, Hadley proposed his ideas in a paper titled "Concerning the Cause of the General Trade Winds." His model was based on the premise that solar heating at the equator causes warm air to rise and flow poleward, while cooler air descends at higher latitudes (Hartmann, 2016). Although Hadley's model did not fully account for all aspects of atmospheric circulation, it laid the groundwork for understanding the large-scale circulation patterns now recognized as the Hadley cell, Ferrel cell, and Polar cell (Vallis, 2017). These cells collectively form the foundation of the general atmospheric circulation pattern.

Later, William Ferrel expanded on Hadley's work, developing a more detailed three-cell model of global atmospheric circulation. Ferrel's model emphasized the importance of angular momentum conservation in generating westerly winds (Peixoto & Oort, 1992). A significant breakthrough came in the 1950s when the role of short waves (horizontal eddies) in transferring energy toward the poles was recognized, providing a solution to the problem associated with strong westerly winds (Held & Hou, 1980). The circulation patterns between the equator and 30°N and 30°S closely resemble the single-cell model originally proposed by George Hadley in 1735, and this region of circulation is still referred to as Hadley cell circulation (Wallace & Hobbs, 2006).

9.2.1 *Three-Cell Circulation Model*

Wind circulation in the atmosphere is driven by the Earth's rotation and the energy provided by the Sun. In each hemisphere, wind circulates through three distinct cells that facilitate the transport of energy and heat from the equator to the poles.

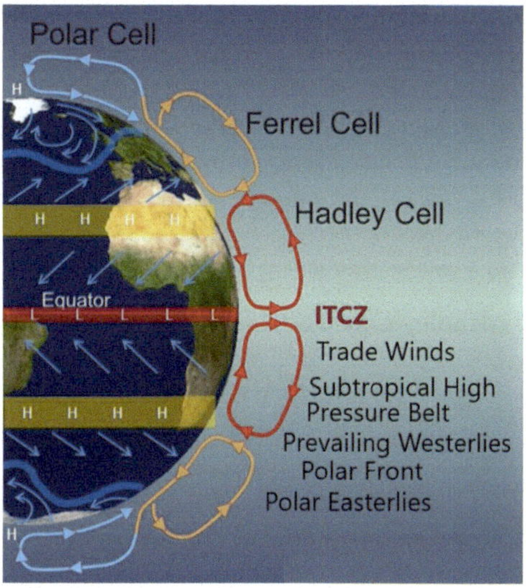

Fig. 9.3 Basic atmospheric circulation model. (1) Hadley's cell; (2) Ferrell's cell; (3) Polar cell

These winds are primarily driven by solar energy at the surface, where warm air rises and cooler air sinks.

The three-cell circulation model, illustrated in Fig. 9.3, describes the fundamental properties of global pressure and wind patterns. The model is based on two key assumptions:

- The Earth's surface has a uniform composition.
- There is an unequal inflow of solar radiation, creating a temperature gradient between the equator and the poles.

Under these conditions, and considering the Earth's rotation, three distinct cells emerge, each characterized by its circulation pattern between the equator and the poles:

1. **Hadley Cell**: Located between the equator and approximately 30° latitude
2. **Ferrel Cell**: Situated in the middle latitudes, between 30° and 60° latitude
3. **Polar Cell**: Found between 60° and 90° latitude

   These cells collectively define the primary mechanisms by which heat and energy are distributed globally, contributing to the overall atmospheric circulation pattern.

9.2.2 *Circulation in a Hadley Cell*

The Earth is enveloped by several major wind belts, which are separated by narrow regions of land or areas where air rises. The direction and positioning of these wind belts are primarily influenced by solar radiation and the Earth's rotation.

The Hadley cell, a key component of the global atmospheric circulation, plays a crucial role in these wind patterns. In the Hadley cell, warm air rises near the equator due to intense solar heating. As this air ascends, it cools and eventually moves poleward at high altitudes. Upon reaching higher latitudes, the cooler air descends around 30° latitude, creating a belt of high pressure. This descending air then flows back toward the equator at the surface, completing the circulation loop. For a detailed visualization of the general atmospheric circulation patterns and the Hadley cell's role within them, refer to Fig. 9.4.

1. **Hadley Cell**: The Hadley cell is an atmospheric circulation pattern characterized by warm air rising near the equator, which then cools and travels poleward at high altitudes. Upon reaching higher latitudes, the air descends as cooler, denser air and moves equatorward near the surface. This cell represents a convective loop dominating subtropical and tropical climates, where air rises in the lower latitudes, moves poleward aloft, and descends around 30° latitude before returning equatorward.
2. **Ferrell Cell**: Named after William Ferrell in 1856, the Ferrell cell operates in the mid-latitudes, between approximately 30° and 60° latitude. In this cell, surface air flows poleward and eastward, while higher-altitude air flows equatorward and westward. This cell is characterized by its role in balancing the heat transport between the tropical Hadley cell and the Polar cell, influencing the prevailing westerlies in the mid-latitudes.
3. **Polar Cell**: The Polar cell is in high latitudes, from about 60° to 90° latitude. In this cell, cold, dense air descends near the poles, creating high-pressure zones. This descending air diverges at the surface, flowing outward toward the lower latitudes. At higher altitudes, the air flows back toward the poles. The surface winds in the Polar cell are easterly, known as polar easterlies.

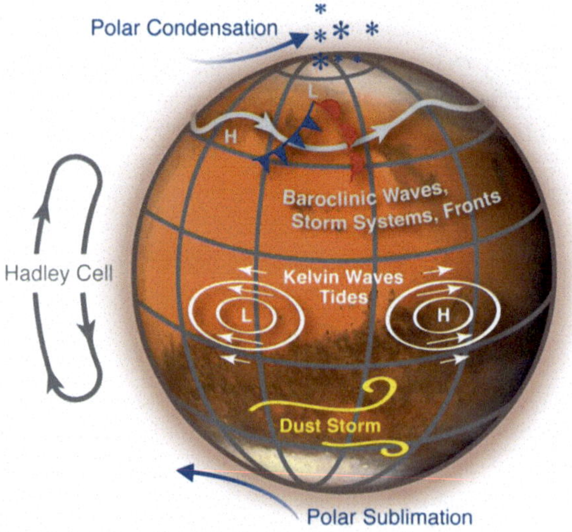

Fig. 9.4 Basic circulation patterns and Hadley cell

A detailed view of the basic cellular circulations is shown in Fig. 9.4. At or near the equator, where solar radiation is intense, the Earth's surface heats the air, causing it to rise. This process creates a zone of low pressure at the equator, known as the Intertropical Convergence Zone (ITCZ). As air from the subtropics converges and rises at the equator, it ascends into the upper atmosphere, reaching a maximum vertical height of about 14 km—the top of the troposphere. This rising air constitutes a segment of the Hadley cell circulation pattern. The Hadley cell transports air poleward at high altitudes and then descends near 30° latitude, creating high-pressure areas known as subtropical high-pressure belts. At these latitudes, the descending air completes the Hadley cell by flowing back toward the equator at the surface.

Surface air flows between the equatorial low-pressure zone and the subtropical high-pressure belts. The pressure gradient force drives the air from high to low pressure, while the Coriolis force deflects the winds. This deflection causes the winds to blow from east to west in the tropics, creating the trade winds (or tropical easterlies). Additionally, air moving toward the subtropical high-pressure zone is deflected to create the westerly winds in the mid-latitudes.

9.2.3 *Intertropical Convergence Zone*

The equatorial region of the globe consistently receives a substantial flow of energy throughout the year. Intense solar heating occurs at lower latitudes due to the high angle of the Sun and the nearly equal length of day and night year-round. The heat absorbed by the Earth's surface is transferred to the atmosphere through radiation and latent heat release. Additionally, when water vapor condenses, it releases heat into the surrounding air.

The resulting warmer, less dense air rises readily into the humid tropical atmosphere. This convective ascent creates an extensive low-pressure zone around the equator, commonly referred to as the equatorial trough. This region is known as the Intertropical Convergence Zone (ITCZ) (Fig. 9.5). As the air rises, it diverges near

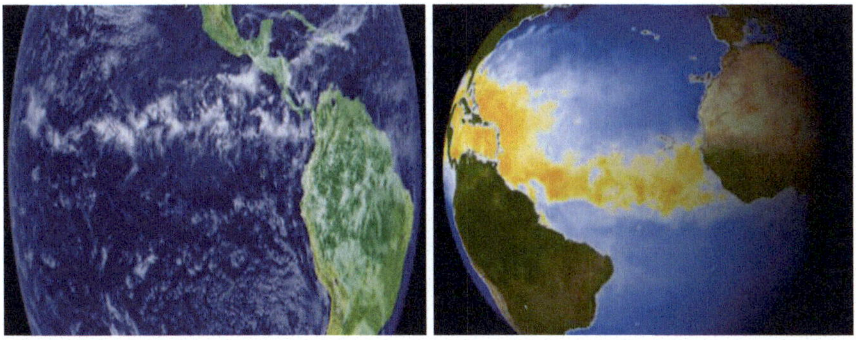

Fig. 9.5 Intertropical Convergence Zone (ITCZ)

the top of the troposphere and flows toward the poles. As it moves poleward, the air begins to converge again at latitudes around 25° to 35° in both hemispheres.

The Intertropical Convergence Zone (ITCZ) is a region encircling the globe near the equator, where the trade winds from the Northern and Southern Hemispheres converge. The intense solar radiation and warm ocean waters in this area heat the air, increasing its humidity and buoyancy. This buoyant air rises, assisted by the converging trade winds. As it ascends, the air expands and cools, leading to the condensation of moisture and the formation of near-constant thunderstorms. The subtropical high-pressure belts, crucial to Earth's climate, drive the general atmospheric circulation in a manner akin to major ocean currents. In these belts, air descends toward the Earth's surface, compressing and warming, which reduces its relative humidity. Major deserts, such as the Sahara, are situated within these subtropical high-pressure zones. As the air descends and diverges from the center of the high-pressure belt, the subtropical winds are generally weak and variable, especially near the equator.

In the Northern Hemisphere, the pressure gradient between the subtropical high at approximately 30°N and the low-pressure zone at the equator generates the northeast trade winds. In contrast, in the Southern Hemisphere, air is deflected to the left, producing southeasterly winds. The convergence of these northeast and southeast trade winds at the equator forms the ITCZ, characterized by calm conditions and light winds without a dominant direction.

Poleward of the subtropical high, between 30° and 60° latitude, the deflected poleward flow creates the westerly wind belts. At the poles, Earth's axial tilt and spherical shape result in a net loss of energy, leading to the formation of cold, dense air masses. These air masses descend, forming polar highs, and move toward lower latitudes. The Coriolis effect causes this air to be deflected eastward, creating the polar easterlies. The collision between these polar easterlies and the westerlies generates rising air and forms a broad low-pressure zone known as the subpolar low, a region marked by frequent storms and migrating pressure systems.

Global winds and pressure patterns are critical to the Earth's climate, largely determining the geographic distribution of rainfall. Low-pressure systems typically bring wet conditions, while high-pressure areas tend to inhibit precipitation.

9.3 Models of Zonal Distribution of the Pressure Field

In the Southern Hemisphere, there exists a single, consistent zonal pressure pattern along the subpolar low-pressure zone. However, in other latitudes, particularly in the Northern Hemisphere, where landmasses dominate over oceans, this zonal pattern is replaced by "semi-permanent" high- and low-pressure cells. In January, the pressure and wind patterns reveal a dominant high-pressure center, known as the Siberian anticyclone, positioned over the frigid terrain of Northeast Asia. This season also sees the presence of two intense "semi-persistent" low-pressure centers, the Aleutian and Icelandic lows, extending over the North Pacific and North Atlantic, respectively. Conversely, during summer, the pressure patterns in the Northern

Hemisphere shift dramatically. The high temperatures over the continents generate depressions that replace the winter high-pressure centers. During this season, the subtropical high-pressure center in the North Atlantic, known as the Azores anticyclone in winter, shifts near Bermuda and is referred to as the Bermuda anticyclone.

9.3.1 Upper Tropospheric Winds and Pressure Distribution Patterns

Circulation in the upper layers of the troposphere differs markedly from that at the surface. Between 15° and 20° latitude north and south, strong easterly winds emerge, considered an extension of the trade winds. For most of the upper troposphere, as one moves poleward, a predominant westerly flow known as the upper westerlies prevails. Above the Earth's friction layer, upper-level winds travel at higher speeds, including extremely powerful winds that encircle the globe, known as jet streams. These jet streams are driven by strong pressure gradients resulting from large temperature differences, particularly at high latitudes where the surface temperature gradient peaks. This maximum coincides with the polar front, the boundary between polar and tropical air masses, and is the location of the polar jet stream corridor of high-speed winds that play a key role in the formation and movement of large pressure systems (baric systems) across the mid-latitudes. Additionally, a subtropical jet stream forms around 30° latitude due to the convergence of air at high altitudes, corresponding to the same latitude discussed earlier. At the poles, extremely cold air descends, forming a high-pressure field at the surface. This downward motion creates a vast area of low pressure at higher altitudes, known as the circumpolar vortex. The polar jet stream lies near the equatorial boundary of this vortex. The polar vortex expands and contracts throughout the year as energy flows shift, affecting the surface. The equatorial edge of the jet stream adopts a wavy pattern, with cold air penetrating southward and warm air moving poleward. This process results in a transfer of cold air to lower latitudes and warm air to higher latitudes, leading to what is known as meridional flow within the polar vortex. At times, this edge flattens, causing the flow to move more directly from west to east, a condition known as zonal flow. Under these circumstances, there is minimal geographic exchange of air and energy, bringing relatively mild maritime polar air masses over the continents and reducing pronounced temperature differences.

9.4 Circulation Models

Wind patterns are categorized based on their horizontal scale of movement, leading to the identification of specific scales of circulation, such as planetary, synoptic, and mesoscale circulations. The Coriolis effect, a phenomenon resulting from Earth's rotation, plays a significant role in shaping these wind patterns (Holton & Hakim, 2013). This effect causes moving objects, including air masses, to be deflected to the right in the Northern Hemisphere and the left in the Southern Hemisphere, due to the conservation of angular momentum (Vallis, 2017). The deflection occurs because the Earth rotates more rapidly at the equator than at the poles, creating a variation in velocity with latitude.

The global wind pattern, also referred to as the "general circulation," is a key component of atmospheric dynamics, reflecting how air is transported across the globe to redistribute heat and energy (Peixoto & Oort, 1992). Each hemisphere's surface winds are organized into three distinct wind belts: the trade winds, the westerlies, and the polar easterlies (Fig. 9.6). These wind belts are separated by major latitudinal pressure zones, including the Intertropical Convergence Zone (ITCZ), subtropical high-pressure belts, and polar front. The interaction of these wind belts with atmospheric pressure systems leads to complex weather patterns, particularly in mid-latitudes where cyclonic activity is common (Trenberth et al., 2018).

Polar Winds (60–90° Latitude): Near the poles, the Coriolis effect weakens, allowing winds to flow predominantly from east to west. These are known as polar easterlies.

Westerly Winds (30–60° Latitude): In the mid-latitudes, between 30° and 60° latitude, the Coriolis effect is stronger, causing winds to deflect from west to east. This results in prevailing westerly winds that flow from west to east.

Trade Winds (0–30° Latitude): Near the equator, where the Coriolis effect is weaker, air moves almost directly from east to west, forming trade winds that blow from east to west in both hemispheres.

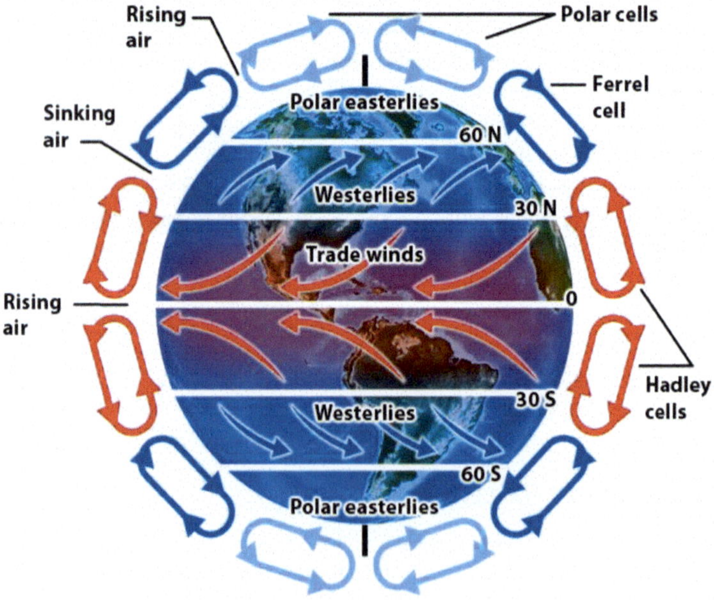

Fig. 9.6 Permanent winds of the general atmospheric circulation, illustrating the primary wind belts that shape global weather patterns

Circulations with a smaller horizontal scale than global circulation are classified as synoptic-scale circulations or weather map scales. Mesoscale circulations, which include phenomena like gale-force winds and onshore and offshore gusts, impact smaller regions and involve more intense vertical motions. The smallest circulations, both in spatial and temporal dimensions, are categorized as micro-scale circulations. These include local and often chaotic winds such as tornadic winds, turbulent eddies, and sandstorms.

9.4.1 East and West Winds

The circulation between 30° and 60° latitude in both hemispheres results in prevailing westerly winds. Air moving from the poles toward the equator creates easterly winds in both hemispheres. The area where cold polar easterly winds collide with the warm westerly flow at high latitudes is known as the polar front, a critical region in meteorology. If Earth's surface were uniform, it would feature two high-pressure belts and two low-pressure belts parallel to the equator. These four belts, starting from the equator, are:

1. The equatorial low-pressure belt, known as the Intertropical Convergence Zone (ITCZ), is a region where the trade winds converge.
2. The subtropical high-pressure zone is located between approximately 20° and 35° latitude on either side of the equator.
3. The subpolar low-pressure zone, around 50–60° latitude.
4. The polar high-pressure zone is situated near the poles.

9.5 Key Drivers of Global Climate: Pressure, Temperature, and Precipitation

Global atmospheric pressure systems significantly influence the geographic distribution of precipitation. This relationship is evident in how atmospheric patterns shape the distribution of precipitation and climate (Fig. 9.7). Generally, climates follow a horizontal pattern extending from the equator toward the poles. The climatic conditions at 30°S, for example, are closely associated with the location of the Intertropical Convergence Zone (ITCZ), the semi-permanent subtropical high-pressure zone (STH), and high-latitude cyclones. Global temperature distribution refers to the spatial pattern of temperatures across the Earth's surface.

This distribution is influenced by various factors, including latitude, altitude, and proximity to large bodies of water, ocean currents, and atmospheric circulation patterns. At the equator, the climate is characterized by some of the highest rainfall on Earth, including tropical rainforest (Af) and tropical monsoon (Am) climates. Recorded data often shows annual rainfall exceeding 2540 mm per year in these

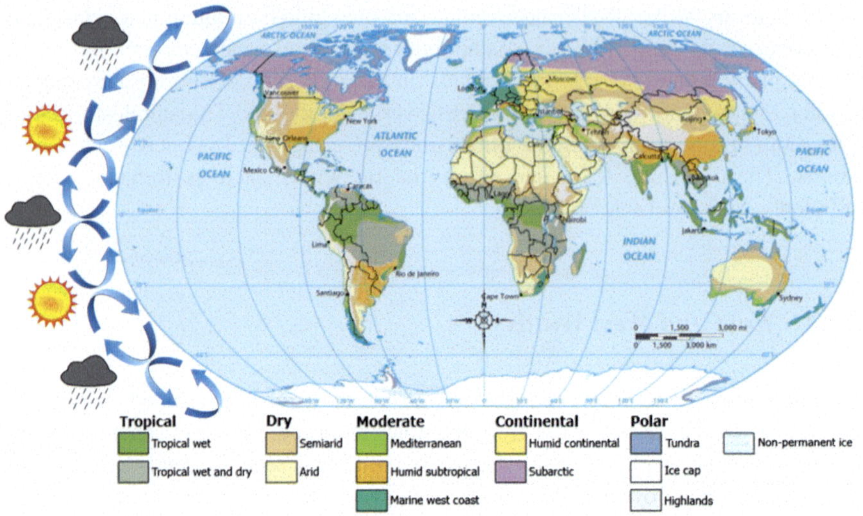

Fig. 9.7 Global climatic regions and their connection to atmospheric convection cells. Rainy and wet conditions occur at 0° and 60° latitudes due to rising air and low pressure, while dry conditions dominate at 30° and 90° latitudes due to sinking air and high pressure. (Credit: Adapted by PW; Map by Waitak, revised by Splette at en.wikipedia; Sun by Inductiveload; Raincloud by Calusarul; [CC BY-SA 3.0], via Wikimedia Commons)

regions. This heavy rainfall results from warm, moist air masses converging in a low-pressure belt over the equator, rising through convection. In Africa and Asia, the monsoon climate is marked by distinct wet and dry seasons. The dry season occurs due to the downwind flow of air from the subtropical high-pressure (STH) belt, while the wet season is influenced by the presence of the ITCZ and the moisture carried by trade winds along the coast. The tropical wet/dry climate (Aw), commonly known as the "savanna" climate, emerges as a transitional zone between the monsoon climate and the rainforest environment. During the low solar season (winter), this climate is dominated by the subtropical high-pressure belt, resulting in extremely dry conditions. The tropical steppe climate (BS) is predominantly controlled by the STH for most of the year. However, during the high solar season, when the ITCZ shifts into the region, the climate receives some rainfall. Moving poleward from the steppes, the tropical desert climate (BW) emerges, characterized by extreme aridity. This climate is the driest on Earth, with some areas experiencing no measurable annual rainfall, largely due to the persistent influence of the subtropical high-pressure zone throughout the year. At around 30°N latitude, a similar tropical steppe climate can be found, though the influence of seasonal pressure systems is reversed. During the low STH season, the region experiences dry conditions, with occasional brief rains from high-latitude winter cyclones. Finally, the Mediterranean region exhibits a dry subtropical climate (Cs), often referred to as a "Mediterranean climate," featuring a dry summer season and a wet winter season. This region receives more annual rainfall than the neighboring steppe climate, despite its distinct summer drought.

9.6 Ocean Circulation

Approximately 70% of Earth's surface is covered by oceans, enabling water to circulate freely across vast distances. This circulation is a fundamental driver of global weather and climate, significantly influencing Earth's energy balance, general atmospheric circulation, and the interactions between the atmosphere, land, and surface waters (Knauss & Garfield, 2016). Oceans transport about half of the heat around the planet, playing a crucial role in regulating Earth's climate system (Talley et al., 2011). Solar energy is unevenly distributed across the Earth's surface, with the highest influx occurring at the equator, creating temperature gradients between the equator and the poles. Both ocean and air currents respond to these temperature differences, facilitating the transfer of heat from the equator toward the poles (Cushman-Roisin & Beckers, 2011).

Sea surface temperature (SST) refers to the temperature of the upper few meters of the ocean and is a critical parameter for understanding and monitoring oceanic and atmospheric processes (Fig. 9.8). SST is crucial for forecasting climate and weather patterns because it directly impacts atmospheric circulation (Trenberth et al., 2018). The large-scale movement of ocean water is referred to as thermohaline circulation, which plays a key role in the exchange of water masses, including the supply of oxygen to the deep ocean. This circulation, also known as the meridional overturning circulation, is a primary mechanism in the redistribution of energy and heat in the oceans (Rahmstorf, 2006). Changes in ocean circulation due to global warming can lead to significant climate shifts, as ocean currents and atmospheric circulation are closely interconnected (IPCC, 2019).

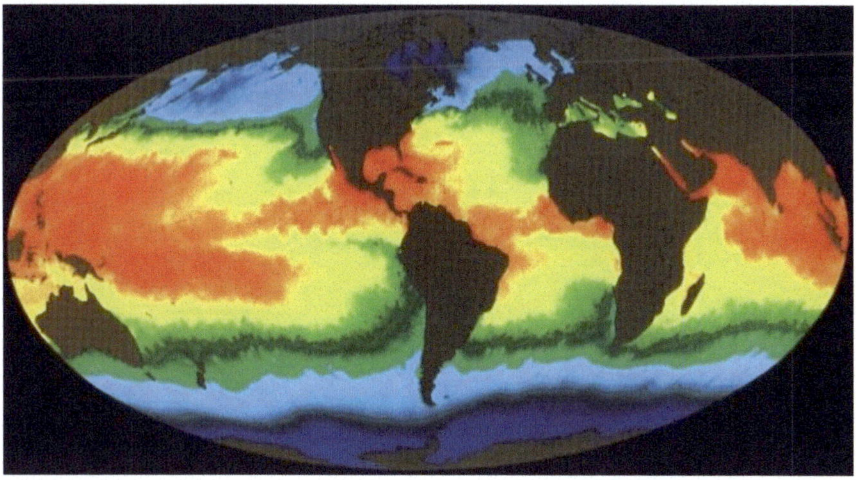

Fig. 9.8 Global sea surface temperature (SST) distribution, illustrating the variation of sea temperatures across different regions of the globe

Wind friction near the ocean's surface drives much of the upper layer motion of the ocean, resulting in surface currents. As wind moves across the ocean, it creates drag, slowing airflow in near-surface layers and initiating turbulent movements at the interface between the ocean and atmosphere (Klinger & Marshall, 1995). The thermohaline circulation is most pronounced during winter when winds are strongest, exhibiting an annual cycle that plays a crucial role in climate regulation. The mixing layer, where wind-driven currents are most active, has an average thickness of about 70 cm. This layer is essential for atmosphere–ocean interaction, influencing heat exchange, carbon absorption, and moisture transport, thereby affecting the climate system (Peixoto & Oort, 1992).

9.6.1 Large Ocean Currents

The primary ocean currents are generated by wind, though some result from changes in water density and salinity. Subtropical high-pressure cells play a significant role in driving many of Earth's major ocean currents. The influence of these currents on climate is most evident in the heating and cooling of coastal regions. Large ocean currents are classified as either warm or cold, and their global distribution is illustrated in Fig. 9.9.

9.6.2 Warm Currents

One of the most significant warm ocean currents in the northern Atlantic is the Gulf Stream (Fig. 9.10). This large, warm current flows northeast between Cuba and southern Florida, runs parallel to the northeastern coast of America almost to

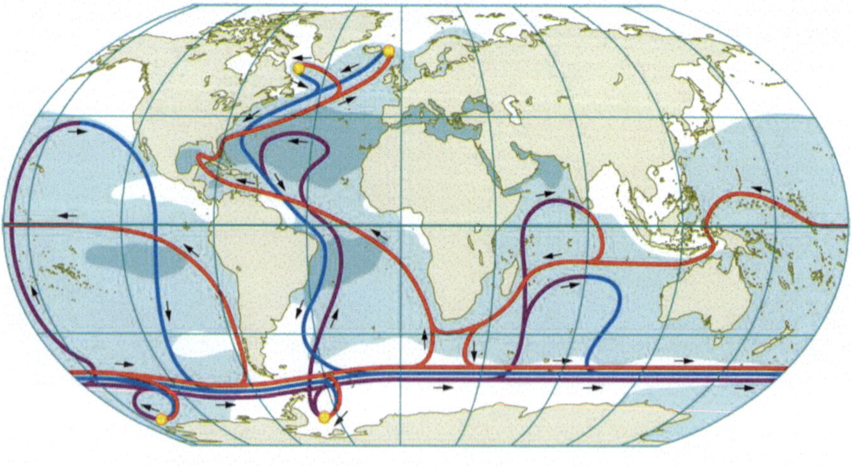

Fig. 9.9 Global ocean currents: this figure illustrates the distribution of major ocean currents around the globe

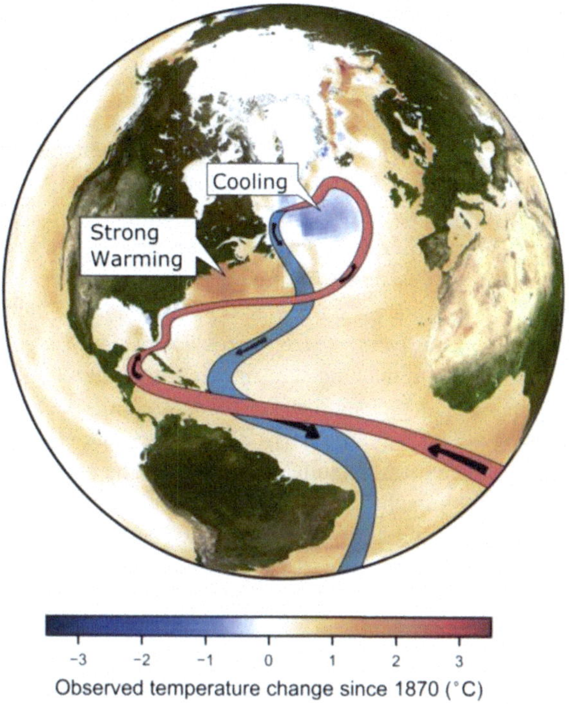

Fig. 9.10 Diagram of the gulf stream system showing the warm surface current and cold deep current, with color shading indicating the temperature trend since the late nineteenth century. (Based on Caesar et al., Nature, 2018)

Newfoundland, and continues as a northeast current. The Northeast Ocean Current then moves eastward, passing the United Kingdom and Norway, and enters the Arctic Barents Sea north of Russia.

The Gulf Stream significantly impacts the climate of much of Europe. The presence of several orographic barriers at these latitudes prevents westerly winds from carrying sea air eastward. In winter, the Gulf Stream raises temperatures in the Norwegian Sea by more than 26 °C above the normal value for this latitude. Consequently, winters in the British Isles, which are situated north of the latitude of the Canadian border, tend to be relatively mild. A similar influence is observed in Georgia, located between 15°N and 20°N latitude. In contrast, the North Pacific Current, a continuation of the Kuroshio Current, does not have as pronounced an impact on the climate of the eastern Pacific as the Gulf Stream does on the eastern Atlantic. This is due to its interaction with the high mountains parallel to the west coast of North America, which obstruct the movement of Pacific sea air across the continent. Additionally, the North Pacific is separated from the Arctic by the narrow, freezing Bering Sea in winter, which prevents the movement of surface water from the North Pacific into the Arctic during mid-to-late winter. In the Southern Hemisphere, warm currents are less prominent. The Indian Ocean currents are primarily influenced by seasonal monsoons, and weak, diffuse warm currents move southwest along the east coast of Australia.

9.6.3 The Gulf Stream and Its Changing Dynamics

The Gulf Stream is a powerful warm ocean current in the North Atlantic that significantly influences the globe's climate system. Originating in the Gulf of Mexico, it flows along the eastern coast of the United States before moving northeastward across the Atlantic Ocean toward Europe. As part of the broader Atlantic Meridional Overturning Circulation (AMOC), the Gulf Stream redistributes heat from the tropics to higher latitudes, shaping climates and weather patterns across continents. The Gulf Stream is essential for moderating temperatures, particularly in Europe, which experiences milder winters than regions at similar latitudes elsewhere. This current also supports diverse marine ecosystems and plays a crucial role in hurricane dynamics by influencing sea surface temperatures in the Atlantic. However, scientific evidence suggests that the Gulf Stream is weakening due to climate change. Melting polar ice, particularly from Greenland, contributes large volumes of freshwater to the North Atlantic. This reduces the salinity and density of seawater, disrupting the mechanisms that drive the circulation. Additionally, rising ocean temperatures and shifting global wind patterns further affect the flow and stability of the Gulf Stream (Fig. 9.10). A weakening Gulf Stream has far-reaching consequences. It may result in more extreme weather, such as harsher winters in Europe, intensified hurricanes in the Atlantic, and prolonged droughts in some regions. Coastal communities along the eastern United States face the added threat of localized sea level rise as ocean circulation slows. Globally, disruptions to the AMOC could amplify feedback mechanisms, exacerbating climate change. Given the Gulf Stream's pivotal role, continued scientific monitoring is vital to understand its dynamics and mitigate the risks associated with its decline. Reducing greenhouse gas emissions and slowing the pace of climate change are critical steps to preserving this vital component of Earth's climate system.

9.6.4 Cold Currents

Cold ocean currents significantly impact the climate, especially in the Southern Hemisphere. Among these, the Peruvian Current is particularly notable. It flows northward along the west coast of South America, starting around 50°S and extending nearly to the equator. This current profoundly influences the climate along most of the western coast of South America, although its effects are concentrated in a narrow coastal strip of the high Andes.

The Peruvian Current, characterized by cold water due to upwelling driven by nearshore winds, creates a strong temperature inversion when combined with subsiding high-pressure subtropical air. These conditions result in minimal rainfall, frequent fog, and low stratus cloud cover along the coast. In the Southern Hemisphere, the Benguela Current flows northward along the southern coast of Africa and the eastern South Atlantic, contributing to cooler conditions in the region.

In the Northern Hemisphere, cold currents are less prominent. The Canary Current moves southwest along the coasts of the Iberian Peninsula and northwest Africa, passing the Canary Islands. Similarly, the California Current flows southwest along the west coast of North America. Notably, during summer, the ocean water offshore north of San Francisco is often the coldest along the North American West Coast. Here, local variations in wave patterns can be more significant than the horizontal movement of water from Alaska to the south. A comparison of surface water temperatures in late August highlights the contrast between cold currents, such as the California Current, and warm currents like the Gulf Stream along the eastern coast.

9.6.5 Ocean Waves

Surface waves in the sea and ocean play a crucial role in ocean circulation. They transfer horizontal momentum to the Ekman layer, which accelerates vertical mixing in the upper water layers. This process is essential for redistributing heat, nutrients, and gases in the ocean, influencing both local and global oceanic systems. Waves also facilitate gas exchange between the ocean and the atmosphere, including the transfer of carbon dioxide and oxygen. Wind blowing over the ocean surface creates waves through friction between the air and water. Initially, this friction causes ripples, which can develop into larger waves as the wind speed exceeds the wave propagation speed. The force of Earth's gravity acts to restore hydrostatic balance, leading to the formation of gravitational waves. These waves are characterized by the crest (highest point) and trough (lowest point), with wave height representing the vertical change between them. The wavelength is the distance between successive crests or troughs.

In deep water, the speed of gravity waves increases with wavelength, meaning longer waves travel faster. For instance, tsunamis with wavelengths of 100 km or more can reach speeds of nearly 900 km/h. In shallow water, wave speed depends on water depth. Strong and persistent winds create the largest waves, particularly in the Indian Ocean's Southern Hemisphere belt between 40°S and 60°S, where average wave heights reach about 7 m, with some waves doubling this height. Near the equator, lighter winds and low waves are common, though large waves can occur during tropical cyclones, typhoons, or hurricanes.

Wave height is a critical metric for understanding ocean dynamics. Figure 9.11 highlights the spatial distribution of average ocean wave heights across the globe, illustrating the regions where wind-driven waves reach their maximum heights, such as the Southern Ocean and areas affected by tropical cyclones. These data emphasize the connection between wave energy, wind patterns, and ocean circulation processes.

Waves on the ocean surface include gravity waves, while other types, such as progressive waves (surface, internal, and tsunamis) and standing waves also exist. Sound waves travel through water approximately 4.5 times faster than through air,

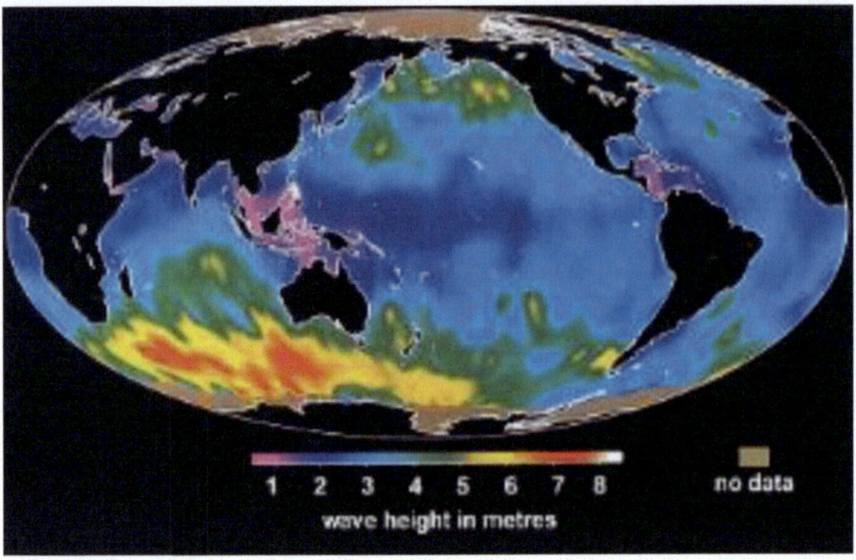

Fig. 9.11 Global distribution of ocean wave heights

reaching the ocean floor at depths of 4 km in a few seconds. The lowest waves are found around the equator, where low wind speeds prevail, but large waves can still form during intense tropical storms.

9.7 Large-Scale Circulation Regimes

In 1920, Sir Gilbert Walker made significant contributions to the understanding of atmospheric dynamics by identifying three major oscillations in the general atmospheric circulation:

- *The Southern Oscillation (SO)*: This oscillation refers to fluctuations in sea level pressure between the western and eastern tropical Pacific Ocean. It is the atmospheric component of the broader El Niño-Southern Oscillation (ENSO) phenomenon, which profoundly impacts global weather patterns. ENSO alternates between warm (El Niño) and cool (La Niña) phases, leading to shifts in precipitation, droughts, and storms across various regions of the globe (Trenberth, 1997).
- *The North Atlantic Oscillation (NAO)*: This oscillation describes changes in the pressure gradient between the Azores High and the Icelandic Low. NAO influences weather patterns, including temperature and precipitation, in the North Atlantic region, impacting Europe, Greenland, and eastern North America. The oscillation has been linked to variations in storm tracks and the intensity of westerly winds (Hurrell, 1995).

- The North Pacific Oscillation (NPO): The NPO is a seesaw pattern of atmospheric pressure variations between a high-pressure area over the North Pacific and the low-pressure Aleutian Low. This oscillation affects weather and climate patterns in the North Pacific region and can influence the weather in North America and parts of Asia, altering jet stream patterns and precipitation distribution (Rogers, 1981).

These oscillations play a critical role in shaping large-scale atmospheric and oceanic interactions, influencing global climate variability and seasonal weather predictions.

9.7.1 North Atlantic Oscillation (NAO) and Arctic Oscillation (AO)

The North Atlantic Oscillation (NAO) is a climate pattern characterized by variations in atmospheric pressure between the Icelandic Low and the Azores High. It significantly affects weather patterns in the Northern Hemisphere and alternates between positive and negative phases. The NAO can impact the strength and behavior of the jet stream, with weakened jet streams changing direction more frequently. The NAO is defined by the pressure differences between the Azores, which typically experiences high pressure, and Iceland, which experiences low pressure. These differences can vary in magnitude. The term "North Atlantic Oscillation" refers to variations in the surface pressure gradient in the North Atlantic region (Fig. 9.12).

Positive NAO Phase: During a positive NAO phase, the pressure difference between the Icelandic Low and the Azores High is stronger than average, leading to intensified westerly winds across the North Atlantic. In Europe, this phase is associated with milder and wetter conditions in northern regions, including the British Isles and Scandinavia. Increased westerly flow brings moist air from the Atlantic, resulting in above-average temperatures and higher precipitation.

Negative NAO Phase: In the negative NAO phase, the pressure difference is weaker than usual, resulting in a weakened jet stream. This phase is often associated with colder and more variable weather in Northern Europe. A weakened jet stream allows cold air masses to penetrate the Arctic, leading to below-average temperatures. Additionally, there is an increased likelihood of storms and blocking patterns, contributing to more extreme and persistent weather events. The negative NAO significantly influences winter weather, with colder and potentially harsher conditions, particularly in Northern and Eastern Europe. It can also affect precipitation patterns, with the positive NAO typically bringing increased precipitation to Northern and Western Europe, while the negative NAO may result in drier conditions. When the pressure difference is large, the NAO is positive, leading to strong westerly winds and more intense and frequent storms across Northwest Europe. Fluctuations in the strength of these features significantly alter the alignment of the jet stream, which ultimately affects temperature and precipitation distribution in the region.

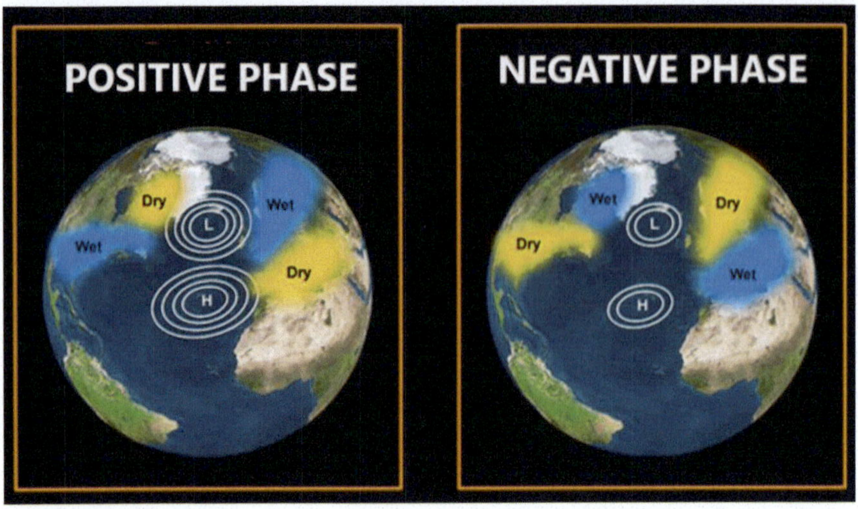

Fig. 9.12 Positive and negative phase of the North Atlantic Oscillation (NAO). (Credit: UCAR)

9.7.2 Polar Vortex

A polar vortex is a large area of low pressure and cold air that surrounds the Earth's polar regions (Fig. 9.13). It is a natural and permanent feature of the atmosphere, typically present in both the stratosphere and troposphere. The polar vortex is strongest during the winter months when the temperature difference between the polar regions and mid-latitudes is most pronounced.

The polar vortex forms and intensifies in winter when the polar regions experience prolonged darkness and extreme cold. The temperature difference between the Arctic (or Antarctic) and mid-latitudes creates strong winds that circulate counterclockwise around the polar region.

Stratospheric Polar Vortex: This is the better-known and more studied part of the polar vortex, found in the stratosphere, the second main layer of the Earth's atmosphere. The stratospheric polar vortex can influence weather patterns in the troposphere below.

Tropospheric Polar Vortex: Often referred to alongside the stratospheric polar vortex, the tropospheric polar vortex exists at lower altitudes. Changes in the stratospheric polar vortex can impact the tropospheric polar vortex and surface weather patterns. Occasionally, the polar vortex weakens, breaks up, or shifts, which can lead to cold air spilling into mid-latitudes, causing extreme cold weather events in regions that do not typically experience such harsh conditions. The term "polar vortex" has gained media attention during periods of extreme cold in North America in recent years. The state of the polar vortex can influence weather patterns, including storm development and the distribution of cold air masses. It is important to note that while the term "polar vortex" is often associated with extreme cold events, it is

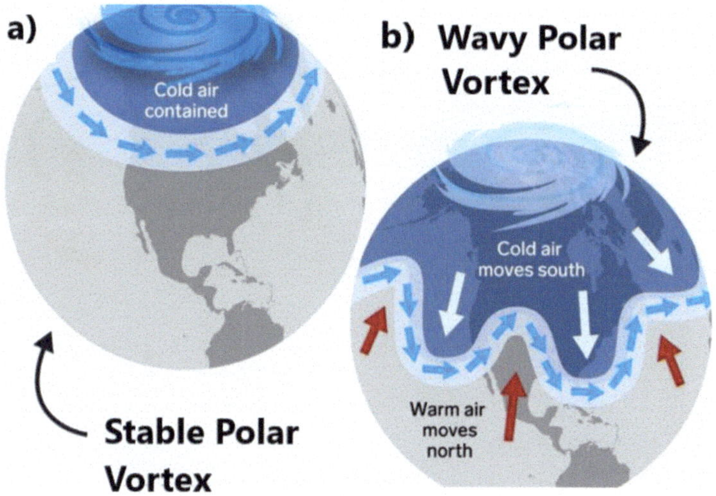

Fig. 9.13 Polar vortex. (**a**) Stable, strong poleward low-pressure center and cold air retention poleward. (**b**) A weak, unstable polar vortex that causes the jet stream to oscillate and meander

a natural and regular feature of Earth's atmospheric circulation. Extreme cold events are not solely caused by the polar vortex but can be influenced by its variability and interactions with other atmospheric features.

9.7.3 Arctic Oscillation

The Arctic Oscillation (AO) is a climate phenomenon characterized by natural variability in atmospheric circulation patterns over the Arctic region. It is a significant component of Earth's climate system and can have considerable impacts on Northern Hemisphere weather patterns, particularly during the winter months.

The Arctic Oscillation (AO) is essentially a climate index that reflects the state of atmospheric circulation over the Arctic (Fig. 9.14). It includes a positive phase, characterized by below-average geopotential heights (also known as negative geopotential height anomalies), and a negative phase, marked by positive geopotential height deviations. In the negative AO phase, the polar center of low pressure (or polar vortex) over the Arctic is weaker, leading to reduced upper-level winds (the westerlies). This weakening allows cold Arctic air to push further south into the Americas and Europe and can also lead to increased storm instability further south. It is important to note that while the AO and NAO are distinct indices, they both describe similar atmospheric phenomena related to pressure gradients in northern latitudes and their effects on temperature and storm patterns. There is ongoing debate about whether one index is a more accurate representation of atmospheric dynamics than the other.

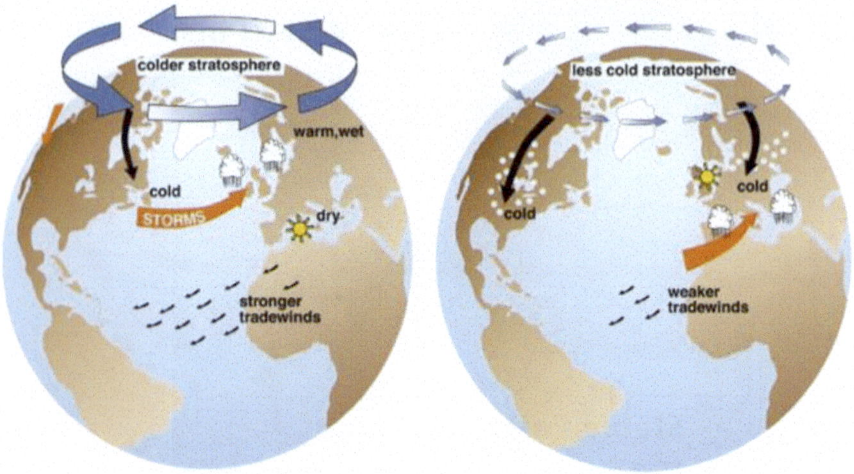

Fig. 9.14 Arctic Oscillation

9.7.4 El Niño-Southern Oscillation

The El Niño-Southern Oscillation (ENSO) is a climate phenomenon characterized by periodic fluctuations in sea surface temperatures (SSTs) across the central and eastern equatorial Pacific Ocean. ENSO is one of the most significant drivers of global climate variability, as its cycles affect temperature, precipitation, and weather patterns worldwide. It is a climate phenomenon characterized by periodic variations in atmospheric pressure and circulation patterns over the equatorial Pacific Ocean. There are three phases of ENSO shown in Fig. 9.15.

El Niño: This phase involves the warming of SSTs in the Central and Eastern Pacific, disrupting normal atmospheric circulation. El Niño typically leads to drier conditions in regions like Southeast Asia and Australia, while bringing increased rainfall and the potential for flooding to parts of South America and the Southern United States. El Niño events also tend to weaken the trade winds, impacting global weather patterns (Philander, 1990). This phase is characterized by abnormal warming of sea surface temperatures in the central and eastern equatorial Pacific Ocean (Fig. 9.15a). This warming disrupts the typical trade winds from east to west, causing changes in atmospheric and oceanic circulation patterns. During El Niño events, normal atmospheric circulation patterns, known as the Walker circulation, weaken or even reverse. This shift causes convection and precipitation to move from the Western Pacific (near Indonesia and Australia) to the Central and Eastern Pacific. Weakened easterly winds allow warm ocean water to spread eastward. El Niño typically brings wetter conditions to the Southern United States and drier conditions to parts of South America, including Peru and Ecuador. It may also lead to increased hurricane activity in the Atlantic. El Niño is associated with drier conditions in Australia and Southeast Asia, potentially causing drought in some regions.

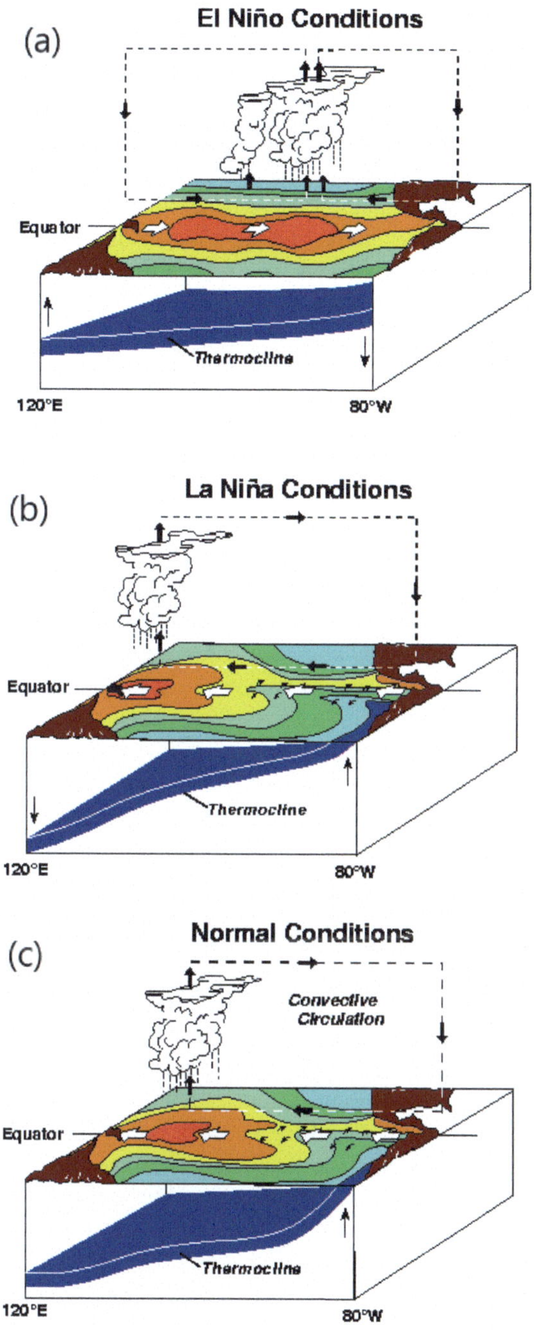

Fig. 9.15 ENSO-Southern Oscillation has three phases: (**a**) El Niño (warm phase), (**b**) La Niña (cold phase), and (**c**) neutral phase. (Credit: Vagaries of the Weather. https://www.vagaries.in)

Conversely, parts of South Asia may experience increased rainfall. East Africa often faces drier-than-average conditions during El Niño events, while Southern Africa may see increased rainfall. Some Pacific Island nations might experience drought during these events. El Niño events are often linked to warmer-than-average temperatures in the Central and Eastern United States, particularly during the winter months, whereas cooler-than-average conditions could occur in parts of the Northern United States and Canada. These events usually last from several months to a year and occur irregularly. El Niño is part of the natural cycle of climate variability, alternating with La Niña and neutral conditions. The frequency and intensity of El Niño events can vary.

La Niña is a climate phenomenon that is part of the broader El Niño-Southern Oscillation (ENSO) climate pattern. La Niña phase is the opposite of El Niño. This climate phenomenon refers to the cooling of SSTs in the Central and Eastern Pacific. This phase often strengthens the trade winds and enhances upwelling in the Pacific, resulting in colder and nutrient-rich waters. La Niña is associated with wetter conditions in Southeast Asia and Australia, while drier and warmer conditions prevail across the southern United States (Trenberth, 1997). It represents the opposite phase of El Niño and is characterized by lower-than-average sea surface temperatures in the central and eastern equatorial Pacific Ocean (Fig. 9.15b). La Niña events significantly impact global weather patterns, affecting precipitation, temperature, and atmospheric circulation. La Niña is marked by cooler sea surface temperatures in the central and eastern Pacific Oceans. This cooling occurs when the normal trade winds from east to west intensify, pushing warm surface waters away from the coast of South America.

The resulting cooler sea surface temperatures lead to changes in atmospheric circulation patterns, including the strengthening of the Walker circulation, which involves the movement of air from east to west along the equator. These changes in atmospheric circulation have widespread effects on global weather patterns. La Niña tends to bring drier and warmer conditions to the Southern United States and wetter-than-average conditions to the Northern United States and parts of Canada. It can also lead to increased hurricane activity in the Atlantic. In Australia and Southeast Asia, La Niña is associated with increased rainfall, which can lead to flooding. Conversely, parts of South Asia may experience drier-than-average conditions. East Africa often experiences increased rainfall during La Niña events, while southern Africa may face drier conditions. Some Pacific Island nations may see increased rainfall, whereas others might face drought. La Niña events are commonly linked to cooler temperatures in the Central and Eastern United States, especially during the winter months, while warmer-than-average conditions may occur in parts of the Northern United States and Canada. La Niña events typically last from a few months to over a year and occur irregularly, alternating with neutral conditions or El Niño events. The frequency and intensity of ENSO events vary and are closely monitored by meteorologists and climatologists.

ENSO-Neutral: As shown in Fig. 9.15c, in this phase, neither El Niño nor La Niña conditions dominate, and SSTs in the equatorial Pacific remain closer to average. However, during neutral phases, ENSO still influences atmospheric and oceanic processes, albeit with less dramatic climate impacts.

ENSO cycles have broad-reaching effects on global weather systems, including shifts in monsoonal patterns, tropical cyclone activity, and the strength of jet streams. This phenomenon can also have socioeconomic impacts, influencing agriculture, fisheries, and water resources in many regions around the world.

9.8 Regional-Scale Winds: Monsoons

The largest seasonal change in Earth's global circulation is the development of monsoons, wind systems that experience significant seasonal shifts in direction. The term "monsoon" refers to winds that change direction with the seasons. Monsoon conditions are found in various regions around the world.

The European atmospheric circulation, influenced by temperature differences between the continent and the North Atlantic Ocean, creates the thermodynamic and climatic conditions that initiate European monsoons. As illustrated in Fig. 9.16, the most notable monsoon pattern is the Indian monsoon over Asia. During the summer, the continent of Asia heats up more than the surrounding ocean due to differences in how land and water absorb heat. This results in a large area of low pressure forming over northern Central Asia and a smaller area over India. This low-pressure system draws moisture-rich air from the Pacific and Indian Oceans. As this moist air moves over the land, it intensifies convection caused by the heated surface. Unstable atmospheric conditions further stimulate air to rise along major mountain systems like the Himalayas, leading to heavy seasonal rainfall and wet monsoons.

In winter, the pattern reverses. The continent cools rapidly, creating a large high-pressure area over northwestern Asia, known as the Siberian anticyclone, and a smaller high-pressure area over India. Drier and colder air from the continent blows

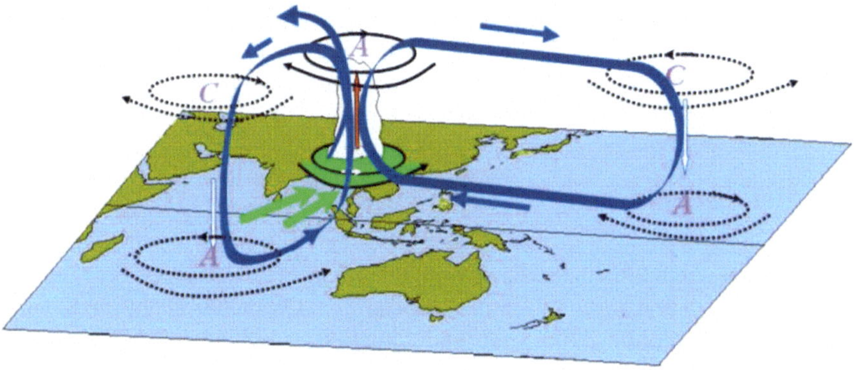

Fig. 9.16 Conceptual model of the characteristic monsoon circulation pattern over the Indian continent during summer. (Credit: Chang (2004) 9.8. Regional-Scale Winds: Monsoons)

Fig. 9.17 Intense and prolonged rainfall during the wet monsoon can lead to flooding in low-lying areas. Riverbanks may overflow, and urban areas can experience waterlogging. The risk of landslides increases, particularly in hilly or mountainous regions, due to soil saturation

toward the coast, creating a dry monsoon season. Monsoon conditions also occur in East Central Africa, where seasonal rains are influenced by the variable pressure systems of the global atmosphere. During the high-sun season, the Intertropical Convergence Zone (ITCZ) moves inland, bringing warm, moist, and unstable air that results in heavy rain (Fig. 9.17). In the low-sun season, the ITCZ shifts outward, and the influence of subtropical high-pressure systems (anticyclones) dominates, preventing precipitation. In the southwestern deserts of North America, pseudomonsoons occur during the summer, often referred to as the Arizona monsoon or the southwest monsoon. This phenomenon involves a relatively small seasonal shift in the wind, resulting in a dry spring followed by a relatively rainy summer in the Southwestern United States and Northwestern Mexico. Intense surface heating forms a heat low over the desert, causing moist cyclonic air from the Gulf to move into the region. This air rises through convergence or orographic lifting, producing rain. In winter, the desert area cools, the low-pressure center dissipates, and dry conditions return.

In Europe, monsoon-like conditions are observed due to temperature differences between the North Atlantic Ocean and the continent. Early summer is rainy in Eastern Europe, while early winter is rainy in Western Europe. Dry periods occur in winter in Eastern Europe and in spring in Western Europe. Major floods are common in some parts of Central Europe due to these monsoon patterns.

9.9 Jet Streams

Jet streams are high-altitude bands of wind that occur in the upper levels of the troposphere, the lowest layer of Earth's atmosphere. Characterized by high wind speeds—often exceeding 100 km/h and sometimes reaching much higher—jet streams are typically found near boundaries between air masses with significant temperature differences, such as between warm tropical air and cold polar air. There

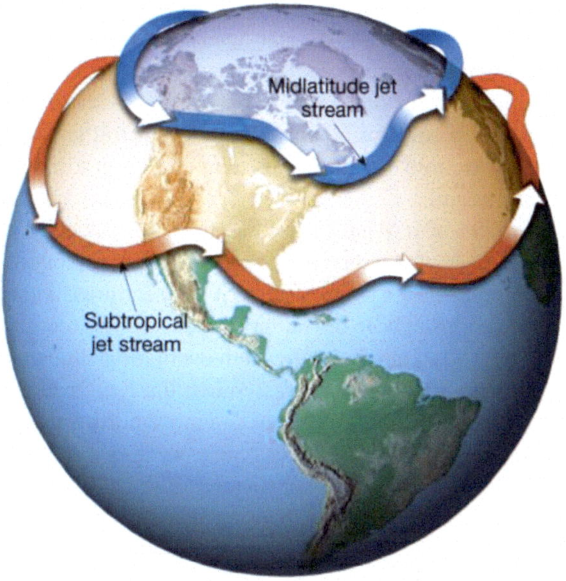

Fig. 9.18 Polar and subtropical jet stream

are usually two main jet streams in each hemisphere: the polar jet stream and the subtropical jet stream (Fig. 9.18).

- **Polar Jet Stream:** Located at high latitudes, typically between 50° and 60° north and south, the polar jet stream marks the boundary between cold polar air and warmer air to the south.
- **Subtropical Jet Stream:** Found at lower latitudes, around 30° north and south, the subtropical jet stream is associated with the transition between warm tropical air and cooler air to the north.

Jet streams are driven by the Earth's rotation and the temperature contrast between different air masses. The Coriolis effect, resulting from the Earth's rotation, contributes to the formation of these high-speed air currents. Jet streams are not constant; they can vary in intensity, location, and shape due to seasonal changes, weather patterns, and other atmospheric phenomena. They play a crucial role in shaping weather patterns and influencing the movement of storms, guiding the development and movement of weather systems, and affecting the distribution of precipitation, temperature, and other atmospheric conditions.

Changes and fluctuations in the intensity of the North Atlantic Oscillation (NAO) and Arctic Oscillation (AO) can significantly alter the alignment of the jet stream, ultimately affecting the distribution of temperature and precipitation. These changes are thought to contribute to the increasing frequency of unusual and extreme weather events across North America, Europe, and Asia.

Various theories propose how the AO might influence mid-latitude weather:

1. **Jet Stream Behavior:** Some scientists suggest that fluctuations in the jet stream, influenced by the AO, can lead to extreme weather conditions. This includes phenomena such as heat waves, severe storms, and floods. The jet stream, a high-altitude wind belt in the troposphere, is driven by the temperature difference between cold Arctic air and milder air to the south. As the Arctic warms faster than the mid-latitudes, this temperature gradient weakens, leading to a weaker jet stream.
2. **Impact of a Weaker Jet Stream:** A weaker jet stream is more susceptible to deviations from its typical west-to-east trajectory due to obstacles such as mountain ranges, varying sea surface temperatures, and energy from dissipating tropical storms. These deviations can result in a more meandering jet stream, which may lead to larger north–south oscillations known as "Rossby waves."
3. **Effects of Rossby Waves:** When the jet stream experiences significant north–south undulations, warm air can move northward, while cold air can move southward. Larger Rossby waves cause weather systems to travel more slowly, making weather patterns more persistent. Prolonged stasis in a weather pattern can result in extreme events, such as prolonged droughts, extended rainfall leading to flooding, persistent cold spells, or prolonged heat waves.

These theories underscore the complex interactions between atmospheric patterns and their influence on regional and global weather extremes.

9.10 Arctic Amplification

Climatologists are increasingly focusing on the potential links between Arctic warming and extreme weather in mid-latitudes. The Arctic is warming at a rate faster than the global surface average, a phenomenon known as "Arctic amplification" (Fig. 9.19). This accelerated warming is partly due to the rapid loss of sea ice cover in the region.

As Arctic sea ice diminishes, the energy from the Sun that would normally be reflected by the bright white ice is instead absorbed by the ocean, leading to further warming. Similarly, the reduction in snow cover over Arctic land areas exacerbates this effect. Arctic amplification is also driven by temperature feedback mechanisms: as the Earth's surface warms, it radiates more energy back into space. However, the Arctic, with its reduced ice and snow cover, reflects and emits less energy compared to lower latitudes, resulting in faster warming in this region. The warming Arctic Ocean inhibits sea ice growth and generates warmer and wetter air masses over the Arctic and adjacent continents. Additionally, the reduction in the temperature gradient between the Arctic and mid-latitudes impacts atmospheric circulation patterns. Another significant factor is "sudden stratospheric warming," which occurs when the stratosphere over the Arctic experiences a rapid increase in temperature. During such events, the temperature in the middle stratosphere above the pole can rise by more than 65 °C compared to conditions with a strong polar vortex. This warming

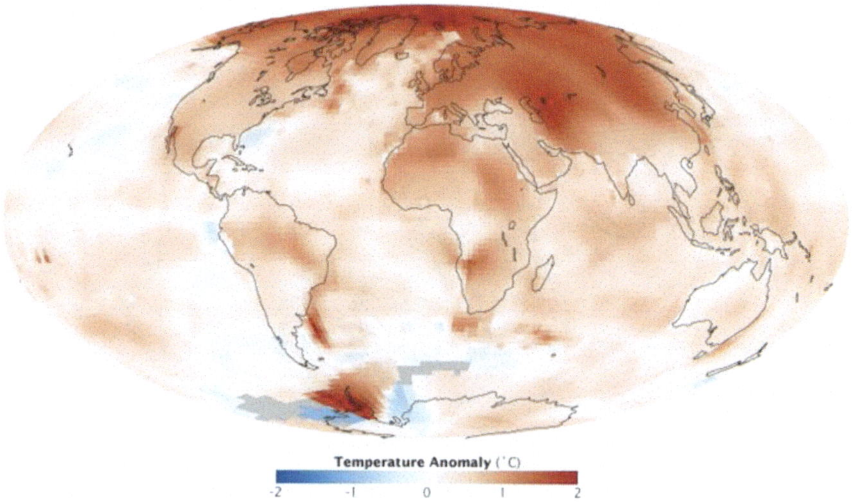

Fig. 9.19 Global temperature anomaly. (Credit: NASA image by Robert Simmon, based on GISS surface temperature analysis data including ship and buoy data from the Hadley Centre. Caption by Adam Voiland)

can weaken the polar vortex and may even cause it to split into two or more smaller vortices. Sudden stratospheric warming events can have substantial weather implications, including the strengthening and persistence of wintry conditions across much of Europe in mid and late January.

9.11 Rossby–Planetary Waves

Oceanic and atmospheric Rossby waves (Fig. 9.20), also known as planetary waves, occur naturally mainly due to the rotation of the Earth. These planetary (oceanic and atmospheric) waves play a significant role in the planet's weather and climate. Slow-moving ocean Rossby waves are significantly different from ocean surface waves. Rossby waves are large, disruptive ocean movements that stretch horizontally across the planet for hundreds of kilometers in a westward direction.

Recent research by de Szoeke and Xie (2023) has provided new insights into how oceanic Rossby waves interact with atmospheric circulation. Their findings suggest that these waves can significantly impact the ocean's heat content and distribution, which in turn affects global weather patterns. They also noted that the interactions between oceanic Rossby waves and atmospheric circulation can amplify climate anomalies, such as those observed during El Niño and La Niña events.

Interaction with Arctic Amplification: The interaction between Rossby waves and Arctic amplification has become a focus of recent studies. As the Arctic warms at an accelerated rate compared to the global average, changes in temperature

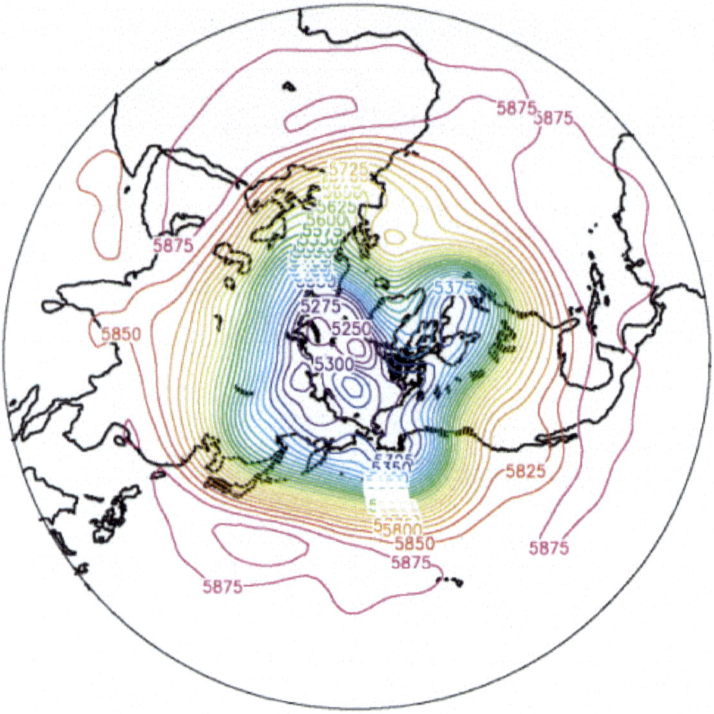

Fig. 9.20 Rossby waves formed by a source with circular circulation

gradients can affect the behavior of Rossby waves. For example, recent research by Cohen et al. (2024) indicates that Arctic warming is altering the typical patterns of Rossby waves, leading to more pronounced and persistent weather anomalies in mid-latitudes. This can contribute to extreme weather events, such as intensified heat waves and severe cold spells.

Future Implications: Understanding the behavior of Rossby waves is crucial for predicting future climate scenarios. The latest models, incorporating satellite observations and advanced climate simulations, are improving our ability to forecast how changes in these waves might impact global and regional climates. Continued research is needed to fully grasp the interactions between Rossby waves, Arctic amplification, and other climatic factors to better predict and mitigate the impacts of extreme weather events.

In summary, Rossby waves are integral to the Earth's climate system, influencing both atmospheric and oceanic conditions. Their interactions with recent climate changes, such as Arctic amplification, underscore the need for ongoing research to understand and predict the complex dynamics of the Earth's climate.

References

Caesar, L., Rahmstorf, S., Robinson, A., Feulner, G., & Saba, V. (2018). Observed fingerprint of a weakening Atlantic Ocean overturning circulation. *Nature, 556*, 191–196.

Chang, C.-P. (2004) East Asian Monsoon (2) (World Scientific Series on Meteorology of East Asia). https://doi.org/10.1142/5482.

Cohen, J., Smith, K., & Yoshida, T. (2024). La Niña and its implications for global climate variability. *Climate Dynamics, 42*(2), 563–581.

Cushman-Roisin, B., & Beckers, J.-M. (2011). *Introduction to geophysical fluid dynamics: Physical and numerical aspects*. Academic Press.

de Szoeke, S. P., & Xie, S. P. (2023). The role of ocean-atmosphere interactions in El Niño dynamics. *Journal of Climate, 36*(4), 1241–1260.

Hartmann, D. L. (2016). *Global physical climatology* (2nd ed.). Elsevier.

Held, I. M., & Hou, A. Y. (1980). Nonlinear axially symmetric circulations in a nearly inviscid atmosphere. *Journal of the Atmospheric Sciences, 37*(3), 515–533.

Holton, J. R., & Hakim, G. J. (2013). *An introduction to dynamic meteorology* (5th ed.). Elsevier Academic Press.

Hurrell, J. W. (1995). Decadal trends in the North Atlantic Oscillation: Regional temperatures and precipitation. *Science, 269*(5224), 676–679.

IPCC. (2019). *Special report on the ocean and cryosphere in a changing climate*. Intergovernmental Panel on Climate Change.

Klinger, B. A., & Marshall, J. (1995). The role of vertical mixing in the dynamics of the thermohaline circulation. *Journal of Physical Oceanography, 25*(5), 775–793.

Knauss, J. A., & Garfield, N. (2016). *Introduction to physical oceanography* (3rd ed.). Waveland Press.

Peixoto, J. P., & Oort, A. H. (1992). *Physics of climate*. American Institute of Physics.

Philander, S. G. (1990). *El Niño, La Niña, and the Southern Oscillation*. Academic Press.

Rahmstorf, S. (2006). Thermohaline ocean circulation. In W. M. Moomaw (Ed.), *Encyclopedia of global warming and climate change* (pp. 215–219). SAGE.

Ramanathan, V., Cess, R. D., Harrison, E. F., Minnis, P., Barkstrom, B. R., Ahmad, E., & Hartmann, D. L. (1987). Cloud-radiative forcing and climate: Results from the Earth radiation budget experiment. *Science, 243*(4887), 57–63.

Randall, D. (2015) An Introduction to the Global Circulation of the Atmosphere. Princeton University Press.

Rogers, J. C. (1981). The North Pacific Oscillation. *Journal of Climatology, 1*(1), 39–57.

Talley, L. D., Pickard, G. L., Emery, W. J., & Swift, J. H. (2011). *Descriptive physical oceanography: An introduction* (6th ed.). Academic Press.

Trenberth, K. E. (1997). The definition of El Niño. *Bulletin of the American Meteorological Society, 78*(12), 2771–2777.

Trenberth, K. E., Fasullo, J. T., & Shepherd, T. G. (2018). The changing character of the general circulation with climate change. *Nature Climate Change, 8*(2), 147–156.

Vallis, G. K. (2017). *Atmospheric and oceanic fluid dynamics* (2nd ed.). Cambridge University Press.

Wallace, J. M., & Hobbs, P. V. (2006). *Atmospheric science: An introductory survey* (2nd ed.). Elsevier Academic Press.

Chapter 10
Complex Interactions of Air Masses and Atmospheric Systems

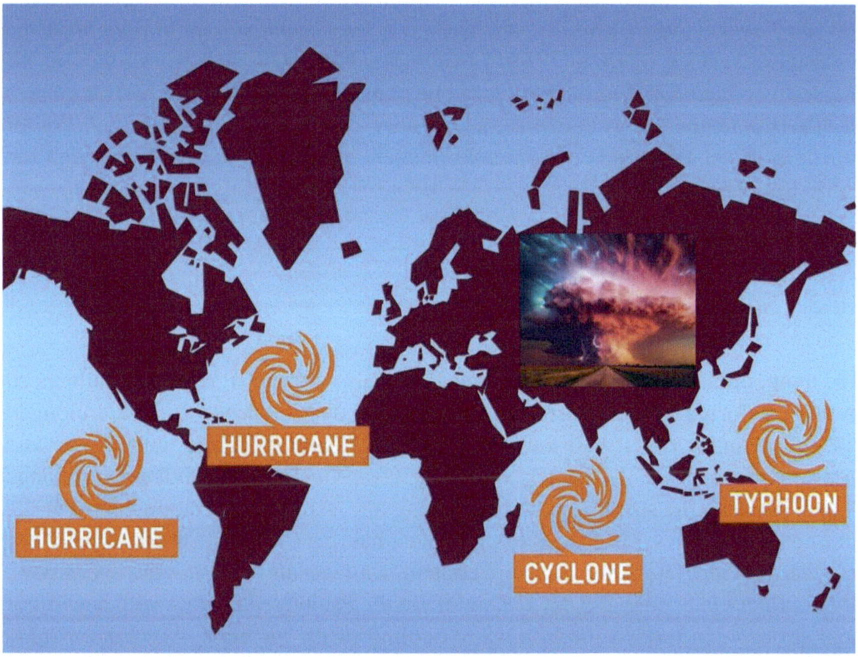

10.1 Definition of Air Masses

The concept of air masses was first identified by Swedish meteorologist Tor Bergeron, who observed that certain meteorological elements exhibited minimal horizontal variation over vast areas. These expansive regions of relatively uniform

© The Author(s), under exclusive license to Springer Nature Switzerland AG 2025

V. Spiridonov et al., *Atmospheric Perspectives*,
https://doi.org/10.1007/978-3-031-86757-6_10

air were termed air masses (Barry & Chorley, 2010). The horizontal extent of an air mass can range from 500 to 5000 km, often spanning entire continents, while vertically, an air mass may encompass the entire troposphere (Bluestein, 1992).

Air masses are characterized by specific attributes, including temperature, humidity, temperature stratification, and the degree of turbidity. While the primary focus here will be on large, synoptic-scale air masses, it is important to note that meteorology also recognizes smaller-scale air masses. An air mass is defined as a widespread body of air with uniform physical properties that develop over a particular region of the Earth's surface, known as the source region (Stull, 2017). This body of air is distinguished by temperature and moisture characteristics that are relatively homogeneous in the horizontal direction across a broad area. Moreover, the vertical variations in temperature and moisture within an air mass tend to be consistent across its horizontal extent (Lamb, 1982). As air masses move away from their source regions, they typically undergo specific modifications. The stagnation or prolonged movement of air over the source region allows the vertical temperature and moisture distribution within the air mass to achieve relative equilibrium with the underlying surface (Harman, 1991). Air masses are separated from adjacent air masses by a boundary region, which can be precisely defined. When two air masses of differing temperatures converge, a boundary known as a front is formed. An air mass can cover thousands of square kilometers and extend vertically through the entire troposphere.

10.2 Source Regions of Air Masses

The temperature and other characteristics of an air mass are significantly influenced by both its origin and its subsequent movement over land or sea. The prolonged contact of an air mass with the surface allows it to acquire the temperature and moisture properties of the underlying region. Over time, heat and moisture exchange occurs between the surface and the overlying air, resulting in a modification of the air mass's properties. For instance, an air mass passing over warm ocean waters may increase in temperature and moisture content, while an air mass moving over a cold, dry continent may cool and dry out significantly. Prolonged contact with a warm or cold surface can lead to the heating or cooling of the air mass. A warm air mass forms when air remains in contact with a warm surface for an extended period, while a cold air mass develops under similar conditions over a cold surface. This heat transfer process occurs gradually, often taking a week or more for the air to warm by 10 °C throughout the atmosphere. For these changes to materialize, a large mass of air must stagnate over a region. The stability and homogeneity of the air mass are enhanced by weak winds and minimal atmospheric disturbances, which allow the air to maintain its temperature and moisture profile across large areas.

Regions of the Earth's surface where air can stagnate and gradually acquire the properties of the underlying surface are known as source regions. These source regions are typically characterized by large, uniform surfaces with little topographical variation, which supports the uniform heating or cooling of the overlying air.

The primary source regions include the high-pressure belts in the subtropics, which give rise to tropical air masses, and the polar regions, which are the source of polar air masses. Additionally, vast continental landmasses and extensive oceanic regions also serve as critical source areas for the formation of air masses. Continental source regions produce dry air masses due to limited moisture availability, while maritime regions generate moist air masses that can significantly influence weather patterns upon moving inland. In the subtropics, for example, the high-pressure zones facilitate the formation of warm, dry tropical air masses (tropical continental, Tc). These air masses often dominate weather conditions in regions like the Sahara Desert and the Southwestern United States, bringing clear skies and hot, dry weather. Conversely, the polar regions give rise to cold polar air masses (polar continental, Pc) that are typically dry and stable. These air masses can extend into temperate latitudes, bringing frigid conditions during the winter months. Over oceanic regions, maritime tropical (mT) and maritime polar (mP) air masses develop, characterized by high humidity and frequent cloud cover. Maritime tropical air masses, originating from subtropical oceans, play a crucial role in generating thunderstorms and heavy rainfall when they encounter cooler landmasses or colder air aloft. The most significant source regions for air masses include:

Polar Regions Cold air masses develop over these high-latitude areas, where the surface is predominantly ice or snow-covered, leading to very low temperatures. Polar air masses (both continental and maritime) influence much of the mid-latitude weather, especially in winter.

Subtropical High-Pressure Zones Found near 30° latitude in both hemispheres, these areas are known for their stable and warm conditions, fostering the formation of tropical air masses. These high-pressure zones are responsible for the development of tropical maritime (mT) air masses over oceans and tropical continental (Tc) air masses over deserts and landmasses.

Continental Landmasses Large land areas, particularly those in mid-to-high latitudes, serve as the source for continental air masses. The interior of continents, far from oceanic influences, produces dry, stable air masses that can be either cold or warm depending on the season. For example, the Siberian High is a well-known source region for extremely cold, dry air masses that influence winter weather patterns across Eurasia.

Oceanic Regions Extensive bodies of water in subtropical and polar regions serve as source regions for maritime air masses. Oceans moderate the temperature of air masses, leading to more temperate and moist conditions. For example, the Gulf of Mexico is a key source region for warm, moist maritime tropical air masses that influence the weather of the Southeastern United States.

As air masses move away from their source regions, they typically undergo modification as they encounter new surfaces. The temperature and moisture characteristics can change significantly, especially when an air mass moves over a surface with contrasting properties. For instance, when a maritime tropical air mass moves inland over a cold, dry continent, it can lose moisture, stabilize, and cool. This interaction between air masses and their environments is critical for understanding weather patterns, such as the development of fronts, storms, and precipitation events.

10.3 Formation Criteria

Air mass source regions are specific geographic areas where air masses originate. For an air mass to form, two primary criteria must be met:

- The region must be large, homogeneous, and physically uniform.
- Air mass stagnation must occur, characterized by the presence of weak surface winds.
- The longer an air mass remains over its source region, the more it acquires the characteristics of the underlying surface.

10.3 Classification of Air Masses

Air masses are classified primarily based on their area of origin and their fundamental temperature and humidity characteristics. The classification also depends on the distribution and nature of the source region's surface, whether it is a continent or an ocean (Barry & Chorley, 2010). The Bergeron classification is the most widely accepted system for categorizing air masses (Bluestein, 1992). This classification system involves three letters: The first letter describes its moisture properties: "c" for continental air masses (dry) and "m" for maritime air masses (moist). The second letter indicates the source region:

T for Tropical
P for Polar
A for Arctic or Antarctic
M for Monsoon
E for Equatorial
S for Superior air

The last refers to the adiabatic drying and heating of air, formed from a descending flow in the atmosphere (Stull, 2017). A generalized map of global air masses based on this classification is illustrated in Fig. 10.1.

10.3.1 Continental Polar and Continental Arctic Air Masses (cP and cA)

Continental polar (cP) and continental arctic (cA) air masses originate from the extensive snow-covered regions of Eastern Europe and Russia. These air masses are primarily a winter phenomenon, occurring between November and April. During the summer months, when the land heats up considerably, these air masses lose their cold characteristics and are typically classified as tropical continental.

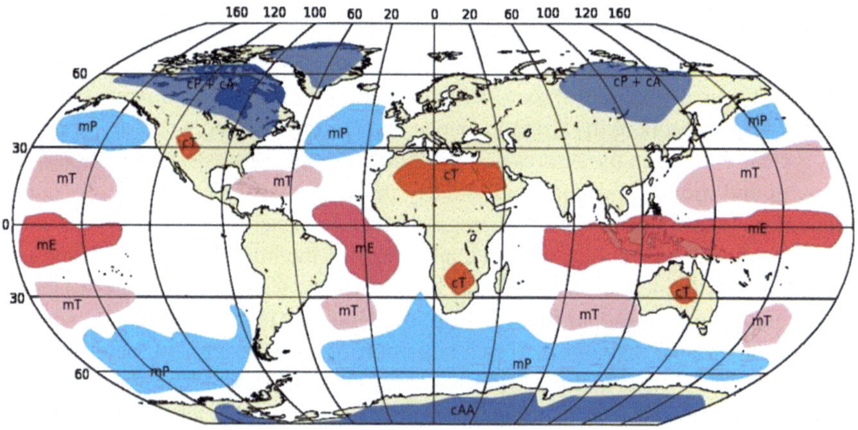

Fig. 10.1 Source regions of air masses (Bergeron classification)

The weather conditions they bring depend heavily on their temperature stability and moisture content. True to their classification, cP and cA air masses are cold and dry, often leading to clear, crisp weather. While cP air masses are generally not associated with heavy precipitation, they can produce significant snowfall when they pass over the Great Lakes in late fall and winter, a process known as lake-effect snow. This occurs when the cold, dry air absorbs moisture and heat from the relatively warmer lake water, leading to localized snow on the downwind shores.

10.3.2 Maritime Polar Air Masses (mP)

Maritime polar (mP) air masses form over the ocean at higher latitudes and are characterized by their moist, cool-to-cold conditions. These air masses originate in Northern Canada and Greenland, traveling to the British Isles on a northwesterly airstream. As they pass over the relatively warm waters of the North Atlantic, they become increasingly unstable, leading to significant weather effects.

In winter, mP air masses bring instability that results in frequent and often intense precipitation. Torrential rain over the sea and persistent rain over the exposed western and northern regions of the British Isles are common. Convection over the sea during these colder months can also lead to hail and thunderstorms, particularly in mountainous areas. While western and northern Britain bear the brunt of these effects, the eastern regions may experience less rainfall due to reduced surface heating.

In contrast, during the summer months, when land temperatures exceed those of the sea, the dynamics shift. The eastern part of the British Isles tends to receive the heaviest rainfall, as the warmer land promotes instability and convection within the mP air mass. Despite these seasonal variations, maritime polar air masses are consistently associated with frequent rainfall throughout the year, reflecting their inherently moist and unstable nature.

10.3.3 Maritime Tropical Air Masses (mT)

Maritime tropical (mT) air masses, which influence much of North America, form over the warm waters of the Atlantic Ocean, the Gulf of Mexico, and the Caribbean Sea, typically between the Azores and Bermuda. These air masses are characterized by warm and moist conditions, particularly in their lower layers. While mT air is inherently unstable above its source region, it tends to stabilize and becomes saturated as it moves over cooler waters. During winter, the influence of mT air masses is limited, as continental polar (cP) air dominates the Central and Eastern United States. However, in summer, mT air from the Gulf of Mexico, the Caribbean, and the adjacent Atlantic expands across a large portion of the continent, persisting for extended periods. This results in widespread moisture and precipitation, particularly in the eastern two-thirds of the United States. The specific characteristics of mT air masses vary, depending on factors such as temperature, moisture distribution, and air stability. These variations contribute significantly to the region's summer weather patterns, making maritime tropical air a key player in the climate of North America.

10.3.4 Equatorial Air Masses (E)

Equatorial air masses (E) are characterized by being very moist, warm, and unstable up to high altitudes. These air masses originate in the equatorial belt, where intense heating and the continuous upward movement of moist air led to the development of towering vertical clouds. This process frequently results in strong storms accompanied by heavy precipitation. Equatorial air masses rarely penetrate temperate latitudes, and when they do, it is usually at higher altitudes, meaning they have minimal impact on the weather in those regions.

10.3.5 Modification of Air Masses

As air masses move away from their source regions, they undergo modifications. In their source region, air masses acquire characteristics reflective of the underlying surface, such as temperature and moisture content, which can be either cold or warm and dry or moist. Modifications to air masses occur due to several factors:

(a) Temperature differences between the air mass and the surface: A cold air mass moving over a warmer surface is heated from below, increasing instability in its lowest layers. Conversely, a warm air mass moving over a colder Earth's surface is cooling from below, becoming more stable in its lower layers.
(b) Vertical movements caused by cyclones and anticyclones: These atmospheric systems can induce vertical air movements that modify the stability and moisture content of the air mass.

(c) Terrain topography: The physical features of the land can influence the characteristics of an air mass, particularly as it moves over mountains or across valleys.

An air mass traveling over the sea follows a maritime path, typically increasing its moisture content through evaporation from the ocean surface, especially in its lowest layers. In contrast, an air mass moving over land follows a continental track, generally remaining relatively dry. These modifications play a crucial role in determining the weather patterns experienced in various regions.

10.4 Atmospheric Fronts

Fronts are boundary surfaces that separate air masses of different densities, typically distinguishing between a warmer, more humid air mass and a colder, drier one. These boundaries are critical in meteorology as they often serve as zones of enhanced weather activity, including cloud formation, precipitation, and storms (Bluestein, 1993). The density differences across a front arise primarily from variations in temperature and moisture content, making fronts essential mechanisms for redistributing energy and moisture across the atmosphere (Holton & Hakim, 2012). The four basic types of fronts are:

Warm Front Occurs when a warm air mass advances and displaces a colder air mass. As the lighter warm air overrides the denser cold air, it rises gradually, leading to cloud formation and steady, prolonged precipitation (Barry & Chorley, 2010).

Cold Front Forms when a cold air mass pushes into a region occupied by warmer air. Cold fronts typically result in rapid uplift of warm air, producing cumulonimbus clouds, thunderstorms, and sometimes intense weather phenomena such as heavy rain or hail (Ahrens, 2013).

Stationary Front Develops when two air masses meet, but neither is strong enough to replace the other. This lack of movement often results in extended periods of cloudiness and precipitation, with the potential for flooding in areas where the front lingers (Bluestein, 1993).

Occluded Front Forms when a cold front overtakes a warm front, lifting the warm air off the surface entirely. Occlusions can lead to complex cloud structures and a mix of precipitation types, such as rain and snow, depending on the atmospheric temperature profile (Holton & Hakim, 2012).

Additionally, boundaries such as lines of instability can form between dry, dense air and moist, less dense air, often resulting in strong convective activity. In regions like the Central United States, dry lines are common and can trigger severe thunderstorms, including tornadoes, particularly during spring and summer (Markowski & Richardson, 2010). Cyclonic and anticyclonic systems, as well as terrain features like mountains, also influence frontal development. For instance, orographic lifting can intensify precipitation along fronts, while cyclones enhance frontal activity due to wind convergence (Ahrens, 2013). A detailed description of each type of front, including their formation processes, associated weather patterns, and key characteristics, follows in the next subsections.

10.4.1 Warm Front

A warm front is a transition zone that forms when a warm air mass advances and replaces a colder air mass (see Fig. 10.2). It is one of the three main types of fronts in meteorology, alongside cold fronts and occluded fronts. As a warm front moves through an area, it typically brings changes in weather conditions. The passage of a warm front is often characterized by gradual increases in temperature and humidity, along with shifts in wind direction and precipitation patterns.

Warm fronts develop when a warm air mass advances into an area occupied by a colder air mass. Because warm air is less dense than cold air, it tends to rise over the cooler, denser air. As the warm air ascends, it cools and condenses, resulting in the formation of clouds and precipitation. Unlike the intense, convective storms typically associated with cold fronts, precipitation from warm fronts is generally more gradual and steadier. Common cloud types associated with warm fronts include stratus, nimbostratus, and altostratus, which often bring overcast skies and extended periods of rain or snow. The passage of a warm front typically leads to a gradual increase in temperature and humidity. As the warm air replaces the cooler air, conditions become milder. Before the arrival of the warm front, winds usually blow from the east or southeast. After the front passes, the winds shift to the south or southwest, reflecting the change in air masses.

On weather maps, warm fronts are depicted as a line with semicircles pointing in the direction of movement, and the symbol is commonly drawn in red. Warm fronts generally move more slowly than cold fronts, with their movement influenced by larger atmospheric patterns and regional topography. Weather associated with a warm front can persist for an extended period, with a gradual onset of precipitation and a more gradual transition from colder to warmer conditions compared to the abrupt changes associated with cold fronts.

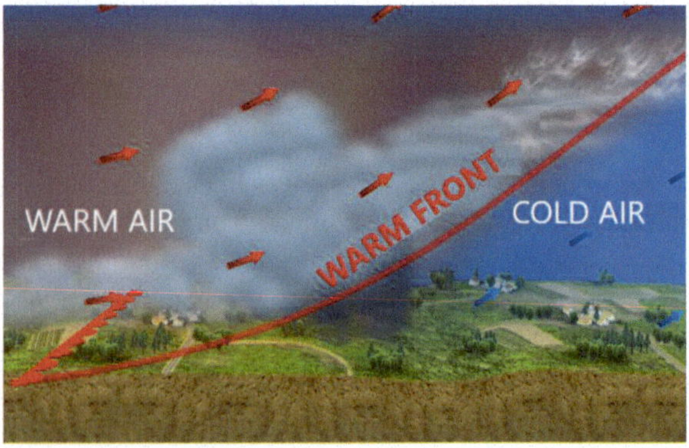

Fig. 10.2 A warm front showing the transition between cold air ahead and warm air rising at the frontal boundary

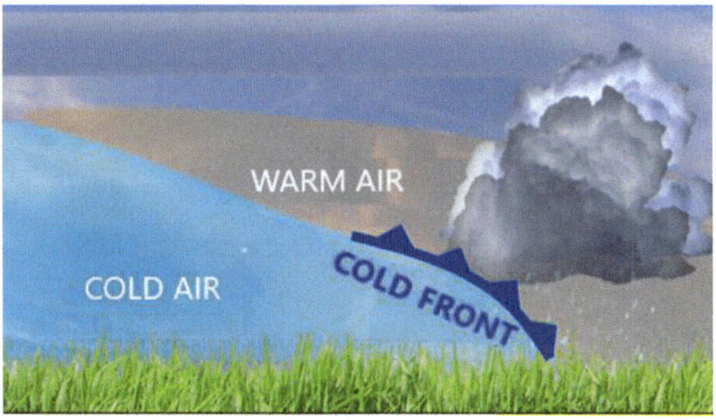

Fig. 10.3 A cold front, depicted as a wedge of cold air, advances into the warm sector, forcing the warm air to rise along the frontal boundary

10.4.2 Cold Front

A cold front is an essential concept in meteorology, representing the boundary between a colder air mass and a warmer air mass. It forms when a cold air mass advances and replaces a warmer air mass (see Fig. 10.3). Cold fronts often lead to rapid changes in weather, characterized by the swift ascent of warm, moist air and the development of intense, convective storms.

Because cold air is denser than warm air, it slides underneath the warmer air, forcing it to rise. This upward movement of warm air can result in the rapid formation of clouds and precipitation. The precipitation associated with cold fronts can vary, including thunderstorms, heavy rain, snow, or a mix of these, depending on the temperature contrast and the season. Cold fronts are typically linked to more intense and sometimes severe weather compared to warm fronts. As warm air is forced upward, it can generate towering cumulonimbus clouds and lead to thunderstorms, lightning, hail, and strong winds. When a cold front passes through an area, there is often a noticeable and abrupt drop in temperature, marking a significant shift from warm to cold conditions.

Before a cold front arrives, winds generally blow from the south or southwest. After the front passes, the winds shift to the northwest or north. On weather maps, cold fronts are depicted as a line with triangles pointing in the direction of movement, and the symbol is usually drawn in blue. Cold fronts typically move faster than warm fronts, resulting in quicker weather changes. The intensity of the weather associated with a cold front depends on various factors, including the temperature difference between the air masses and the level of atmospheric instability.

10.4.3 Stationary Front

A stationary front is a boundary between two air masses with different temperatures that remains largely stationary or moves very slowly (see Fig. 10.4). It is one of the three main types of fronts in meteorology, alongside warm fronts and cold fronts. A stationary front forms when the forces pushing warm and cold air are roughly equal, preventing significant movement of the front. This occurs when a warm air mass and

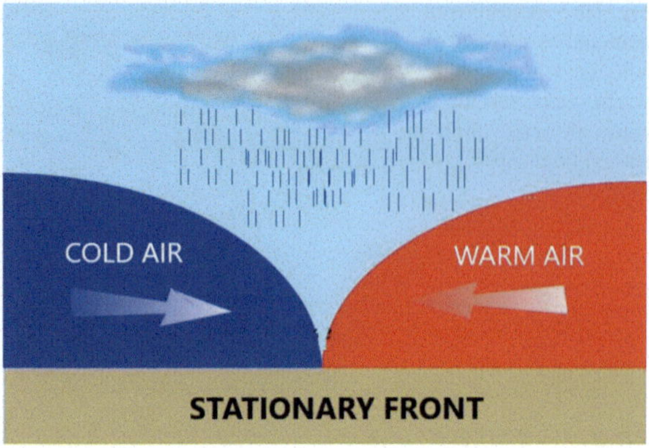

Fig. 10.4 A stationary front forms when two slow-moving air masses of different temperatures meet, with neither air mass strong enough to displace the other

a cold air mass converge, but neither is strong enough to displace the other, resulting in a front that stays nearly in place. The weather along a stationary front can be varied but is typically characterized by extended periods of cloudy and wet conditions. Precipitation associated with stationary fronts tends to be more continuous and steady compared to the intense, sporadic precipitation often linked with moving fronts. Temperature changes on either side of a stationary front are usually small and gradual because the air masses are not actively moving relative to each other. Nevertheless, noticeable differences in temperature and humidity can still occur.

Winds along a stationary front can vary in direction and intensity, with no clear or consistent shift as seen with warm and cold fronts. On weather maps, a stationary front is depicted by alternating red semicircles and blue triangles on opposite sides of the front line. The direction of these symbols indicates the movement of air on each side of the front. Although a stationary front generally remains in place or moves slowly, it may meander or oscillate over time. It can eventually transition into a warm or cold front depending on changes in atmospheric conditions. While the weather associated with a stationary front is often less dramatic than that of moving fronts, it can persist for longer periods, leading to extended cloudy skies and precipitation.

10.4.4 Occluded Front

An occluded front forms when a faster-moving cold front overtakes a slower-moving warm front, lifting the warm air mass off the ground and creating a complex weather system (see Fig. 10.5). Occluded fronts are among the more dynamic and intricate types of weather fronts and typically occur in the mature stages of extratropical cyclones (areas of low pressure). There are two main types of occluded fronts:

1. Cold Occlusion: This occurs when the cold air behind the cold front is colder than both the cold air ahead of the warm front and the warm air ahead of the warm front.

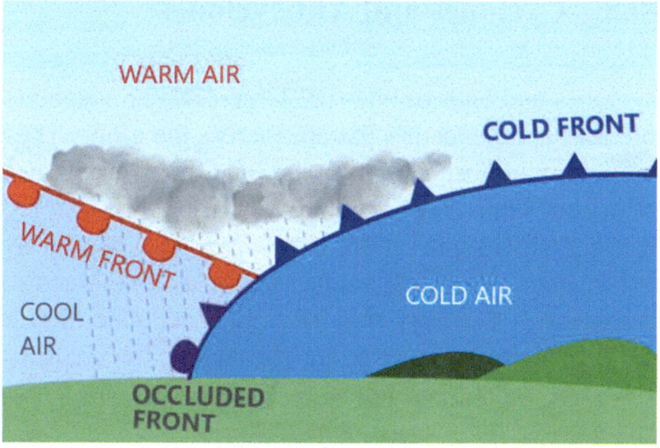

Fig. 10.5 Schematic representation of an occluded front, where a fast-moving cold front overtakes a slower-moving warm front. This collision forces the warm air to rise, resulting in a complex weather system with varied clouds and precipitation

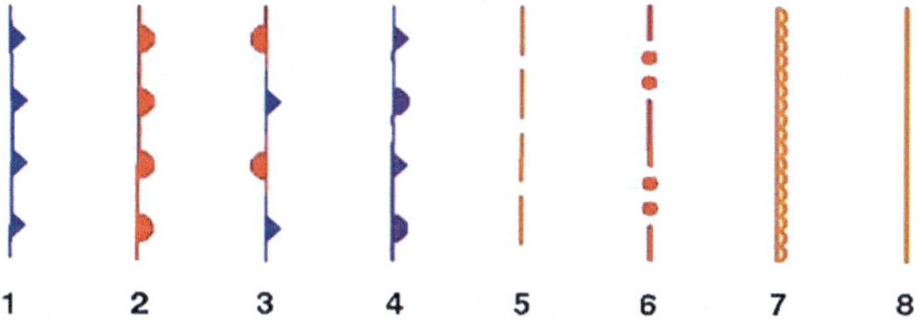

Fig. 10.6 Symbols and characteristics of various surface fronts and meteorological features: (1) cold front; (2) warm front; (3) stationary front; (4) occluded front; (5) valley; (6) line of instability; (7) dry line; (8) tropical wave

2. Warm Occlusion: This happens when the cold air behind the cold front is not as cold as the cold air ahead of the warm front.

Occluded fronts bring complex and varied weather conditions. The rising warm air can lead to the development of different types of clouds, such as nimbostratus and stratocumulus, which produce precipitation.

The temperature changes associated with occluded fronts depend on the type of occlusion and specific conditions. In cold occlusions, temperatures may decrease, while in warm occlusions, temperature changes might be less pronounced. The wind direction around occluded fronts is variable and depends on the type of occlusion. Generally, winds associated with occluded fronts can be strong and gusty. On weather maps, occluded fronts are represented by a combination of symbols used for cold and warm fronts: a line segment with triangles (pointing in the direction of movement) and semicircles (on the side of the warmer air mass). Occluded fronts are commonly found in the later stages of extratropical cyclones and are linked with low-pressure systems. They can bring a mix of weather conditions, including precipitation and temperature variations, often resulting in more persistent weather patterns compared to warm or cold fronts alone. A symbolic representation of the main characteristics of surface fronts and their corresponding features is shown in Fig. 10.6.

10.5 Cyclones and Anticyclones

Cyclones and anticyclones are large-scale atmospheric systems that play a crucial role in shaping weather patterns across the globe. These systems are driven by differences in atmospheric pressure, and their formation and evolution are influenced by factors such as temperature contrasts, the Coriolis effect, and surface features like landmasses and oceans. Cyclones are typically associated with stormy, unsettled weather, while anticyclones are linked to fair and stable conditions. The interaction between these systems forms the backbone of synoptic-scale meteorology and is essential for understanding global weather dynamics (Barry & Chorley, 2010; Bluestein, 1993). Cyclones, or low-pressure systems, often develop along fronts where contrasting air masses meet.

They are characterized by inward-spiraling winds, rising air, and the potential for cloud formation and precipitation. Anticyclones, or high-pressure systems, in contrast, feature outward-moving winds and descending air, which inhibit cloud formation and are typically associated with clear skies (Ahrens, 2013; Holton & Hakim, 2012).

10.5.1 General Characteristics

Cyclones and anticyclones are atmospheric systems characterized by organized circulation around centers of low and high pressure, respectively. In the Northern Hemisphere, cyclones exhibit counterclockwise wind circulation around their centers due to the Coriolis effect, while anticyclones show clockwise circulation. In the Southern Hemisphere, this pattern reverses, with cyclones having clockwise circulation and anticyclones featuring counterclockwise rotation (Bluestein, 1993; Holton & Hakim, 2012). The general structure and circulation patterns of cyclones and anticyclones are illustrated in Fig. 10.7.

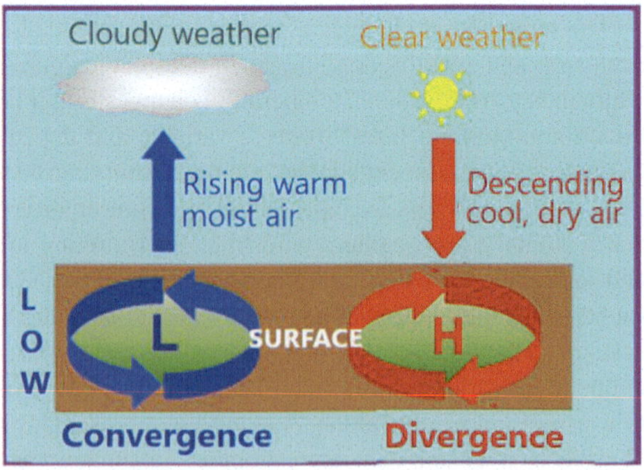

Fig. 10.7 Diagram of low-pressure (cyclone) and high-pressure (anticyclone) systems, showing their characteristic circulation patterns and pressure centers

In a cyclone, the pressure is lowest at the center, known as depression. The pressure gradient force directs air toward the center, where the air rises, leading to cloud formation and, often, precipitation. Cyclones are generally associated with stormy weather, including rain and sometimes severe weather conditions such as thunderstorms or even tornadoes (Ahrens, 2013).

Conversely, an anticyclone features the highest pressure at its center. Here, the pressure gradient force drives air outward from the center, leading to descending air. This descending motion suppresses cloud formation, resulting in clear, dry, and calm weather conditions. Anticyclones are often associated with fair weather but can also contribute to heat waves or prolonged dry spells (Barry & Chorley, 2010).

When a cyclone or anticyclone interacts with a frontal boundary, it is referred to as a wave cyclone or mid-latitude frontal storm. These systems are more complex and can significantly influence local and regional weather, producing varied conditions such as snow, rain, or strong winds depending on the characteristics of the fronts involved (Bluestein, 1993). Cyclones are associated with unstable atmospheric conditions and vertical air movements. In a cyclone, air near the Earth's surface converges inward toward the low-pressure center and then rises. As it ascends, the air spreads outward and away from the center. One key factor influencing the formation of cyclones and anticyclones is the development of irregularities in the jet stream. When the jet stream—a band of high-altitude winds—oscillates along an east–west axis, it can impact cyclonic or anticyclonic systems in the lower troposphere. This interaction can lead to the development of systems that move across continents in an easterly or northeasterly direction. In rare cases, these systems can intensify significantly, resulting in severe weather conditions such as storms, tornadoes, and hurricanes, which can cause extensive damage to life and infrastructure.

10.5.2 Low-Pressure Atmospheric Systems: Cyclones

Cyclones are characterized by a low-pressure system where air converges toward the center, creating counterclockwise rotation in the Northern Hemisphere and clockwise rotation in the Southern Hemisphere. These low-pressure systems, also known as depressions, form under unstable conditions that cause air to rise, creating a surface low-pressure area.

The rising air cools and condenses, leading to cloud formation and often resulting in cloudy and humid weather. In the Northern Hemisphere, winds circulate counterclockwise around the low-pressure center, with isobars—lines of equal atmospheric pressure—being densely packed around the center, which leads to strong winds. Depressions can be identified on weather maps as areas with densely spaced isobars, typically appearing roughly circular. Within these depressions, isobars around the center of the lowest pressure form troughs.

Cyclones are often associated with fronts. The pressure gradient force in a cyclone acts from the periphery toward the center, following the direction of

Fig. 10.8 A satellite image of a cyclone in the Northern Hemisphere, overlaid with a schematic illustration depicting the circulation patterns and forces acting within the low-pressure system, including the wind flow (V), Coriolis effect (F_d), pressure gradient force (**PGF**), and centrifugal force (F_c)

decreasing pressure. As air flows into the low-pressure area from all directions, the Coriolis force deflects the cyclonic flow to the right (Fig. 10.8). This deflection creates a counterclockwise rotation around the low-pressure center in the Northern Hemisphere, resulting in convergence near the center of the system.

Convergence occurs at the surface as air from different directions meets and is forced to rise, while divergence happens at higher altitudes as the air moves away from the center. Near the Earth's surface, friction also affects airflow, causing it to veer and creating an angular movement that results in a counterclockwise wind pattern around the cyclone. The convergence at the surface leads to rising moist air, which, in turn, drives divergence aloft. Consequently, surface winds around the Northern Hemisphere cyclone blow counterclockwise and converge toward the center, promoting rising air and continued cyclonic activity.

10.5.3 High-Pressure Anticyclone System

A high-pressure system, commonly referred to as an anticyclone, is characterized by an area of elevated atmospheric pressure on the Earth's surface. Anticyclones are distinguished by high surface pressure at the center due to the descending motion of air within the system. In the Northern Hemisphere, the wind around an anticyclone circulates clockwise, whereas, in the Southern Hemisphere, it circulates counterclockwise. The pressure gradient force in an anticyclone acts from the center toward the periphery, with the Coriolis force and centrifugal force causing air parcels to be deflected outward (Fig. 10.9).

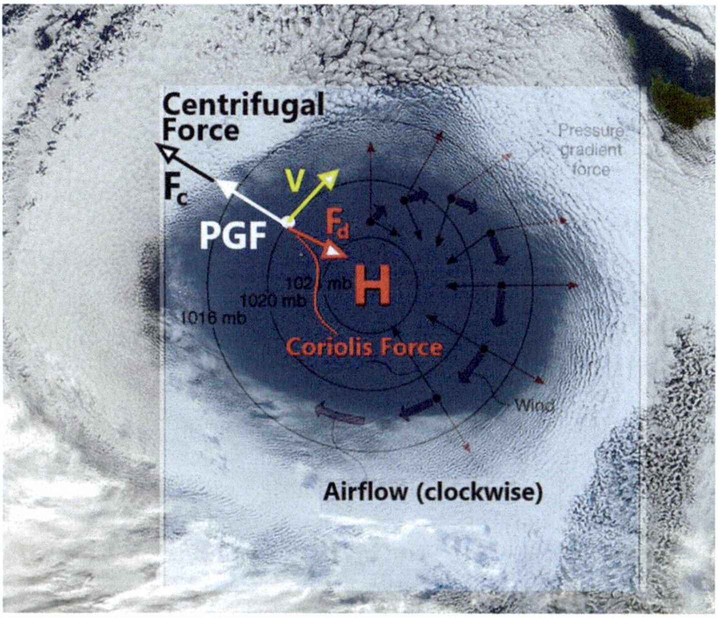

Fig. 10.9 A satellite image of an anticyclone in the Northern Hemisphere, overlaid with a schematic illustration depicting the circulation patterns and forces acting within the high-pressure system, including the wind flow (V), Coriolis effect (F_d), pressure gradient force (**PGF**), and centrifugal force (F_c)

Anticyclones are marked by descending air at their centers, which warms as it descends. This warming suppresses cloud formation and promotes stable atmospheric conditions. Consequently, anticyclones are often associated with clear skies and dry weather. They contribute to prolonged periods of favorable weather, which can include temperature inversions where a layer of warmer air traps cooler air near the surface. This phenomenon can lead to extreme temperature variations, such as cold nights and hot days in winter and hot, dry conditions in summer. During winter, anticyclones can result in cold, clear weather that fosters ice and fog formation, while in summer, they can bring hot, dry conditions and occasional heat waves. At the surface, anticyclones exhibit divergent airflow, with air moving outward from the center, whereas convergence occurs aloft. This divergence prevents the system from filling with air, thereby maintaining its high-pressure characteristics and stability. Figure 10.9 illustrates an anticyclone with high surface pressure (1031 hPa) and clear skies visible from satellite imagery. Unlike cyclones, which generally move northeast, anticyclones typically advance southwest. Their gradual pressure increase toward the center leads to stable weather conditions. Large, persistent anticyclones, known as blocking anticyclones, are often found at high latitudes during winter. They generally drift in a zonal flow from west to east and move poleward. These stagnant systems can cause one region to experience prolonged dryness, while another remains under cyclonic influence. Additionally, the subsidence associated with large anticyclones can lead to temperature inversions and increased air pollution levels. In winter, the clear, calm conditions often result in frost and fog due to radiational cooling and light winds. In summer, anticyclones bring long, sunny days and warm temperatures, though very high temperatures can occasionally trigger thunderstorms.

10.5.4 Formation of Cyclones

Cyclones begin their formation in mid-latitudes where a prominent temperature gradient and forced upwelling, often associated with a stationary front like the polar front, are present. In the initial phase, air converges and rises from the Earth's surface within the cyclone. This convergence, coupled with the temperature gradient, leads to the development of fronts as warm air moves northeastward and cold air moves southwestward (Bluestein, 1993). The warm front advances in a northeasterly direction, while the cold front moves southeastward, creating a warm sector between them. Concurrently, the low-pressure system intensifies, resulting in widespread precipitation ahead of the warm front and a narrow band of snow along the cold front (Ahrens, 2013). This intensification of weather activity is accompanied by an increase in wind speed and typically occurs over a 12–24-h period.

The **Norwegian cyclone model**, also known as the polar front theory or the Bergen school model, provides a framework for understanding the formation and life cycle of extratropical cyclones. Developed by Norwegian meteorologists, including Vilhelm Bjerknes, in the early twentieth century, this model elucidates the processes involved in cyclone development, intensification, and dissipation (Barry & Chorley, 2010). The stages of cyclone development according to this model are illustrated in Fig. 10.10:

1. Polar Front (a, b): The model begins with the presence of a temperature gradient along the polar front, where cold polar air meets warm tropical air. A stationary front separates these contrasting air masses.
2. Wave Development (c): Small disturbances, or waves, along the polar front lead to the development of cyclones. A low-pressure trough at high altitudes amplifies these disturbances (Bluestein, 1993).
3. Cyclogenesis (d): These disturbances evolve into frontal waves along the polar front. As the waves develop, a low-pressure system forms at the surface.
4. Cyclonic Development (e): The system continues to develop cyclonically, with warm air rising over cold air along the front. Eventually, the cyclone undergoes occlusion as the cold front catches up with the warm front (Ahrens, 2013).
5. Mature Cyclone (f): A mature cyclone exhibits a characteristic triangular structure with a warm front, a cold front, and an occluded front. Intensification occurs due to the release of latent heat and other dynamic processes (Holton & Hakim, 2012).
6. Dissipation (g, h): The cyclone dissipates as it loses its energy source, which is the temperature contrast along the polar front. The system moves away from the polar front and gradually disintegrates (Barry & Chorley, 2010).

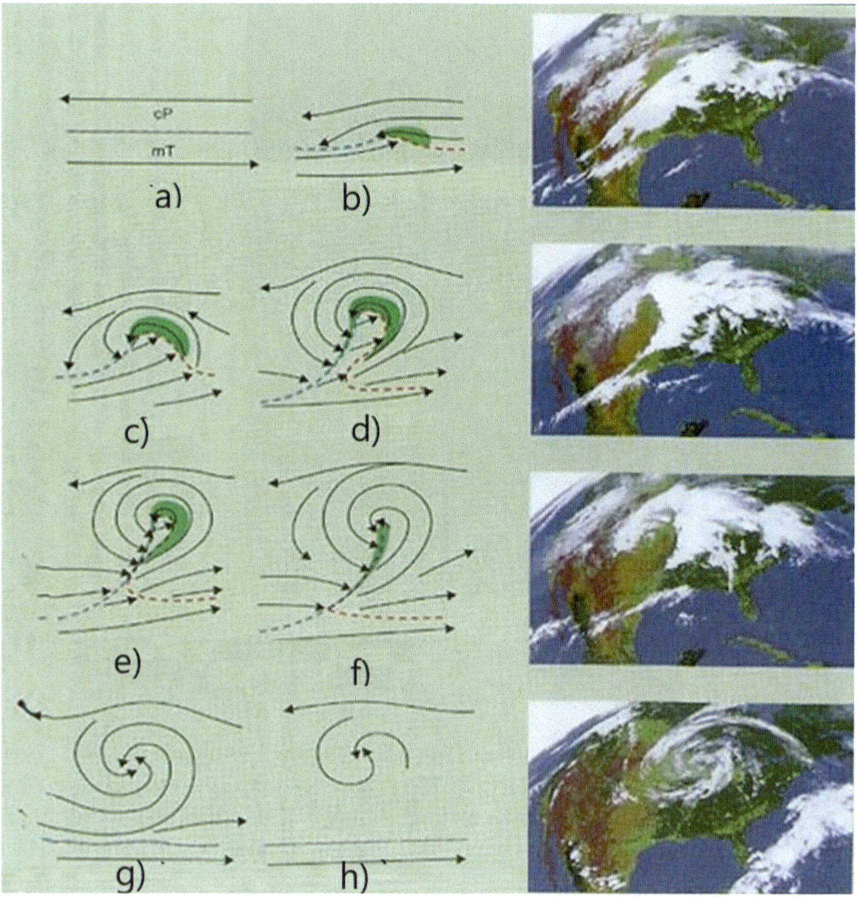

Fig. 10.10 Life cycle of an extratropical cyclone (Norwegian model)

10.5.5 Weather Conditions Associated with Cyclones and Anticyclones

The weather associated with cyclones and anticyclones is closely tied to their dynamic structure and thermodynamic properties. Cyclones, characterized by low-pressure systems, are synonymous with unstable and variable weather conditions. These systems typically bring overcast skies, enhanced cloud cover, and frequent precipitation, which can range from light rain to intense thunderstorms and even severe weather events such as hail or tornadoes. The convergence of moist air and the lifting mechanism associated with cyclones foster the development of extensive cloud cover and prolonged periods of rainfall. Cyclones also generate strong winds due to the pressure gradient force, which contributes to turbulent and gusty conditions. The weather in cyclonic systems is marked by its variability and can change rapidly, often resulting in dramatic shifts in temperature and humidity. In contrast,

Fig. 10.11 Weather conditions associated with cyclones and anticyclones

anticyclones, or high-pressure systems, are associated with stable, calm, and often favorable weather conditions (Fig. 10.11). These systems are characterized by descending air that inhibits cloud formation, leading to clear skies and dry weather. The subsiding air in anticyclones suppresses vertical motion, which contributes to stable atmospheric conditions. During the day, anticyclones can bring warm temperatures and sunshine, while at night, clear skies can lead to significant cooling and potential temperature inversions. This can result in frost formation and fog, particularly in winter. The high-pressure center of an anticyclone also causes gentle winds that can enhance the stability of the weather, although these winds are generally less intense compared to those associated with cyclones. Additionally, anticyclones can lead to prolonged periods of dry weather and may contribute to heat waves in summer or cold spells in winter. During winter, anticyclones can create clear and frosty conditions, often leading to the development of fog and temperature inversions. In summer, they can produce extended spells of hot and dry weather, which can sometimes lead to drought conditions. Overall, the contrasting weather patterns associated with cyclones and anticyclones highlight the influence of atmospheric pressure systems on local and regional weather, with cyclones bringing dynamic and unstable conditions and anticyclones fostering calm and stable weather.

10.6 Tropical Cyclones

Tropical cyclones are among the most intense and destructive weather systems on the planet, known for their ability to produce severe wind, rain, and storm surges that can devastate coastal regions. These storm systems are driven by heat from warm ocean waters and are primarily located between the Tropics of Cancer and Capricorn. Tropical cyclones have significant implications for climate and weather patterns, influencing precipitation and temperature distributions across vast areas. Understanding their formation, structure, and behavior is crucial for forecasting and mitigating their impacts on human societies and natural environments (Kossin et al., 2018).

In recent years, research has focused on how climate change may alter the frequency, intensity, and distribution of tropical cyclones. Some studies suggest that warmer ocean temperatures could lead to more intense storms, while others indicate that increased atmospheric moisture may enhance rainfall associated with these cyclones (Emanuel, 2005; Knutson et al., 2010). The dynamics of tropical cyclones are complex and influenced by various atmospheric conditions, making them a significant area of study within meteorology.

10.6.1 General Overview

Tropical cyclones are powerful storm systems characterized by low-pressure centers, intense winds, and organized thunderstorms that form over warm ocean waters (Fig. 10.12). They are among the most destructive natural phenomena, capable of

Fig. 10.12 Satellite view of a tropical storm, showing the storm's organized structure, including the central low-pressure area, spiral rainbands, and the eyewall

causing extensive damage to coastal areas and loss of life. The term “tropical cyclone” encompasses several categories of storms, including tropical depressions, tropical storms, hurricanes, and typhoons. In their mature phase, tropical cyclones are among the most intense tropical storms, with wind speeds exceeding 90 m/s and torrential rainfall. Their diameters typically range from 100 km in the developing stage to over 2000 km in the fully mature stage. A tropical cyclone features a warm core at its center, very steep pressure gradients, and strong spiral cyclonic winds near the Earth’s surface. These winds become nearly circular near the cyclone’s center. The pressure gradients and wind speeds in tropical cyclones are almost always significantly stronger than those found in extratropical storms. The cloud and precipitation patterns vary; however, there are typically spiral bands in the outer vortex, with the most intense precipitation and winds occurring in the cyclone’s eyewall (Rappaport et al., 2009). Tropical cyclones are classified based on their wind speeds:

- Tropical Depressions: Maximum wind speeds less than 60 km/h
- Tropical Storms: Maximum wind speeds between 60 km/h and 110 km/h
- Hurricanes or Typhoons: Maximum wind speeds exceeding 110 km/h

The nomenclature of these storms varies by region. In the Atlantic Ocean and the Northeastern Pacific Ocean, they are referred to as “hurricanes” (Fig. 10.13). In the Northwestern Pacific Ocean, they are called “typhoons,” while in the Indian Ocean, they are known as severe cyclonic storms. For example, Fig. 10.14 shows a satellite image of Typhoon Soulik near Japan on August 20, 2018, captured by the Moderate Resolution Imaging Spectroradiometer (MODIS) aboard NASA’s Terra satellite (Wang and Lee 2019; Knutson et al., 2010).

Fig. 10.13 (**a**) Hurricane Katrina (August 2005), a large and extremely powerful storm that caused extensive destruction and significant loss of life. (**b**) Satellite view of Typhoon Soulik near Japan on August 20, 2018, showcasing its well-defined structure and intensity

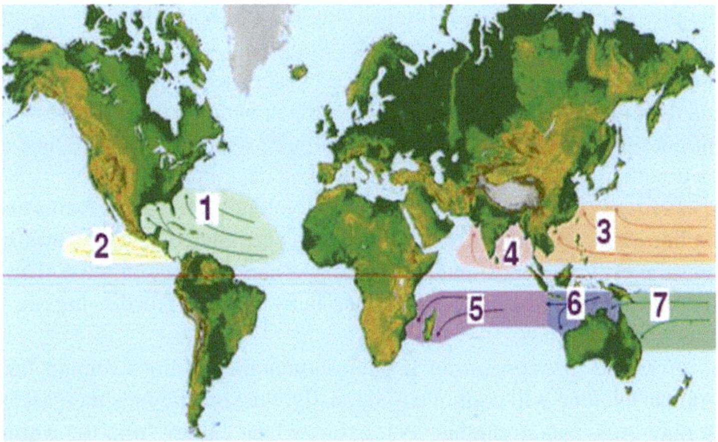

Fig. 10.14 Source regions of tropical cyclone formation: This map illustrates the seven primary basins where tropical cyclones commonly develop. The areas are marked as follows: (1) The Atlantic Basin, (2) the Northeast Pacific Basin, (3) the Northwest Pacific Basin, (4) the North Indian Basin, (5) the Southwest Indian Basin, (6) Southeast Indian/Australian Basin, and (7) Australian/Southwest Pacific Basin. Each region is characterized by specific climatic conditions and sea surface temperatures that contribute to tropical cyclone genesis. The figure highlights the spatial distribution and geographical extent of tropical cyclone activity across these basins

10.6.2 Formation of Tropical Cyclones

The formation of tropical cyclones necessitates several key factors and processes:

1. *Warm Ocean Waters*: Tropical cyclones require ocean water temperatures of at least 26.5 °C to a significant depth, typically 50 m or more. This warmth provides the necessary heat and moisture to fuel the cyclone. The heat from the ocean surface is crucial for the cyclone's energy, enabling the evaporation of water that later condenses into clouds, releasing latent heat that further drives the storm's development.
2. *Atmospheric Instability*: A pre-existing weather disturbance, such as a cluster of thunderstorms, is essential for cyclone formation. The atmosphere must be unstable, allowing warm, moist air near the surface to rise. As this air ascends, it cools and condenses, creating a low-pressure area. This instability promotes continued vertical motion and helps sustain the cyclone's development.
3. *Coriolis Effect*: The Coriolis effect, resulting from the Earth's rotation, is crucial for cyclonic rotation. It causes moving air to be deflected to the right in the Northern Hemisphere and the left in the Southern Hemisphere. This effect is necessary for the formation of cyclonic circulation. Tropical cyclones generally do not form within about 5° latitudes of the equator because the Coriolis effect is too weak in this region to initiate rotation.
4. *Low-Pressure System*: As warm, moist air rises from the ocean surface, it creates a low-pressure area. The release of latent heat during condensation further low-

ers the pressure at the surface, intensifying the cyclone. This low-pressure system acts as a focal point for cyclonic development.

5. *Cyclonic Circulation*: The Coriolis effect causes the rising air to begin rotating around the low-pressure center. As more air is drawn into this low-pressure zone, it continues to rise and spiral, creating a circular wind pattern. Cyclonic circulation is essential for the cyclone's structure and development.
6. *Eye Formation*: In the center of a mature cyclone, a relatively calm and clear region known as the "eye" forms. The eye is surrounded by the eyewall, which features the cyclone's most intense winds and heaviest precipitation. The formation of the eye is a result of the balance between outward-flowing air and the inward-spiraling winds.
7. *Maturation and Intensification*: If environmental conditions remain favorable, the tropical cyclone will continue to intensify. Sustained winds increase, and the storm may grow and strengthen as it extracts heat energy from the warm ocean waters. The cyclone's structure becomes more organized, with a well-defined eye and eyewall.
8. *Flooding or Dissipation*: Upon making landfall, a tropical cyclone can cause widespread damage due to strong winds, heavy rainfall, and storm surges, leading to significant flooding. Alternatively, if the storm moves over cooler waters or encounters strong vertical wind shear, it may weaken and eventually dissipate. The storm's intensity decreases as it loses its primary energy source from the warm ocean.

The detailed sequence of these processes illustrates the complex interplay of oceanic and atmospheric conditions required for the development, intensification, and eventual dissipation of tropical cyclones.

10.6.3 Basic Factors

The formation of tropical cyclones is influenced by several key atmospheric factors and disturbances. These include:

- Easterly Waves: Also known as tropical waves, represent large undulations or troughs in the tropical easterly wind belt. They move westward across the tropics and are characterized by low-pressure valleys. Easterly waves are often precursors to tropical cyclones, as they create areas of low pressure where convection can intensify, leading to the formation of cyclones. These waves are a primary mechanism for cyclogenesis in tropical regions.
- Western Pacific Line Disturbances: These disturbances manifest as lines of convection, like lines of instability, and typically originate over the West African continent. They then move westward into the Atlantic Ocean, often traveling at a faster pace than tropical waves. Their presence can enhance convection and trigger the development of tropical cyclones.
- Upper Trough with Tropical Upper Tropospheric Trough: This refers to a trough or cold core low-pressure system in the upper atmosphere that contributes to the

development of tropical cyclones. The upper trough creates favorable conditions for convection by allowing warm, moist air from the surface to rise more readily. This process facilitates the formation of tropical storms and cyclones as it enhances vertical wind shear and instability.

- Old Frontal Boundaries: Remnants of old polar fronts or frontal boundaries can serve as a starting point for tropical cyclone formation. These boundaries, which may still have residual temperature gradients and convection lines, can trigger the development of tropical storms when they interact with warm, moist air in tropical regions. Such disturbances are particularly noted in the Atlantic Ocean, where they can lead to early or late-season hurricanes, especially in the Gulf of Mexico or the Caribbean Sea.

These basic factors interact in various ways to create the conditions necessary for tropical cyclone formation. Each contributes to the development of cyclones by inducing instability, enhancing convection, or modifying atmospheric conditions to favor cyclogenesis. Understanding these factors is crucial for predicting and analyzing tropical cyclone behavior and impact.

10.6.4 Initiation and Formation of Tropical Cyclones

The formation of tropical cyclones, encompassing hurricanes, typhoons, tropical depressions, and tropical storms, depends on several critical factors. A key element is sea surface temperature (SST), as warm ocean waters provide the heat and moisture necessary for cyclogenesis. SSTs typically need to exceed 26 °C for cyclone development, with higher temperatures increasing evaporation rates and fueling the storm through latent heat release. Seasonal changes and climatic patterns like El Niño and La Niña also influence SST variations, affecting the frequency and intensity of tropical cyclones.

Geographical location plays a significant role, as tropical cyclones primarily form over tropical and subtropical oceans, between 5° and 20° latitude. They rarely develop near the equator due to the weak Coriolis effect. The distribution of SST and atmospheric conditions determines preferred formation zones. Additionally, the presence of pre-existing atmospheric disturbances, such as clusters of thunderstorms or low-pressure systems, provides the initial perturbations for cyclone development. These disturbances may originate from mesoscale instabilities or interactions with broader atmospheric processes.

Wind conditions are another essential factor. Low vertical wind shear, defined as minimal differences in wind speed and direction between the lower and upper atmosphere, is crucial for cyclone formation. High wind shear can disrupt the storm's structure, while favorable wind patterns facilitate the organized rotation needed for cyclone growth. Thermodynamic conditions also contribute significantly, as the atmosphere must be unstable enough to allow warm, moist air to rise and cool, forming a low-pressure system. High mid-level moisture content supports cloud formation and intensifies the storm.

Lastly, the vertical stratification of the atmosphere is vital. A well-defined structure with a warm core at the center and cooler air aloft is characteristic of tropical cyclones, helping to maintain their strength and organization throughout their development.

10.6.5 *Areas of Formation*

Tropical cyclones typically form in regions where sea surface temperatures meet the necessary threshold of approximately 27 °C. However, their formation is rare within 5 degrees latitude of the equator due to the insufficient Coriolis effect. The primary areas of tropical cyclone formation are shown in Table 10.1.

Table 10.1 Basic characteristics of tropical cyclones

The Atlantic Basin, which includes the North Atlantic Ocean, the Gulf of Mexico, and the Caribbean Sea	In the North Atlantic Basin, the hurricane season officially begins on June 1 and runs through November 30. Activity is generally low from mid-September. A broad peak in activity starts in late May or early June and continues until late October or early November, with the maximum storm occurring in late August
Northeast Pacific Basin	The Northeast Pacific Basin experiences regular cyclone activity throughout the year, with minimal intensity in February and the first half of March. The main season extends from July to November, with peak intensity in late August
Northwest Pacific Basin, including the South China Sea	In the Northwest Pacific Basin, cyclone activity shows a double peak: one from May to November and another from April to December. Strong storm surge cyclones, with winds exceeding 119 km/h, typically occur from April to June and again from late September to early December
North Indian Basin (including the Bay of Bengal and the Arabian Sea)	The North Indian Basin sees activity beginning in late October or early November, with a double peak in cyclone activity occurring in mid-January and mid-February. This period of heightened activity continues until early march and then ends in May
Southwest Indian Basin (from Africa to about 100 E)	For the Southwest Indian Basin, activity also starts in late October or early November, with a double peak like the North Indian Basin, in mid-January and mid-February, continuing until early March and ending in May
Southeast Indian Australian Basin (100°E to 142°E)	In the Southeast Indian Australian Basin, the delay in peak activity in February is more pronounced compared to the southwest Indian Basin. Cyclone activity begins in late October or early November, with peaks in mid-January and mid-February, lasting until early March and ending in May
The Atlantic Basin, which includes the North Atlantic Ocean, the Gulf of Mexico, and the Caribbean Sea	In the North Atlantic Basin, the hurricane season officially begins on June 1 and runs through November 30. Activity is generally low from mid-September. A broad peak in activity starts in late May or early June and continues until late October or early November, with the maximum storm occurring in late August

The seven standard basins where tropical cyclones typically form are illustrated, highlighting the distribution and geographical extent of cyclone activity. Each region has distinct characteristics and seasonal patterns of cyclone formation, influenced by local and regional climatic factors. Table 10.1 provides a summary of the areas of tropical cyclone formation, including their main characteristics and periods of activity.

10.6.6 Classification of Tropical Cyclones

Tropical cyclones are classified based on their organized low-pressure systems, associated thunderstorms, and wind speeds. A tropical cyclone with a well-defined circulation and sustained winds of about 60 km/h or less is termed a "tropical depression." When the cyclonic circulation becomes more organized and sustained winds range from 60 to 120 km/h, it is classified as a "tropical storm."

The terminology for tropical cyclones varies by region:

1. Hurricanes occur in the North Atlantic Ocean, Northeast Pacific Ocean, and the South Pacific Ocean east of 160° E. The term "hurricane" derives from "Huracan," a term used by Caribbean Indians to describe a storm. This term was adapted from the ancient Taino language, where Hurakán was their god of evil.
2. Typhoons are found in the Northwest Pacific. A "super typhoon" is characterized by maximum sustained winds of at least 241 km/h.
3. Tropical storms are present in the Southwest Pacific, west of 160°E, and the Southeast Indian Ocean, east of 90°E.
4. Cyclones refer to storms in the northern part of the Indian Ocean.
5. Tropical storms in the southwestern part of the Indian Ocean are also classified as such.

10.6.7 Tropical Storm Structure

The structure of a tropical storm is characterized by key features such as the eye and the eyewall. As depicted in Fig. 10.15, the air rotates toward the cyclone's center in a counterclockwise direction in the Northern Hemisphere (and clockwise in the Southern Hemisphere), then rises and flows outward at higher altitudes. In the center, a calm, clear area known as the "eye" forms, typically devoid of clouds. The strongest winds and heaviest precipitation occur in the surrounding eyewall.

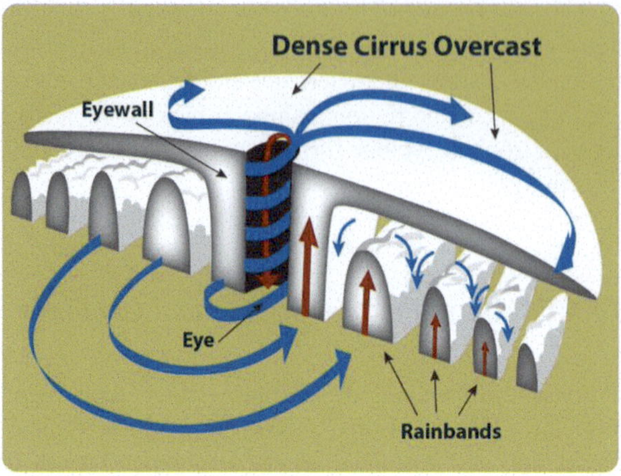

Fig. 10.15 Structure of a tropical storm

10.6.8 Cyclone Eye

The eye of a cyclone is the calm center of a strong tropical cyclone, typically circular and ranging from 40 km to 65 km in diameter. It is surrounded by the eyewall, a vertical band of intense thunderstorms where the cyclone's most severe weather occurs. The eye is perhaps the most recognizable feature of tropical cyclones. In well-developed storms, the eye is marked by light winds, clear skies, and warmer temperatures. In contrast, weaker tropical cyclones may have a less defined eye, sometimes obscured by central dense overcast, which is a region of high, thick clouds that appear prominently in satellite images. In some cases, disorganized storms may have an incomplete eyewall or an eye characterized by heavy rain. However, the eye is consistently the location of the storm's lowest barometric pressure at sea level. Despite the intense activity surrounding it, the air temperature near the ocean's surface within the eye remains relatively uniform throughout the storm. The formation of a tropical cyclone's eye is driven by the conservation of angular momentum and the resulting centrifugal force. As air accelerates toward the storm's center, it is deflected outward by centrifugal force, a phenomenon that occurs due to the wind's angular momentum. This outward force increases with the speed of the cyclone's rotation. In powerful storms, the rapid rotation of air around the center, with wind speeds of approximately 119 km/h, causes air to ascend rapidly, forming the eyewall at heights of 16–32 km. This process also creates a vacuum at the storm's center, drawing air downward, which clears the eye of clouds. During nighttime, the passage of the eye can often be recognized by the visibility of stars in the sky. The abrupt transition from the calm conditions of the eye to the violent winds of the surrounding eyewall is hazardous, particularly for those unfamiliar with the structure of a tropical cyclone. People might mistakenly believe the storm has passed when encountering the eye's calm, only to be caught off guard by the storm's resumption as the eyewall returns, bringing intense winds from the opposite direction.

10.6.9 Eyewall

The eyewall is a tropical cyclone's most dangerous and destructive part, as shown on Fig. 10.16. Within the eyewall, the winds are strongest, precipitation is heaviest, and deep convective clouds rise from near the Earth's surface to altitudes of up to 15 km. The strong winds are primarily driven by rapid changes in atmospheric pressure near the eye, creating a significant pressure gradient force. Wind speeds are highest at around 300 meters above sea level, where friction is minimal, but closer to the surface, friction slows them down. Above 300 meters, the wind speeds decrease due to the weakening horizontal pressure gradient force, which is influenced by the storm's temperature structure.

In the core of a tropical cyclone, the air is warmer, causing the atmospheric pressure in the center to decrease more slowly with altitude than in the surrounding atmosphere. This reduced contrast in atmospheric pressure with height weakens the horizontal pressure gradient, leading to a decrease in wind speed. The surface friction not only reduces wind speed but also causes the wind to turn inward toward the area of the lowest pressure. As the air flows into the eye at low pressure, it cools by expansion, extracting heat and water vapor from the sea surface. The areas with maximum heating exhibit the strongest updrafts, and the eyewall is where the largest vertical wind speeds—ranging from 5 to 10 m/s—are found within the storm. Although these updraft speeds are much lower than the horizontal winds, they are crucial for sustaining the towering convective clouds in the eyewall. A significant portion of the heavy rainfall associated with tropical cyclones is produced by these clouds.

The upward movement of air in the eyewall also causes the eye to be wider at higher altitudes than at the surface. As the air ascends, it maintains its angular momentum, which depends on the distance from the cyclone's center and the wind speed near the center. Since wind speed decreases with height, the air must move away from the storm's center as it rises. When the updrafts reach the stable tropopause, the air flows outward. The Coriolis force then deflects this outward flow,

Fig. 10.16 Wall structure at the eye of a tropical storm

creating a broad anticyclonic circulation aloft. Thus, the horizontal circulation in the upper levels of the tropical cyclone is opposite to that near the surface.

10.6.10 Precipitation Belts

In addition to the deep convective cells (compact areas of vertical air movement) that surround the eye of a tropical cyclone, secondary cells often form in bands around the storm's center. These bands, commonly known as rainbands, spiral inward toward the storm's core. In some instances, the rainbands remain stationary relative to the storm's center, while in other cases, they appear to rotate around it. The rotating cloud bands are often linked to the apparent wobbling of the storm's path. This wobble can cause significant differences between the predicted and actual landfall positions of the storm. Sometimes there are gaps between the spiral rainbands, leading to intermittent periods of rain or intense wind. As a tropical cyclone makes landfall, surface friction increases, which in turn enhances the convergence of airflows within the eyewall and the associated vertical air movement. This increased convergence and the ascent of moisture-laden air are responsible for the torrential rains associated with tropical cyclones, which can exceed 250 mm within 24 hours. In some cases, the storm may stall, allowing heavy rain to persist for several days in a single area. In extreme situations, rainfall totals of up to 760 mm have been observed over 5 days.

10.6.11 Tropical Cyclone Dimensions

Tropical cyclones typically span large areas, with diameters ranging from 100 km to 2000 km, although their size can vary considerably. It is important to note that the diameter of a cyclone is not necessarily an indicator of its intensity. For example, Hurricane Andrew in 1992, which was the second most powerful and destructive hurricane to strike the North American continent after Hurricane Katrina in 2005, was relatively small. Conversely, the largest recorded typhoon occurred in the Northwest Pacific on October 12, 1979, with storm-force winds extending up to 1087 km in radius. In contrast, the smallest tropical cyclone on record, Cyclone Tracy, struck Darwin, Australia, on December 24, 1974, with gale-force winds extending only 48 km in radius. Despite their varying sizes, the destructive winds and rains of a hurricane can cover a wide area. For example, in the Atlantic Basin, the most intense storm recorded was Hurricane Wilma in 2005, which had a central pressure of 882 hPa and sustained wind speeds of approximately 280 km/h.

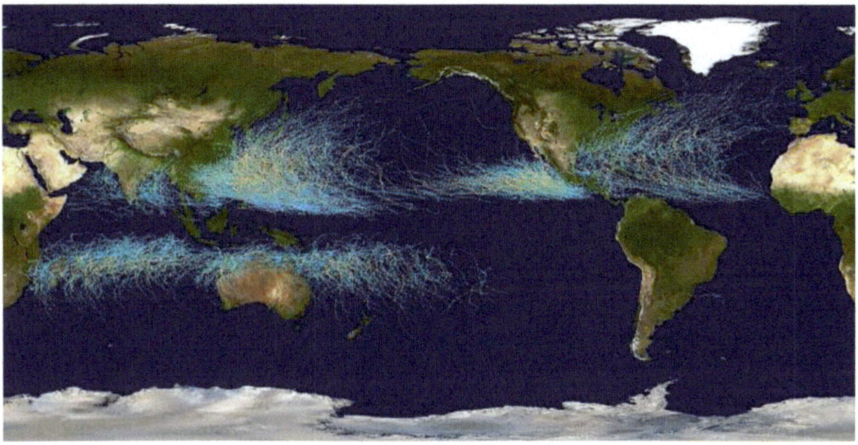

Fig. 10.17 Global tropical cyclone tracks between 1985 and 2005

10.6.12 Tropical Cyclone Life Cycle and Trajectories

Most tropical cyclones last only a few days, though some may persist significantly longer, while others dissipate within 12 hours. Despite the strong winds within them, tropical cyclones themselves move relatively slowly, typically around 5 m/s. As they mature, their speed increases to about 20 m/s. When a tropical cyclone makes landfall, it weakens rapidly as the energy supply from the warm ocean surface ceases. These systems rarely retain their tropical characteristics at temperate latitudes, often transforming into extratropical cyclones.

Typical trajectories of tropical cyclones are illustrated in Fig. 10.17. Although the regions where tropical cyclones form are well-known, and they generally move westward, their exact paths can vary significantly and are often challenging to predict. The movement of hurricanes also depends on atmospheric winds that are perpendicular to their path. For example, a cyclone forming in the tropical region of the eastern Atlantic typically moves westward with the trade winds. In the Gulf of Mexico, it tends to turn northwest, posing a potential threat to the U.S. East Coast and Gulf region.

10.6.13 Weather Associated with Tropical Cyclones

Tropical cyclones are invariably accompanied by heavy rainfall, with a single storm capable of delivering up to 3000 mm (l/m^2) of rain. This intense rainfall can continue for days after landfall, often leading to severe and widespread flooding. The steep pressure gradient within the storm causes strong winds, which increase rapidly from near zero in the cyclone's eye to a maximum at a radius of 10–100 km from the center. The most intense winds occur near the leading edge of the storm, making this region particularly dangerous.

10.6.14 Tropical Cyclone Disasters

Every year, starting from June 1, countries along the Gulf and East Coasts face a heightened risk of tropical cyclones. While many are aware of the destructive winds associated with these storms, fewer realize that tropical cyclones can also lead to a variety of other disasters, both directly and indirectly, that threaten life and property (Collins and Walsh, 2019).

To mitigate these risks, it's essential to increase public awareness and education regarding the formation, behavior, and destructive power of tropical cyclones. This knowledge can lead to better preparedness and preventive measures, reducing the potential for disaster.

10.6.15 Storm Surge

The destructive power of tropical cyclones is not only due to strong winds but also to the large storm surges and associated waves they generate. A storm surge is composed of ocean water pushed toward the coast by the intense winds swirling around the cyclone (see Fig. 10.18). This surge, combined with high waves, can cause

Fig. 10.18 Hurricane wave. (Credit: Photo by Ray Bilcliff: https://www.pexels.com/photo/crashing-waves-1494707/)

significant damage, including severe beach erosion and the destruction of coastal infrastructure.

A storm surge is characterized by a rapid rise in sea level, which can exceed 6 m in some cases. Many of the fatalities and accidents associated with tropical cyclones are due to the combined effects of storm surges and the accompanying winds. These large leading waves, when combined with normal waves, create a powerful and destructive hurricane storm surge.

The intensity of the storm surge is influenced by the slope of the continental shelf. A gently sloping coast allows the surge to inundate vast coastal areas, leading to widespread flooding. In contrast, regions with a steeper continental shelf may experience less extensive flooding during severe storms, though the large, destructive waves that accompany the surge can still pose serious threats.

10.6.16 Winds and Floods

Hurricanes are notorious for their destructive winds, with intensity measured by wind speed. The National Hurricane Center provides advisories based on this scale, though it does not account for storm surges. Wind gusts, which are brief but intense bursts of wind, can cause extensive damage to both maritime traffic and land-based infrastructure. For example, hurricanes can lift roofs off buildings by creating lower pressure on the roof's exposed side. This pressure differential can cause the roof to detach, and once it is removed, the hurricane's winds can more easily push the walls. In addition to strong winds, tropical cyclones bring heavy rainfall that can lead to severe flooding (Fig. 10.19). The risk of flooding remains even as winds diminish. Tropical cyclones can cause various types of flooding, including:

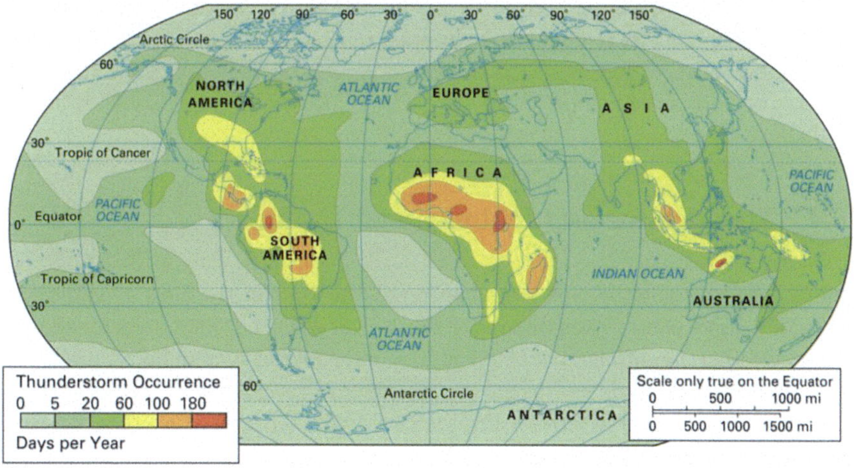

Fig. 10.19 Global patterns of thunderstorm frequency. (Credit: https://www.britannica.com/)

- **Flash Floods**: These occur rapidly, often within minutes or hours of intense rainfall. Water levels can rise quickly, reaching up to 10 m, and can carry away trees and destroy buildings and bridges.
- **Urban Floods**: Although less severe than flash floods, urban flooding can still cause significant damage, such as street flooding and inundation of basements and other structures.

Heavy rainfall from tropical cyclones can result in widespread flooding that persists for days or even weeks, compounding the impact of the storm.

10.6.17 Names of Tropical Cyclones

Tropical cyclones are assigned names to facilitate clear communication between meteorologists and the public regarding weather forecasts, observations, and warnings. The practice of naming tropical storms began in the early twentieth century with an Australian forecaster, who proposed naming them after political figures. In the Pacific, the naming convention evolved significantly over time. From 1950 to 1952, tropical cyclones in the North Atlantic were identified using a phonetic alphabet. In 1953, the system was updated to include female names. The World Meteorological Organization (WMO) and the US National Oceanic and Atmospheric Administration (NOAA) subsequently developed lists for tropical cyclones in the Pacific Northwest. Starting in 1945, cyclones were named using female names, with male names introduced in 1979 (Pielke and Landsea 2000; Landsea and Franklin 2004).

On January 1, 2000, a new naming system was implemented for the Pacific Northwest, featuring a diverse set of names selected by the WMO Typhoon Committee. This new list includes names from various Asian cultures and territories, differing significantly from other global naming conventions. While the list includes both male and female names, it predominantly features names of flowers, animals, birds, trees, and foods. Names are organized alphabetically by their contributors rather than sequentially (Hsu et al., 2012; Kossin and Camargo 2014).

10.7 Thunderstorms, Lightning, and Tornadoes

Thunderstorms, lightning, and tornadoes represent some of the most dynamic and hazardous atmospheric phenomena on Earth. These events are characterized by intense weather conditions that can lead to significant damage and pose threats to life and property. Thunderstorms are the precursor to various severe weather conditions, including lightning strikes, flash floods, hail, and tornadoes. They develop from the complex interactions between warm, moist air and cooler, drier air, leading to strong vertical motions within the atmosphere (Doswell & Brooks, 2018; McNulty & Dyer, 2014).

The study of these phenomena is crucial not only for understanding weather patterns but also for developing effective forecasting techniques to mitigate the risks associated with severe storms. The impact of climate change on the frequency and intensity of thunderstorms and associated severe weather is an area of ongoing research, highlighting the importance of understanding these systems in a changing global environment (Thompson and Edwards 2017). Thunderstorms most frequently occur in tropical latitudes over land, where rapid surface heating leads to strong updrafts (Fig. 10.19). As such, exploring the mechanisms behind thunderstorms, the nature of lightning, and the dynamics of tornado formation provides insights into these potent weather phenomena and their implications for society.

10.7.1 Basic Definition

A storm is a dynamic atmospheric phenomenon characterized by adverse weather conditions, which can manifest in various forms such as thunderstorms, hurricanes, typhoons, tornadoes, blizzards, and more. These events are driven by Earth's complex atmospheric processes involving the interaction of air masses, moisture, and other atmospheric components. Storms typically arise from disturbances in the atmosphere that impact on the Earth's surface, leading to destructive weather conditions. They generally involve clouds with significant vertical development, where convection plays a dominant role (Miller, 2018).

Convection refers to the vertical transport of heat and moisture in the atmosphere, driven by buoyancy forces or external mechanical factors. It can be categorized as free convection, which is a self-sustaining flow driven by temperature gradients, and forced convection, generated by mechanical factors such as forced airlift or horizontal mass convergence. Convection can be further divided into dry convection, which occurs without cloud formation or precipitation, and moist convection, which produces clouds and precipitation. In disorganized convection, there is an irregular distribution of cloud formation due to differential heating and rising warm air. Organized convection features a more regular distribution of convective cloud development over a broader area, typically occurring through shallow convection within the surface boundary layer (1–2 km), where cumulus (Cu) and stratus (Sc) clouds form. Shallow convection is important for cloud cover, vertical moisture transport, and surface precipitation and serves as a precursor to more intense convection. Deep convection affects the entire tropospheric layer, with vertical lifting extending above 500 hPa. It is driven by factors such as frontal boundaries, shear lines, tropical cyclones, and tropical waves or by large-scale air descents associated with polar troughs or front block ridges. The general conditions for storm formation include low-level convergence, upper-level divergence, relative humidity exceeding 70%, and an unstable atmospheric boundary layer (ABL) or planetary boundary layer (PBL).

Thunderstorms, specifically, are localized storms caused by cumulonimbus clouds, always accompanied by lightning and thunder. They often feature strong wind gusts, heavy rainfall, and sometimes hail. Thunderstorms have relatively short life cycles, typically lasting up to 60 minutes, but can persist longer depending on atmospheric conditions. Key characteristics of thunderstorms include strong convective updrafts during their initial stages and strong downdrafts in the vertical precipitation zone, indicating the storm's breakup phase.

10.7.2 Formation of Thunderstorms

The formation of thunderstorms relies on three key factors: moisture, instability, and an updraft mechanism (Cotton et al., 2011).

1. Moisture: Moisture is typically sourced from large bodies of water such as the Atlantic and Pacific Oceans, the Gulf of Mexico, and the Indian Ocean. Sea surface temperatures play a critical role in determining the amount of moisture available for storm formation.
2. Instability: Atmospheric instability occurs when warm, moist air near the surface rises and encounters cooler, drier air aloft. This condition is often triggered by surface heating or the intrusion of cold air over a heated surface. As the warm air rises, it cools, leading to condensation of water vapor and the formation of cumulonimbus clouds, which are characteristic of thunderstorms.
3. Updraft Mechanism: For thunderstorms to develop, there must be a mechanism to initiate and sustain upward vertical motion (Murillo, 2018). Key mechanisms include:

   - Front Boundaries: The interface between cold and warm air masses can initiate upward motion as it provides a region of atmospheric instability.
   - Dry Line Boundaries: Differences in moisture between two air masses can create conditions conducive to vertical air movement.
   - Topographic Effects: Wind blowing over mountains or other obstacles can be forced to rise, contributing to storm formation.

10.7.3 Life Cycle of a Storm

The life cycle of a thunderstorm typically progresses through several stages:

1. Cumulus Stage: This initial stage features the formation of cumulus clouds as rising air causes water vapor to condense. These clouds are usually associated with surface heating or topographic uplift. During this stage, the precipitation is light and short-lived.
2. Mature Stage: The storm reaches its full development during the mature stage. The cloud top reaches the tropopause, where it spreads out into an "anvil" shape.

This stage is characterized by intense weather phenomena, including heavy rainfall, hail, frequent lightning, strong winds, and possibly tornadoes. The storm may exhibit a dark green appearance, indicating severe weather conditions. The mature stage typically lasts between 10 min and 20 min, though storms can persist longer if wind shear (variations in wind speed with height) is minimal.

3. Dissipating Stage: In the final stage, the storm begins to weaken. Downdrafts dominate as the precipitation intensity decreases. Despite the reduction in overall storm activity, strong downdrafts can still occur near the surface. Lightning remains a potential hazard during this phase.

Figure 10.20 illustrates the numerical simulation of a thunderstorm's life cycle using a cloud-resolving model, as demonstrated by Telenta and Aleksic (1988) and Spiridonov et al. (2021). This figure showcases the evolution of the storm through various stages, highlighting the microphysical elements within the cloud. Additionally, the wind flow associated with the storm is represented by arrows, providing a clear depiction of the dynamic processes involved in storm development.

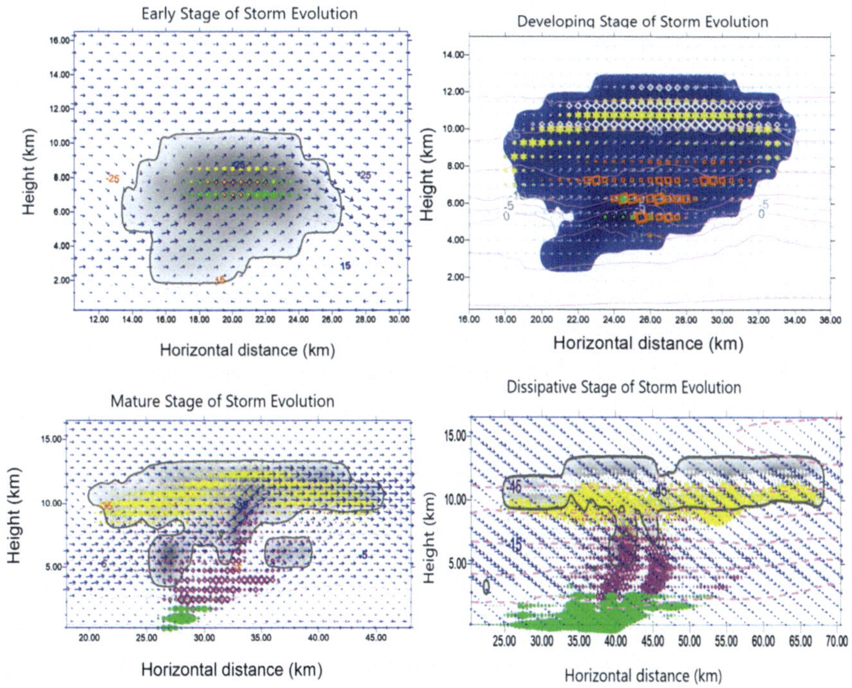

Fig. 10.20 Life cycle of a thunderstorm simulated by a numerical cloud model, illustrating the various stages of development, from initial cumulus formation through to the mature stage and eventual dissipation

Fig. 10.21 Classification of thunderstorms: single cell, multicell, and supercell

10.7.4 Classification of Thunderstorms

Thunderstorms can be classified into three main types based on their convective structure, environmental thermodynamics, and microphysical characteristics: single-cell, multicell, and supercell storms (Fig. 10.21). Each type arises from different combinations of these conditions (Marwitz, 1972).

Single-Cell Storms

Single-cell storms consist of a single convective core, often referred to as a "cell." These storms typically form in isolation and are often triggered by surface heating during a summer afternoon. They generally produce hail, rain, and lightning, with a typical lifespan of 20–30 minutes. The storm's evolution follows three stages: cumulus, maturity, and dissipation. Single-cell storms usually exhibit low wind shear and minimal organization (Wakimoto, 2001).

Multicell Storms

Multicell storms consist of a cluster of cells, each at a different stage of development. These storms feature multiple convective updrafts and downdrafts. They can be further categorized into ordered, weak, and disordered multicellular types based on their structural and evolutionary properties. Multicell storms can produce moderate hail, flash floods, and weak tornadoes (Weiss & Jacobs, 2005).

Supercell Storms

Supercells are a specialized type of single-cell storm that can last several hours. They are characterized by strong vertical rotation, with a rotating updraft known as a mesocyclone. Supercells have well-organized and vigorous updrafts and downdrafts, strong wind shear, and often a rounded, large top. Key factors contributing to supercell formation include high wind shear, a significant storm helicity index, and differential heating due to localized atmospheric conditions. Supercells are known for their potential to produce severe weather, including strong tornadoes. They can generate strong downdrafts, large hail, flash flooding, and a range of tornadoes from weak to violent, making them highly destructive (Brotzge & Morgan, 2003). These storms are particularly dangerous due to their potential for severe weather impacts, including heavy rainfall and significant hail. Thunderstorms, as a whole, are responsible for a variety of severe weather phenomena. They are typically accompanied by strong winds and can produce heavy rain, hail, flash floods, wind gusts, tornadoes, and atmospheric electrical discharges (lightning) (Glickman, 2000).

Based on thermodynamic characteristics, convective instability parameters, and physical processes, thunderstorms are also classified into air mass thunderstorms, instability lines, convective belts, mesoscale convective complexes (MCCs), convective clusters, and mesoscale convective systems (MCSs) (Fig. 10.22).

Air Mass Thunderstorms Also known as local thunderstorms, these systems consist of mass thunderclouds that are generally vertical in structure, short-lived, and typically do not produce severe weather conditions. These storms are composed of one or more convective cells, each progressing through a well-defined life cycle.

Early in the cell's development, air movements are predominantly upward, with a series of rising vortices. As the cell grows, cloud and precipitation particles form.

Fig. 10.22 Sequence of the general appearance of (**a**) an instability line, (**b**) a convective belt, (**c**) a mesoscale convective system (MCS), and (**d**) a mesoscale convective complex (MCC)

When the water and ice load become excessive, downdrafts begin, intensifying as cloud particles evaporate and cool the air. In the mature stage, the cell contains both updrafts and downdrafts and in the later stages, the downdraft spreads through the cell, weakening as precipitation falls. Isolated thunderstorms contain convective cells at various stages of evolution.

Instability Lines These are multicell storms arranged in a line, like mesoscale convective complexes but with a linear rather than spherical structure. Instability lines often form along or ahead of cold fronts by hundreds of kilometers. When the air is unstable, moist, and warm, thunderstorms develop along these lines, bringing heavy rainfall, strong winds, small hail, and lightning, with tornadoes forming in rare cases.

Mesoscale Convective Complexes (MCCs) MCCs are large clusters of thunderstorms that can span extensive regions. They typically develop in areas of warm, moist, and unstable air, where sufficient heat and moisture lift the air to form storms. These complexes are organized with clusters of thunderclouds, often associated with convective vertical flow near mountain slopes or well-defined cold fronts. Multicellular clusters within MCCs consist of a group of cells that move together, constantly forming and dispersing in an environment conducive to thunderstorm formation. Although individual cells are short-lived, the cluster retains its general characteristics and often produces strong storms.

Mesoscale Convective Systems (MCSs) MCSs are organized thunderstorm systems characterized by excessive rainfall over large mesoscale areas and a life cycle lasting several hours or more. They have distinctive shapes visible on satellite images, with regions of both convective and stratiform precipitation. MCSs develop mesoscale circulation in their mature stage, with upward motion in the form of deep-layer ascent due to latent heating and cooling in the convective region. A mid-level mesoscale convective vortex (MCV) often develops in the stratiform region, extending the MCS and potentially contributing to tropical cyclone development. MCS propagation may involve discrete components and may also be influenced by waves and disturbances both generated by and external to the MCS.

10.8 Atmospheric Electricity

Atmospheric electricity refers to the electrical phenomena that occur within Earth's atmosphere, primarily due to the presence of charged particles like ions and electrons. The study of atmospheric electricity encompasses various electrical processes, including lightning, thunder, and the global electrical circuit.

10.8.1 General Characteristics of Lightning

Lightning is a transient electrical discharge of a high current that extends over several kilometers. It results from the formation of a strong electric field within a thunderstorm cloud, typically between ice crystals in the negatively charged upper regions of the cloud and negatively charged hail near the cloud base. The most common source of lightning is the separation of electrical charges within cumulonimbus clouds or thunderclouds. Approximately half of all lightning discharges occur within the thunderstorm cloud itself, known as intra-cloud discharges.

The exact mechanisms behind lightning are not fully understood, including whether it is influenced by sunlight affecting Earth's upper atmosphere or the electromagnetic field. Lightning can also be observed during volcanic eruptions, intense forest fires, severe snowstorms, major hurricanes, and even surface nuclear detonations. It is most commonly associated with thunderstorms, where lightning and the subsequent thunder are integral components of the storm's dynamics (Fig. 10.23) (Rakov & Uman, 2003; Krehbiel, 1986; National Research Council, 1986).

10.8.2 Origin of Atmospheric Currents

The Earth's surface is continuously charged negatively, and this negative charge is replenished by large currents flowing from the atmosphere's "top" to the surface. Thunderstorms act as a primary source for this negatively generated charge, effectively serving as batteries that deliver a significant amount of negative charge to the

Fig. 10.23 (**a**) Atmospheric electricity, lightning. (**b**) Global annual lightning flash density rates. The colors represent flash rates in flashes/km^2/year, ranging from 0.1 (magenta) to 70 (black). Data is sourced from OTD and TRMM-LIS (1995–2003). (Image adapted from NASA by Marit Jentoft-Nilsen, with data provided by the Global Hydrology and Climate Centre Lightning Team)

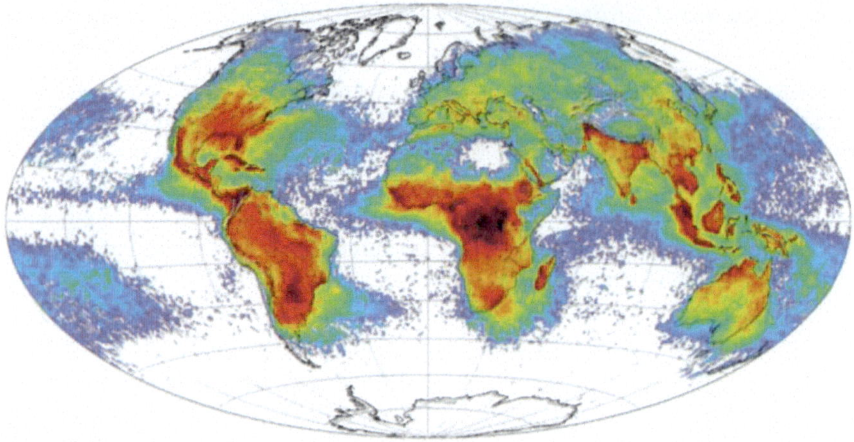

Fig. 10.23 (continued)

Earth. On average, thunderstorms contribute around 1800 A to this process, which is then discharged across regions experiencing fair weather.

With approximately 40,000 thunderstorms occurring globally each day, they continuously pump electricity into the upper layers of the atmosphere, maintaining the voltage difference between the Earth's surface and the atmosphere. However, estimating the exact number of thunderstorms is challenging due to the lack of comprehensive observations overseas and oceans. Nonetheless, evidence suggests that there are about 100 lightning strikes per second worldwide, with peak activity around 19:00 GMT.

10.8.3 Mechanism of Charge Separation

Understanding the mechanism of charge separation within thunderstorms is crucial. Based on numerous experiments, including aircraft flying through thunderstorms, the charge distribution within a thunderstorm cell is depicted as shown in Fig. 10.24. A typical storm is characterized by turbulent regions and strong updrafts and downdrafts (Cotton et al. 2011). Vertical updrafts transport small liquid water droplets from the lower regions of the storm to the higher layers of the cloud, while downdrafts carry hail and cloud ice from the frozen upper regions down through the storm.

When these particles collide, the water droplets freeze and release heat, sometimes causing the surface of ice particles to remain slightly warmer than the surroundings, leading to the formation of "wet hail."

As this hail collides with additional water droplets and ice particles, a critical phenomenon occurs. Electrons are transferred from ascending particles to descending particles. Because electrons carry a negative charge, this process results in a

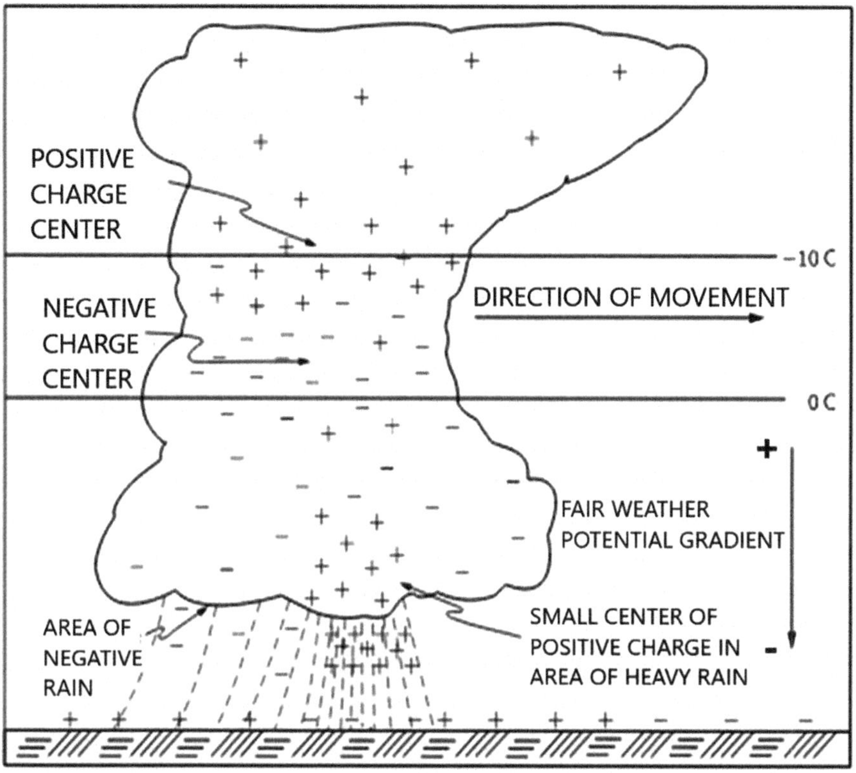

Fig. 10.24 Formation of charge and mechanism of charge separation in a conceptual storm model

negatively charged base and a positively charged top of the storm cloud, creating a distinct charge separation. In addition to a small local region of positive charge near the cloud base, this charge configuration drives the storm's electrical activity. The main negative charge at the cloud's base and the positive charge at the top create the potential difference required to transfer the negative charge to the Earth. Positive charges typically accumulate at altitudes of 6-7 km, where temperatures are around −20 °C, while the negative charges are found between 3 km and 4 km, in a temperature range of −10 to 0 °C. The resulting charge at the cloud's base can create a potential difference of 20–100 MV between the cloud and the Earth, vastly surpassing the 0.4 MV present in fair-weather conditions. These high voltages cause the air to break down, leading to large-scale electrical discharges—lightning. When lightning occurs, it transports the negative charges from the cloud's base to the Earth. This process is rapid; the thundercloud can recharge in just about 5 s after each lightning strike. Another key factor is the influence of the electric field on droplets, which can significantly affect their movement. This movement occurs irregularly, but it generally takes about 5 s for the cloud to return to its pre-strike state, maintaining approximately 4 A of current.

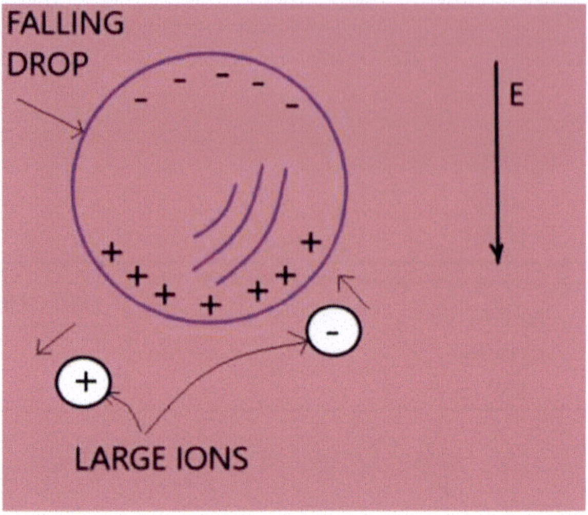

Fig. 10.25 Formation of charge on a falling drop

There are two main theories to explain charge separation in a storm: the drop theory and the Wilson theory. Both theories suggest that precipitation particles carry specific charges, and as these particles move through the air, electric charges are separated, leading to the observed distribution within the cloud (Fig. 10.25).

10.8.4 Creating an Electric Field

The most common source of lightning is the electrical charge released in thunderclouds (cumulonimbus clouds). The atmosphere is a very effective insulator, which prevents electricity from flowing freely. As a result, a significant amount of electrical charge must accumulate before lightning can occur. Once this charge reaches a critical threshold, the strength of the electric field surpasses the insulation capacity of the atmosphere, leading to a lightning discharge.

In a thunderstorm, the electric field develops due to the accumulation of charges. The base of the storm cloud carries a negative charge, which induces a positive charge on the Earth's surface beneath it. This positive charge follows the storm's movement, facilitating lightning from the cloud to the ground. Despite this, the electric field within the storm is generally much stronger than the field between the cloud base and the Earth's surface. Consequently, approximately 75–80% of lightning discharges occur within the storm itself.

As the storm moves, it continues to induce positive charges in the region below. These positively charged particles tend to accumulate on tall structures such as trees, buildings, and power lines. Meanwhile, the storm releases negative charges in a series of rapid, invisible pulses toward the Earth's surface. As the negative charge approaches the ground, the positive charge accumulates on surfaces and objects.

When these charges meet, the result is a visible electrical discharge known as lightning.

10.8.5 Mechanism of Atmospheric Electrical Discharge

The mechanism of atmospheric electrical discharge involves ionization and polarization processes that lead to the formation of charged suspended particles and water droplets in the atmosphere, which subsequently causes cloud charging. In a calm sky, electric charges are evenly distributed, resulting in a neutral atmosphere. However, during a thunderstorm, the presence of various hydrometeors causes significant charge separation. Ice crystals, which carry a positive charge, are lifted to the top of the thunderstorm cloud by updrafts, while hailstones, which carry a negative charge, are moved to the base of the storm by downdrafts (Fig. 10.26). This separation results in a positive charge at the cloud top, a negative charge at the cloud base, and a corresponding positive charge at the ground level.

As the electric field intensifies, it can ionize air molecules, creating a conductive path through the atmosphere known as a leader channel. Step leaders propagate downward in a series of rapid steps, while positive charge currents ascend from the Earth's surface. When these downward leaders and upward currents connect, a conductive path is established, and lightning occurs. The lightning discharge, or flash, is the visible and intense flash of light experienced as lightning. The lightning channel, created by the stepped leaders and streamers, rapidly heats and expands the surrounding air, generating shock waves that produce the sound of thunder. Thunder

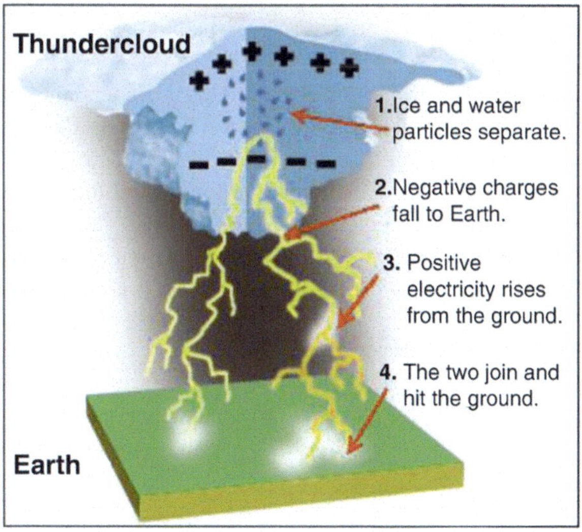

Fig. 10.26 Formation of lightning, illustrating the distribution of charges within a thunderstorm cloud. (Credit: Pearce (1989))

typically follows the flash of lightning, as it travels along the path established by the electrical discharge.

10.8.6 Lightning Strike

A lightning strike is a powerful and sudden discharge of electrical energy between the atmosphere and an object on Earth. This phenomenon involves a rapid flow of electric charge between areas of opposite electrical polarity. Lightning may occur within a single cloud (intra-cloud lightning), between two clouds (cloud-to-cloud or CC lightning), or between a cloud and the ground (cloud-to-ground or CG lightning), which is the most observed type. Lightning occurs when an electrical imbalance builds up between charged regions in the atmosphere or between the atmosphere and the Earth's surface. This imbalance is often caused by turbulent air movement, such as within a thunderstorm, where updrafts and downdrafts cause collisions among ice particles and water droplets, generating static charges. When the electrical potential difference becomes large enough to overcome the insulating properties of air, a rapid discharge of electricity occurs, producing a bright flash and an intense sound wave known as thunder.

A typical lightning strike consists of multiple components, including stepped leaders, return strokes, and dart leaders. The stepped leader, a channel of ionized air, travels downward in a series of steps, creating a path for the electric charge to flow. When this path connects with an upward streamer from the ground, a powerful return stroke occurs, completing the circuit and releasing immense energy in the form of heat, light, and sound. Cloud-to-ground lightning is particularly hazardous as it can cause severe damage to structures, ignite fires, and pose risks to human life. Intra-cloud and cloud-to-cloud lightning, though less likely to impact the surface directly, still play a crucial role in the atmospheric electrical balance and can influence weather dynamics.

The study of lightning strikes helps us understand their contribution to the global electric circuit, their role in nitrogen fixation in the atmosphere, and their potential implications for climate systems. Advanced detection systems and forecasting technologies have been developed to monitor lightning activity and mitigate its risks to people and infrastructure.

10.8.7 Tornado

Cyclones can produce tornadoes, which enhance the storm's destructive power (Fig. 10.27). Tornadoes are often found within the rain zones of the right frontal quadrant, away from the tropical storm's center of motion. A thunderstorm tornado is a violent wind, manifesting as a rotating vortex extending downward from a cumulonimbus cloud.

Fig. 10.27 Tornado over land

A tornado is a prominent rotating column of air, typically rotating counterclockwise in the Northern Hemisphere, that descends from a thunderstorm cloud and touches the ground. Although tornadoes are usually short-lived, lasting only a few minutes, they can occasionally last over an hour and travel several miles, causing significant damage. Tornadoes consist of water droplets, sand, and dust lifted from the ground by strong winds. They frequently occur in the United States and Mexico, generally moving at speeds of up to 50 km/h, with the vortex rotating at speeds of around 160 km/h. When a tornado occurs over water, it is called a waterspout and usually lasts from a few minutes to an hour. Tornadoes cause severe damage and can leave large areas of destruction. The United States experiences more tornadoes than any other country, with approximately 1000 tornadoes annually. The peak tornado season is from April to June. Tornadoes are characterized by continuous vertical rotation and strong vertical wind shear in the surrounding air. Wind shear refers to changes in wind speed and/or direction with height. Vertical movement raises a rotating column of air generated by the horizontal change in wind speed. This creates two types of rotation in a supercell storm: cyclonic (counterclockwise) and anticyclonic (clockwise). Wind shear intensifies cyclonic rotation and reduces anticyclonic rotation (rotation to the right of vertical motion). A supercell storm is a rotating thunderstorm, and the most dangerous product of supercell thunderstorms is the tornado. Tornadoes develop from supercell clouds that have rotating vertical velocities and a characteristic mesocyclone. They usually appear as visible condensation funnels, with the narrow end touching the ground and often surrounded by debris and dust. Tornadoes can cause massive destruction, with wind speeds exceeding 300 km/h, capable of overturning vehicles, uprooting trees, and destroying buildings.

10.8.8 Classification of Tornadoes

Tornadoes are classified using the Enhanced Fujita (EF) Scale, which assesses tornado intensity based on the damage they cause. The scale ranges from EF0 (the weakest) to EF5 (the most destructive). The Enhanced Fujita Scale, introduced in

Table 10.2 Classification of tornadoes according to the EF Scale

EF Scale Class	Wind speed (*km/h*)	Description
EF0	105–137	Weak
EF1	138–177	Moderate
EF2	178–217	Significant
EF3	218–266	Severe
EF4	267–322	Devastating
EF5	> 322	Catastrophic

Fig. 10.28 Waterspout: A tornado occurring over water

1971 and updated to reflect more precise damage assessments, is used to estimate tornado wind speeds (see Table 10.2) (Fujita, 1971; Fujita, 1981; Brooks et al., 2003; Doswell et al., 2009).

10.8.9 Types of Tornadoes

Multivortex Tornado This type of tornado features multiple rotating columns of air, or vortices, that are separated yet contribute to a single tornado. These multi-vortex tornadoes exhibit complex vortex dynamics and can be highly destructive due to the intense winds generated by the individual vortices.

Companion Tornado A smaller, weaker tornado that forms near a larger, more powerful tornado. The companion tornado may rotate around the stronger tornado or be influenced by its presence.

Waterspout A tornado occurring over water (Fig. 10.28). Waterspouts can be further classified into two types:

- Fair-Weather Waterspout: Typically observed in calm weather conditions, usually forming over warm water.

- Tornado Waterspout: Associated with severe weather systems, like land-based tornadoes, and can exhibit strong winds and significant damage potential.

References

Ahrens, C. D. (2013). *Meteorology today: An introduction to weather, climate, and the environment* (10th ed.). Cengage Learning.

Barry, R. G., & Chorley, R. J. (2010). *Atmosphere, weather, and climate* (9th ed.). Routledge.

Bluestein, H. B. (1992). *Synoptic-dynamic meteorology in Midlatitudes: Observations and theory of weather systems* (Vol. 1). Oxford University Press.

Bluestein, H. B. (1993). *Synoptic-dynamic meteorology in midlatitudes: Observations and theory of weather systems* (Vol. 1). Oxford University Press.

Brooks, H. E., Lee, J. W., & Heintz, F. (2003). *The enhanced Fujita scale*. Severe Weather Research, Inc.

Brotzge, J. A., & Morgan, R. G. (2003). The relationship between radar and damage reports from supercell thunderstorms. *Weather and Forecasting, 18*(5), 832–845. https://doi.org/10.1175/1520-0434(2003)018<0832>2.0.CO;2

Collins, D. C., & Walsh, K. J. E. (2019). Climate change and tropical cyclones: Recent advances in research and understanding. *Weather and Climate Extremes, 24*, 100192. https://doi.org/10.1016/j.wace.2019.100192

Doswell, C. A., & Brooks, H. E. (2018). The severe thunderstorm of April 25, 2011: A summary of events and analysis of the processes involved. *Weather and Forecasting, 33*(3), 929–952. https://doi.org/10.1175/WAF-D-17-0197.1

Doswell, C. A., Brooks, H. E., & Maddox, R. A. (2009). Dawn of the Tornado: The importance of the enhanced Fujita scale. *Bulletin of the American Meteorological Society, 90*(3), 263–269.

Cotton, W. R., Bryan, G. and Van den Heever, S. C. (2011) Storm and Cloud Dynamics. International Geophysics Series, Vol. 99, Academic Press, 820 pp.

Emanuel, K. A. (2005). Advances in hurricane modeling. *Nature, 438*(7068), 274–275. https://doi.org/10.1038/438274a

Fujita, T. T. (1971). *Recommendations for the development of the enhanced Fujita scale*. National Weather Service.

Fujita, T. T. (1981). Tornadoes and downbursts in the context of generalized planetary scales. *Journal of the Atmospheric Sciences, 38*, 1511–1534.

Glickman, T. S. (2000). *Glossary of meteorology*. American Meteorological Society.

Harman, J. R. (1991). *Synoptic climatology of the westerlies: Process and patterns*. Routledge.

Holton, J. R., & Hakim, G. J. (2012). *An introduction to dynamic meteorology* (5th ed.). Academic Press.

Hsu, H. H., & Raghavan, K. (2012). Global warming and tropical cyclone names: An examination of the naming convention. *Climate Change, 112*(3–4), 725–738. https://doi.org/10.1007/s10584-011-0272-4

Knutson, T. R., Sirutis, J. J., Garner, S. T., & Held, I. M. (2010). Simulation of the 20th century Atlantic hurricane activity. *Geophysical Research Letters, 37*(2), L02706. https://doi.org/10.1029/2009GL041124

Kossin, J. P., & Camargo, S. J. (2014). The impact of the changing climate on tropical cyclone naming conventions. *Bulletin of the American Meteorological Society, 95*(12), 2097–2103. https://doi.org/10.1175/BAMS-D-13-00162.1

Kossin, J. P., Hall, T. C., & Knutson, T. R. (2018). Elevated tropical cyclone activity in the North Atlantic. *Nature, 554*(7693), 126–130. https://doi.org/10.1038/nature25476

Krehbiel, P. J. (1986). *Lightning. In the Earth's electrical environment* (pp. 25–56). National Academies Press. This work provides insights into the mechanisms of lightning and its effects on the Earth's atmosphere.

Lamb, H. H. (1982). *Climate, history and the modern world.* Methuen.

Landsea, C. W., & Franklin, J. L. (2004). Atlantic hurricane database uncertainty and control of the intensity of tropical cyclones. *Journal of Climate, 17*(24), 4243–4260. https://doi.org/10.1175/JCLI-3249.1

Markowski, P., & Richardson, Y. (2010). Mesoscale meteorology in midlatitudes. *Wiley-Blackwell.*

Marwitz, J. D. (1972). The structure and motion of severe hailstorms. *Journal of Applied Meteorology, 11*, 166–179.

McNulty, S. G., & Dyer, J. (2014). A new paradigm for the role of thunderstorms in the production of lightning and the contribution of lightning to climate change. *Bulletin of the American Meteorological Society, 95*(11), 1707–1721. https://doi.org/10.1175/BAMS-D-13-00178.1

Miller, J. A. (2018). Fundamentals of thunderstorm formation: The role of moisture and instability. *Journal of Atmospheric Sciences, 75*(2), 673–688. https://doi.org/10.1175/JAS-D-17-0144.1

Murillo, J. (2018). Understanding thunderstorm dynamics: The role of surface conditions and environmental factors. *Meteorological Monographs, 57*, 15.1-15.24. https://doi.org/10.1175/AMSMONOGRAPHS-D-17-0001.1

National Research Council. (1986). *Earth's electrical environment.* https://doi.org/10.17226/898. This report examines the electrical phenomena of the atmosphere, including lightning, and the mechanisms involved in its occurrence.

Pearce, Q. L. (1989) Lightning and Other Wonders of the Sky (Amazing Science), *Julian Messner*, 64 pp. ISBN-13: 978-0671685348.

Pielke, R. A., & Landsea, C. W. (2000). Normalizing losses from natural disasters: A global perspective. *Natural Hazards Review, 1*(1), 37–50. https://doi.org/10.1061/(ASCE)1527-6988(2000)1:1(37)

Rakov, V. A., & Uman, M. A. (2003). Lightning: Physics and effects. Cambridge university press. This book offers an in-depth examination of the physical processes associated with lightning, including its formation and effects on the environment.

Rappaport, E. N., Goni, G. J., & Schmit, T. J. (2009). Tropical cyclone intensity change: A comparison of observations and model simulations. *Weather and Forecasting, 24*(4), 1007–1023. https://doi.org/10.1175/2009WAF2222204.1

Spiridonov, V., Curic, M., Velinov, G., & Jakimovski, B. (2021). Numerical simulation of a violent supercell tornado over Vienna airport initialized and initiated with a cloud model. *Atmospheric Research, 261*. https://doi.org/10.1016/j.atmosres.2021.105758

Stull, R. B. (2017). *Practical meteorology: An algebra-based survey of atmospheric science.* University of British Columbia.

Telenta, B., & Aleksic, N. (1988). A three-dimensional simulation of the 17 June 1978 HIPLEX with observed ice multiplication. 2nd Int. Cloud Modelling Workshop, Toulouse, WMO/TD, No. 268, 277–285.

Thompson, G., & Edwards, R. (2017). The changing nature of severe storms: Implications for forecasting. *Meteorological Applications, 24*(2), 209–222. https://doi.org/10.1002/met.1685

Wakimoto, R. M. (2001). The role of shear in the development of severe convective storms. *Monthly Weather Review, 129*(5), 943–958. https://doi.org/10.1175/1520-0493(2001)129<0943>2.0.CO;2

Wang, C., & Lee, S. K. (2019). Tropical cyclone intensity changes in response to global warming. *Nature Climate Change, 9*(11), 866–871. https://doi.org/10.1038/s41558-019-0635-0

Weiss, S. J., & Jacobs, J. M. (2005). Characteristics of severe thunderstorm environments. *Weather and Forecasting, 20*(3), 399–415. https://doi.org/10.1175/WAF860.1

Chapter 11
Exploring Natural Hazards: From Earthquakes, Floods, and Beyond

11.1 Natural Hazards: Basic Features

Natural disasters are catastrophic, abrupt phenomena driven by environmental forces that inflict profound devastation on human life, infrastructure, and ecosystems (Mokhtari et al., 2022; Smith et al., 2021). These events often strike with

© The Author(s), under exclusive license to Springer Nature Switzerland AG 2025
V. Spiridonov et al., *Atmospheric Perspectives*,
https://doi.org/10.1007/978-3-031-86757-6_11

minimal or no advance notice, posing significant challenges to preparedness and effective response (Sharma et al., 2020). Natural disasters manifest in diverse forms, categorized according to their origin and characteristics. Among the most prevalent types are earthquakes, hurricanes, typhoons, cyclones, tornadoes, floods, volcanic eruptions, wildfires, heat waves, droughts, tsunamis, landslides, blizzards, avalanches, ice storms, and epidemics (Khakzad, 2020; Nedaei, 2021). These occurrences, often called natural hazards, represent severe and adverse weather or climate-related phenomena (Gonzalez et al., 2021; Wang & Liu, 2022). While such hazards are ubiquitous globally, certain regions exhibit heightened vulnerability to specific threats (Richardson Varas, 2020; Smith et al., 2022). A natural hazard escalates into a disaster when it destroys lives and livelihoods (Mokhtari et al., 2022). Natural hazards are broadly classified into four primary categories based on their underlying causes (Gonzalez et al., 2021):

1. Geological hazards
2. Meteorological hazards
3. Hydrological hazards
4. Biological hazards

11.1.1 Geological Hazards

Geological hazards arise from the dynamic processes within the Earth's crust, including tectonic plate movements and volcanic activity (Fig. 11.1). These natural events or phenomena, driven by geological and tectonic forces, can lead to catastrophic outcomes for human societies, ecosystems, and infrastructure (Schwartz et al., 2021). The impacts of such hazards are often profound, resulting in significant loss of life, displacement of communities, and widespread destruction. Key examples of geological disasters include earthquakes, volcanic eruptions, and landslides, each presenting unique challenges and consequences based on their nature and magnitude. In addition to the immediate threats posed by these hazards, secondary effects, such as tsunamis, triggered by undersea earthquakes, and lahars, formed by volcanic ash mixing with water, can exacerbate the damage and extend the range of affected areas. For example, earthquakes not only cause ground shaking but may

Fig. 11.1 Geological disasters: earthquakes, volcanic eruptions, and landslides

also lead to soil liquefaction, where the soil temporarily behaves like a liquid, further destabilizing structures and causing extensive ground failure (Elliott & Oppenheimer, 2019). Human activity can sometimes intensify these hazards. For instance, mining operations and reservoir-induced seismicity—where the filling of large dams creates pressure on fault lines—can trigger or worsen seismic activity. Similarly, deforestation and poor land-use practices in mountainous regions can increase the frequency and severity of landslides. Furthermore, advancements in geotechnical and geophysical monitoring have improved hazard forecasting, allowing for better preparedness and mitigation strategies (Patton et al., 2022). Early warning systems, especially in earthquake-prone and volcanic regions, have become vital tools in reducing the human toll of such events, though complete prediction remains elusive. As populations expand and urbanize in areas vulnerable to geological hazards, the challenge of managing and mitigating the risks posed by these natural phenomena continues to grow. Recognizing the interplay between natural forces and human development is key to enhancing resilience and reducing future losses.

11.1.2 Meteorological Hazards

Meteorological hazards originate from processes within the Earth's atmosphere. These events, also known as atmospheric or weather-related disasters, are driven by atmospheric dynamics and weather phenomena (Fig. 11.2). They can often lead to substantial and often detrimental impacts on human societies, ecosystems, and economies (Smith et al., 2020). The causes of these hazards are linked to natural variability in atmospheric conditions as well as anthropogenic influences, such as

Fig. 11.2 Meteorological disasters: consequences of weather, water, and climate extremes

climate change, which amplifies the frequency and intensity of some extreme weather events (Jones & McCabe, 2021). Meteorological hazards encompass a wide range of extreme weather events, including extreme temperatures, heat waves, cold spells, hurricanes, tornadoes, droughts, and severe storms. Each of these phenomena presents unique challenges, with the potential to disrupt communities and inflict widespread damage. For instance, heat waves can increase mortality, particularly among vulnerable populations, and place stress on energy systems and water resources, while hurricanes and tropical storms can cause devastating flooding, wind damage, and coastal erosion (Williams & Burch, 2022). In addition to immediate destruction, meteorological hazards often have long-term consequences. Droughts, for example, can severely impact agriculture, leading to food shortages, economic losses, and mass displacement in the most affected regions (Taylor et al., 2019). Similarly, cold spells can disrupt transportation networks, cause energy shortages, and increase mortality rates due to hypothermia and respiratory illnesses. Advances in meteorology and climate science have improved early warning systems for these hazards, allowing for better preparedness and risk mitigation (Davis & Hayes, 2023). However, the growing unpredictability of weather patterns due to climate change poses a significant challenge for forecasting and managing these events. Extreme meteorological events are becoming more frequent and severe, especially with rising global temperatures, changes in precipitation patterns, and shifts in jet stream behavior, all of which can intensify the occurrence of hurricanes, droughts, and storms (Smith et al., 2020). Furthermore, urbanization and population growth in areas vulnerable to meteorological hazards—such as coastal regions susceptible to storm surges and inland regions prone to extreme heat—exacerbate the risk. As human-induced climate change continues to accelerate, the need for adaptive strategies, improved infrastructure, and international cooperation to mitigate the impacts of meteorological hazards is becoming increasingly critical (Jones & McCabe, 2021).

11.1.3 Hydrological Disasters

Hydrological disasters are water-related natural events or processes that have severe impacts on human societies, ecosystems, and economies. These disasters are closely tied to the Earth's hydrological cycle, which regulates the movement of water between the atmosphere, land, and oceans. Hydrological disasters can take many forms, including floods, droughts, and landslides, and they can affect regions worldwide, often with devastating consequences (Kundzewicz et al., 2020; Blöschl et al., 2019). Hydrological disasters are influenced by various factors, including climate variability, land-use changes, and human activities such as deforestation and urbanization (Dai et al., 2021). Climate change is expected to exacerbate the frequency and intensity of hydrological disasters, making effective management and mitigation strategies increasingly critical (IPCC, 2021).

Floods: Floods are among the most common and destructive hydrological disasters, resulting from excessive rainfall, snowmelt, or storm surges. They can lead to significant loss of life, displacement of communities, and extensive damage to infrastructure and property (Jongman et al., 2015). Conversely, droughts are characterized by prolonged periods of below-average precipitation, which can severely impact agriculture, water supply, and natural ecosystems. The economic implications of droughts can be profound, leading to food shortages and increased prices, as well as long-term ecological degradation (Mastrorillo et al., 2016). Typically caused by heavy rainfall, storms, rapid snowmelt, dam failures, or a combination of these factors, flooding results in the inundation of land. It leads to widespread destruction of infrastructure, displacement of populations, loss of crops, and damage to ecosystems.

Droughts: Defined by extended periods of below-average rainfall, high temperatures, and elevated evaporation rates, droughts create water shortages, agricultural failures, food insecurity, and economic disruptions and can even provoke social unrest in affected regions.

Landslides: Triggered by excessive rainfall, seismic activity, volcanic eruptions, or human activities that destabilize slopes, landslides can cause significant destruction of infrastructure, loss of life, alteration of landscapes, and disruption to transportation networks.

Storm Surges: Generated by powerful storms like hurricanes, typhoons, or tropical cyclones, storm surges occur when strong winds drive seawater onto coastal areas. These surges lead to coastal flooding, severe erosion, damage to buildings, and destruction of coastal ecosystems.

Flash Floods: Resulting from intense rainfall over a short period, especially in mountainous or urban regions with poor drainage, flash floods rapidly inundate low-lying areas, often with little warning. These events can cause high casualties and widespread damage.

Riverbank Erosion: Increased water flow, human interventions that alter river dynamics, or shifts in sediment transport can result in riverbank erosion. This process leads to the loss of agricultural land, displacement of communities, and changes in the shape and flow of rivers.

Tsunamis: Triggered by underwater earthquakes, volcanic eruptions, or landslides, tsunamis displace massive amounts of water, generating powerful waves that can devastate coastal areas. The resulting flooding causes widespread infrastructure damage, loss of life, and long-term economic disruption. To enhance resilience to hydrological disasters, early warning systems and comprehensive disaster risk management frameworks are essential. These measures can help communities prepare for and respond to water-related hazards, ultimately reducing their impacts on lives and livelihoods (López & Azevedo, 2022; WMO, 2020).

11.1.4 Biological Hazards

Biological hazards arise from the biological processes occurring on Earth and are predominantly associated with the spread of diseases and pests (Fig. 11.3). These hazards encompass a range of events, including epidemics, pandemics, and airborne diseases. Biological disasters—also referred to as biologically induced disasters—occur due to the emergence or proliferation of infectious diseases, harmful biological agents, or other biological threats that endanger human health and the environment (Patterson et al., 2017; Mackenzie et al., 2019). These disasters can stem from natural processes, human activities, or a combination of both. Examples include pandemics, epidemics, the emergence of new infectious diseases, bioterrorism, zoonotic diseases, vector-borne diseases, foodborne disease outbreaks, and environmental contamination. Each of these presents significant risks to public health and ecological stability (World Health Organization, 2020; Jones et al., 2021). The rise of global trade and travel, along with environmental changes, has facilitated the spread of biological hazards, making them more challenging to control and manage (Heymann et al., 2015). Climate change further exacerbates these risks, as it can alter the distribution of pathogens and vectors, leading to increased incidences of disease (Smith et al., 2014; Intergovernmental Panel on Climate Change, 2021). To mitigate the impacts of biological hazards, robust surveillance systems, public health initiatives, and international cooperation are essential. These measures are critical in detecting outbreaks early, implementing effective responses, and ultimately protecting public health (Bavari et al., 2019; Gollust et al., 2022).

Natural disasters inflict substantial material damage annually across various regions of the planet Earth. Figure 11.4 illustrates the global distribution of natural hazards based on statistical data from 1980 to 2005. According to the World Meteorological Organization (WMO), the hazard distribution is as follows: floods (32%), storms (22%), epidemics and famine (17%), droughts (9%), earthquakes (9%), landslides (5.3%), extreme temperatures (2.5%), fires (1.9%), volcanic eruptions (1.8%), and tsunamis (0.5%). This data indicates that nearly 90% of all natural disasters are of meteorological origin, driven by weather, climate, and water-related factors.

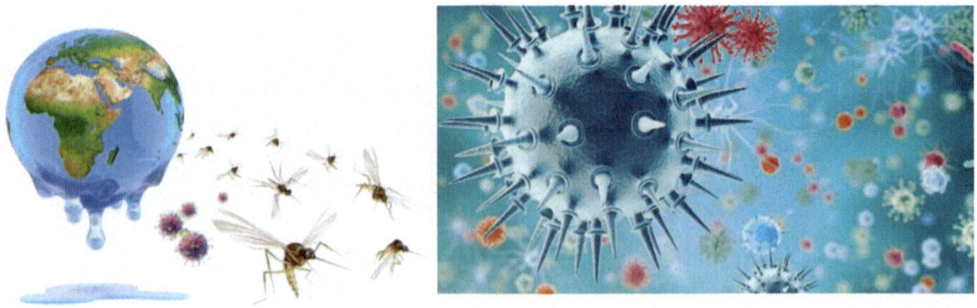

Fig. 11.3 Biological disasters: a spectrum of threats including pandemics, epidemics, emerging infectious diseases, bioterrorism, zoonotic outbreaks, vector-borne diseases, foodborne illnesses, and environmental contamination

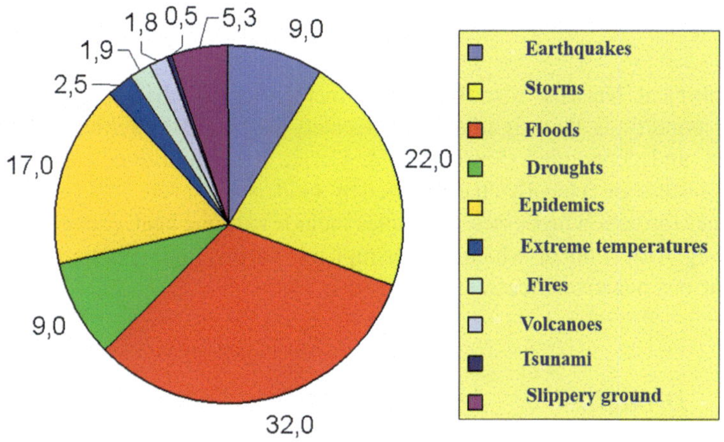

Fig. 11.4 Distribution of disasters resulting from weather, climate, and water (1970–2019) by category. (Source: WMO)

Human and material losses resulting from natural disasters present a significant barrier to sustainable development. Accurate forecasting and clear, understandable warnings, coupled with public education on preparedness, are crucial for mitigating the impact of these hazards before they escalate into full-blown disasters. Investing in disaster preparedness yields substantial benefits; for instance, every dollar spent on preparedness can prevent up to seven dollars in economic losses—a remarkable return on investment. Recognizing the ongoing global impact of natural disasters, the United Nations General Assembly designated the 1990s as the International Decade for Natural Disaster Reduction (IDNDR). The primary goal of this initiative was to reduce the loss of life, property destruction, and socioeconomic disruption caused by natural disasters, including earthquakes, tsunamis, floods, landslides, volcanic eruptions, droughts, and locust infestations. The decade focused on highlighting the importance of these phenomena and emphasized the need for targeted international action, especially in developing countries, to minimize disaster-related damage. This involved a systematic approach: experts were encouraged to study specific regions to develop procedures for continuous monitoring, setting disaster thresholds, and establishing early warning systems. These procedures aim to reduce fatalities and mitigate material damage. The approach distinguishes between passive and active measures of protection—where passive measures involve monitoring and reporting, and active measures involve proactive interventions to reduce adverse effects. In alignment with these goals, many countries have taken significant steps toward disaster risk reduction. For instance, Mexico, Canada, and the United States collaborated to form a team of 50 scientists tasked with developing a project to mitigate disaster damage over 3 years. More recently, as signatories to the Sendai Framework for Disaster Risk Reduction 2015–2030, member countries have committed to preventing new and reducing existing disaster risks through integrated and inclusive measures. These efforts focus on reducing hazard exposure, enhancing preparedness for response and recovery, and ultimately strengthening resilience.

11.2 Meteorological Hazards

Meteorological hazards, also known as meteorological disasters, are extreme weather conditions that significantly adversely impact the environment, human societies, and economies (Spiridonov & Ćurić, 2011; Spiridonov et al., 2013). These disasters are typically characterized by sudden and severe changes in weather or weather formation processes. Examples include extreme heat, excessive rainfall, and strong winds, all of which can disrupt the Earth's atmospheric balance and cause various negative effects.

11.2.1 Droughts

Droughts are extended periods of abnormally low precipitation that can occur anywhere in the world. This slow-onset phenomenon is characterized by a significant deficit of rainfall over an extended period. When drought leads to shortages of water and food, it can severely impact public health, increasing morbidity and potentially leading to fatalities. The primary cause of drought is a prolonged deficit in precipitation (Fig. 11.5), which results in insufficient water availability for various uses, including agricultural, ecological, and municipal needs. Droughts are often accompanied by other climatic factors such as high temperatures, strong winds, and low

Fig. 11.5 Precipitation deficit and the occurrence of drought in a specific area

relative humidity. Unlike sudden disasters, droughts develop gradually, sometimes over months or years, and are influenced by climate feedback mechanisms. Drought differs from aridity in that it is a temporary condition that may occur intermittently, whereas aridity is a persistent characteristic of low-rainfall regions. The impact of drought highlights the vulnerability of human communities to this natural hazard. Drought can be defined through two main approaches:

1. Conceptual Definition: This approach emphasizes the importance of drought and its effects on agriculture, food production, and other sectors.
2. Technical Definition: This approach focuses on identifying the onset and conclusion of droughts, assessing the severity, and determining deviations from average rainfall within a climatic period.

Several types of drought are recognized, each with distinct characteristics:

- Meteorological Drought: Defined by the degree of dryness relative to average conditions and the duration of the dry period. This definition must be region-specific, as atmospheric conditions influencing precipitation deficits vary by region.
- Agricultural Drought: Relates meteorological drought to agricultural impacts, including the lack of precipitation, discrepancies between actual and potential evaporation, water deficits in the soil, and reduced levels of underground or surface water.
- Hydrological Drought: Associated with prolonged low water flow and/or reduced water volume in watersheds and reservoirs. It may persist for months or years and can be exacerbated by human activities such as land-use changes and degradation.
- Socioeconomic Drought: Connects the availability of economic goods to meteorological, hydrological, and agricultural droughts. It occurs when the demand for goods like water, food, and energy exceeds supply due to weather-related disruptions. This type of drought reflects the impact of climate variability on economic resources and highlights the strain on supply systems during periods of inadequate water availability (Fig. 11.6).

Fig. 11.6 Socioeconomic drought and its impact on energy supply

11.2.2 Massive Floods from Tropical Cyclones

As was discussed in Chap. 10, tropical cyclones are intense weather systems characterized by low central pressure and a highly organized circulation pattern that forms over tropical and subtropical ocean waters (Fig. 11.7). These systems can span from 200 to 2000 km in diameter. A defining feature of tropical cyclones is their warm core, accompanied by a steep pressure gradient and powerful cyclonic winds. These winds, which rotate counterclockwise in the Northern Hemisphere and clockwise in the Southern Hemisphere, can reach speeds of up to 300 km/h.

One of the most severe impacts of tropical cyclones is their ability to generate large storm surges—significant increases in sea level caused by strong winds pushing seawater toward the coast. When combined with high waves, these surges can lead to catastrophic flooding in coastal regions. The heavy rainfall associated with tropical cyclones can also exacerbate flooding, leading to both flash floods and prolonged riverine or urban floods. The destructive force of tropical cyclones extends beyond flooding. The intense winds can cause widespread damage to infrastructure, disrupt maritime traffic, and result in significant property damage, including the destruction of homes. The combined effects of storm surges, heavy rainfall, and strong winds make tropical cyclones particularly devastating to coastal and inland areas alike.

11.2.3 Flash Floods

Flash floods are sudden, intense floods that occur within a short period, often within 6 h of heavy rainfall. They are typically triggered by weather systems such as cyclones, storms, fronts, or other cloud formations under unstable atmospheric conditions.

Fig. 11.7 Satellite image of a tropical cyclone. (Credit: Opportunity Desk)

While a single extreme weather event cannot be directly linked to climate change, increased intensity and frequency of such events are anticipated consequences of climate change, leading to more frequent and severe floods and droughts worldwide.

Flash floods result from a combination of intense rainfall, surface runoff, snowmelt, and local topography. The impact of heavy rainfall is exacerbated in areas where the ground is already dry, as the soil's limited absorption capacity can lead to rapid runoff and flooding. Flooding can range from small-scale flash floods to extensive inundation of large areas. It poses significant risks, including loss of life, damage to buildings and infrastructure, and destruction of crops and livestock.

Extreme weather events, such as severe storms and flash floods, also have broader health implications. In addition to direct injuries and fatalities from drowning and accidents, such events can lead to secondary health issues, such as increased incidence of gastrointestinal diseases and mental health challenges, including depression and post-traumatic stress disorders.

For example, the supercell storm that struck Skopje on August 6, 2016, resulted in catastrophic flooding, with 23 fatalities, extensive property damage, and significant economic losses (Fig. 11.8). Beyond immediate impacts, flash floods can contaminate drinking water sources, leading to waterborne diseases and creating breeding grounds for disease-carrying insects.

Inadequate response and preparation for such events can exacerbate their effects. Levees and dams may fail under the pressure of floodwaters, and the disruption of essential services, including medical care, can further compound the disaster's impact. Effective flood management requires addressing direct and indirect consequences to enhance resilience and preparedness for future extreme weather events.

11.2.4 Landslides

Landslides, also known as mass wasting or slope failures, are gravity-driven processes where materials such as rock, debris, sand, soil, or sediment are transported down a slope (Fig. 11.9). These events can occur suddenly and pose significant

Fig. 11.8 A scene from the catastrophic flood in Skopje on August 6, 2016, highlighting the severe impact on the city. (Source: https://mvr.gov.mk/)

Fig. 11.9 Landslides and slippery ground

threats and damage, especially in localized areas. Although landslides are not typically classified as meteorological disasters, they can be triggered or exacerbated by meteorological factors such as heavy rainfall, flash floods, rapid snowmelt, and associated warming and precipitation.

Weather events that contribute to landslides include intense rainfall, which increases the weight of the slope and reduces friction, making it easier for materials to move downward. Landslides are particularly common in regions with steep terrain, such as mountain slopes, river valleys, and areas near dams. These locations are more susceptible to landslide complexes, debris flows, and flash floods. The likelihood of a landslide occurring is heightened during the rainy season and is especially frequent during severe storms. The increased water content from rainfall can significantly weaken the stability of slopes, leading to the sudden downslope movement of materials. Landslides can cause extensive damage to infrastructure, disrupt transportation networks, and pose serious risks to human safety and property.

11.2.5 Avalanches

An avalanche is a rapid and often violent descent of snow, ice, and debris down a mountain slope (Fig. 11.10). These events can reach speeds exceeding 150 km/h and carry with them rocks and other debris, causing significant destruction. The force of the moving snow generates a powerful wind that can devastate infrastructure, buildings, forests, and mountain resorts. Each year, thousands of avalanches occur globally, resulting in approximately 500 fatalities. The destructive potential of avalanches underscores the need for effective monitoring and early warning systems in avalanche-prone areas.

Fig. 11.10 A snow avalanche cascading down the slopes of the Himalayas near Mount Everest, illustrating the immense power and scale of such natural events

Fig. 11.11 Desert landscape and sandstorm event. (Source: WMO)

11.2.6 Desert and Dust Storms

Desert and dust storms (Fig. 11.11) are phenomena where strong, turbulent winds lift large amounts of dust or sand into the atmosphere, creating dense, low-visibility conditions. These storms are most common in arid and semi-arid regions such as parts of Africa, Australia, China, and the United States. They pose significant health risks to individuals exposed to the dust, particularly those working outdoors without adequate protection. Additionally, dust storms severely impact transportation by reducing visibility to just a few meters, making travel hazardous and disrupting daily activities.

11.2.7 Heat Waves and Cold Spells

Extreme temperature events, such as heat waves and cold spells, pose significant challenges to both human health and infrastructure. These phenomena are characterized by prolonged periods of abnormally high or low temperatures, each bringing its own set of impacts and risks (Fig. 11.12).

Heat Waves: Heat waves represent a significant threat in regions of high latitudes, characterized by prolonged periods of extreme temperature and humidity. These conditions can persist for several days, particularly during the peak summer months. Heat waves typically occur under anticyclonic conditions, marked by stable, high-pressure systems that inhibit cloud formation and lead to intense solar heating. The advection of warm air into the region exacerbates the heat, resulting in exceptionally high temperatures. Additionally, stagnant air masses, especially over urban areas, can give rise to "heat islands," where localized temperatures remain elevated, often resulting in "tropical nights" with minimum temperatures not falling below 20 °C. The health impacts of extreme heat are profound, posing significant risks to vulnerable populations such as the young, elderly, and those with pre-existing health conditions. For instance, the summer of 2003 witnessed severe heat waves across Western Europe, with countries like France, Italy, the Netherlands, Portugal, Spain, and the United Kingdom experiencing a tragic loss of approximately 40,000 lives due to extreme temperatures.

Cold Spells: Conversely, cold spells bring about a range of health risks associated with hypothermia and exacerbation of circulatory and respiratory ailments. These phenomena are characterized by sustained periods of significantly lower-than-average temperatures, often accompanied by other adverse weather conditions such as strong winds and snow. Cold spells can lead to severe physiological stress, particularly for vulnerable groups, including the elderly, young children, and individuals with pre-existing health conditions. The impact of extreme cold is multifaceted, influencing not only human health but also infrastructure and energy demands, further compounding the challenges posed by such events.

Fig. 11.12 Heat waves and cold spells. Illustrations of extreme temperature events, depicting the characteristics and impacts of prolonged periods of intense heat and cold

11.2.8 Severe Storms, Lightning, and Tornadoes

Severe storms represent some of the most hazardous atmospheric phenomena, characterized by intense weather conditions and sudden changes in the environment. These storms are often associated with atmospheric electrical discharges, most notably lightning, which is a dramatic and powerful manifestation of the storm's electrical activity. Lightning is accompanied by thunder, a sound wave produced by the rapid expansion of air heated by the lightning strike. In addition to these electrical phenomena, severe storms typically bring heavy rain, strong winds, and in some cases, snow. The severity of these storms can vary widely, from localized heavy showers to widespread, destructive events. Tornadoes, a particularly violent and destructive type of storm, occur with notable frequency in certain regions, such as the Great Plains of North America. These rotating columns of air, extending from the base of a thunderstorm to the ground, can achieve wind speeds exceeding 500 km/h, causing catastrophic damage to structures, vegetation, and landscapes. While tornadoes are most common in the United States, particularly in "Tornado Alley," they can also occur in temperate latitudes and other parts of the world. The formation of tornadoes is often associated with severe thunderstorms and can lead to the development of flash floods, which exacerbate the damage by overwhelming drainage systems and inundating areas with rapid, high-volume runoff. Severe storms also contribute to widespread lightning strikes, which can be particularly dangerous during dry periods. In such conditions, lightning becomes a critical factor in igniting wildfires in forests and grasslands. These wildfires can spread quickly, fueled by dry vegetation and strong winds, leading to significant environmental and property damage.

In summary, severe storms, lightning, and tornadoes represent a complex interplay of atmospheric forces that can result in various forms of destructive weather. Understanding these phenomena is crucial for preparing for and mitigating the impacts of these potentially devastating events.

11.2.9 Forest Fires

Forest fires, also known as wildfires, represent a significant and often catastrophic natural hazard that can be triggered by both natural and anthropogenic causes. These fires can be ignited by lightning strikes or human activities, particularly during and after extended periods of drought, which can occur in almost any region of the world (Fig. 11.13). The conditions conducive to fire formation include extremely high temperatures, low relative humidity, and prolonged dry spells, which together create a highly flammable environment.

Recent advancements in fire science have highlighted the role of climate change in altering fire regimes. Increasing temperatures and shifting precipitation patterns are contributing to more frequent and intense fire events. For example, prolonged

Fig. 11.13 Forest fires: conditions, spread, and impacts

drought conditions linked to climate change can extend the fire season and increase the likelihood of severe fires. Moreover, the accumulation of dry vegetation, often exacerbated by changing land management practices, can provide ample fuel for these fires, making them more challenging to control. Wind is another critical factor that influences the behavior of forest fires. High winds can rapidly spread flames, carrying embers to new areas and creating spot fires far from the original blaze. This phenomenon, known as “fire whirls” or “fire tornadoes,” can result in explosive fire behavior, complicating firefighting efforts and increasing the risk to communities and ecosystems. The impact of forest fires extends beyond immediate destruction. They can lead to significant ecological changes, including the loss of habitat for wildlife and the alteration of soil properties. Post-fire environments often experience increased erosion and sedimentation, which can affect water quality in nearby rivers and streams. Additionally, the release of carbon dioxide and other greenhouse gases from burning vegetation contributes to global warming, creating a feedback loop that exacerbates the risk of future fires. Human health and safety are also critically impacted by forest fires. Smoke from fires can lead to air quality issues, affecting respiratory health and increasing the risk of cardiovascular problems. Communities near fire-prone areas may face evacuation orders, property loss, and long-term economic disruptions. In summary, forest fires are a complex and multifaceted hazard influenced by a range of natural and human factors. Advances in understanding the dynamics of fire behavior, coupled with the impacts of climate change, underscore the need for effective fire management and mitigation strategies to protect ecosystems, communities, and public health.

11.2.10 Heavy Rain, Snowstorms, and Strong Winds

Heavy rain, snowstorms, and strong winds are severe weather conditions that pose significant risks to both human safety and infrastructure. These phenomena can disrupt daily life, impact transportation, and cause extensive property damage.

Fig. 11.14 Heavy snowfall: impacts on infrastructure and safety

Heavy Rain: Heavy rainfall can lead to a range of problems, particularly for vulnerable groups such as the elderly, young children, and those with pre-existing health conditions (Fig. 11.14). Intense downpours can overwhelm drainage systems, leading to flash floods and prolonged flooding in urban and rural areas. This can result in significant damage to homes, infrastructure, and agricultural land. Flooding can also disrupt transportation networks, impede communication systems, and pose health risks due to contaminated water.

Snowstorms: Snowstorms, characterized by heavy snowfall and often accompanied by strong winds, present a serious challenge to infrastructure and daily activities. Accumulation of snow can lead to the collapse of building roofs and structural damage, particularly in areas not designed to handle large amounts of snow. Snowstorms can also cause major disruptions to transportation systems, with roads becoming impassable and airports closing for takeoffs and landings. This can lead to widespread travel delays and economic impacts.

Strong Winds: Strong winds are a dangerous weather phenomenon affecting various sectors, including aviation, maritime activities, and construction. High winds can pose serious risks to pilots, sailors, and fishermen, as well as cause damage to tall structures such as towers, cranes, and skyscrapers. Wind speeds capable of reaching hurricane force can lead to the overturning of vehicles, uprooting of trees, and damage to power lines, resulting in power outages and disrupted services.

Blizzards: Blizzards are severe snowstorms characterized by extremely low temperatures, high winds, and heavy snowfall. The combination of these factors can create whiteout conditions, drastically reducing visibility and making travel extremely hazardous. Blizzards pose significant dangers to both people and livestock, leading to potential hypothermia and frostbite for those caught outdoors. In addition to their immediate impact, blizzards can lead to long-term disruptions in transportation and supply chains.

11.2.11 Extreme Air Pollution

Extreme air pollution is a significant environmental and public health concern, encompassing a range of harmful substances and gases emitted from industrial processes, vehicles, and various human activities. Atmospheric pollutants include particulate matter, volatile organic compounds (VOCs), nitrogen oxides (NO_x), sulfur dioxide (SO_2), and carbon monoxide (CO), all of which contribute to the degradation of air quality.

In addition to emissions from industrial and vehicular sources, natural events such as forest fires, agricultural practices like grain clearing, and volcanic eruptions also release pollutants into the atmosphere. During periods of stable atmospheric conditions, these pollutants can accumulate, leading to smoke and haze that significantly reduce visibility (Fig. 11.15). The persistence of these pollutants in the air can have severe health impacts, particularly in urban and industrial areas where pollution levels are highest. Air pollution affects human health in numerous ways, including respiratory and cardiovascular problems. Exposure to high levels of pollutants can lead to chronic conditions such as asthma, bronchitis, and other respiratory diseases.

Vulnerable populations, such as children, the elderly, and individuals with preexisting health conditions, are particularly at risk. In addition to health impacts, air pollution has broader environmental consequences. It can lead to the formation of smog, a mixture of smoke and fog that reduces visibility and poses health risks. Acid rain, resulting from the reaction of sulfur dioxide and nitrogen oxides with water vapor, can damage ecosystems, soil, and water bodies. Ground-level ozone, a harmful air pollutant formed by the reaction of VOCs and NO_x in the presence of sunlight, contributes to respiratory issues and damages crops. The depletion of the ozone layer, driven by pollutants such as chlorofluorocarbons (CFCs), allows increased ultraviolet (UV) radiation to reach the Earth's surface, impacting human health and ecosystems. Moreover, air pollution plays a role in enhancing the greenhouse effect, leading to global warming. Greenhouse gases such as carbon dioxide and methane trap heat in the Earth's atmosphere, contributing to climate change and

Fig. 11.15 Air pollution in urban areas and industrial zones: sources and effects

its associated impacts. A persistent high-pressure system, often associated with stable atmospheric conditions and temperature inversions, can exacerbate pollution episodes. Temperature inversions occur when a layer of warmer air traps pollutants close to the ground, leading to periods of extreme air pollution with high concentrations of harmful substances. This is particularly problematic in urban areas with specific topographic features, such as valleys or basins, where pollutants can become trapped and accumulate. The combination of these factors results in episodes of poor air quality and significant health risks for residents.

In summary, extreme air pollution is a complex issue with widespread implications for health, the environment, and visibility. Addressing this challenge requires comprehensive strategies for pollution control, monitoring, and public awareness to mitigate its effects and improve air quality.

11.2.12 Tsunami

A tsunami is a powerful and potentially devastating wave phenomenon often generated by a variety of meteorological or atmospheric disturbances. The term "tsunami" originates from the Japanese words "Tsu," meaning "Harbor," and "Nami," meaning "Wave," reflecting the historical impact of such waves on coastal harbors and communities. Unlike typical ocean waves, tsunamis are characterized by their long wavelengths and high energy, allowing them to travel across entire ocean basins. Tsunamis can be triggered by several types of atmospheric disturbances, including atmospheric gravity waves, pressure oscillations, and frontal passages. These disturbances are often associated with severe weather events such as typhoons, hurricanes, and intense precipitation systems. The atmospheric pressure changes induced by these disturbances generate barotropic ocean waves in the open ocean. Barotropic waves, which are influenced by the pressure and density of the water column, can propagate across vast distances with minimal loss of energy. In addition to atmospheric sources, tsunamis can also be caused by underwater seismic activity, such as earthquakes, volcanic eruptions, and underwater landslides. These events displace large volumes of water, creating waves that travel at high speeds across the ocean. As tsunamis approach shallow waters near the coast, their speed decreases, and their wave heights increase due to the process of wave shoaling. This phenomenon is often amplified by coastal features such as bays, estuaries, and coastal basins, where the wave energy can become concentrated. Resonance effects, where the natural frequency of the coastal basin matches the frequency of the incoming waves, can further amplify the wave heights. This results in dramatic and potentially catastrophic waves that are often associated with tsunami events. The impact of a tsunami can be devastating, causing widespread damage to coastal infrastructure, flooding, and loss of life. The waves can inundate large areas of land, destroying buildings, infrastructure, and ecosystems. The force of the water can also lead to significant changes in coastal geomorphology, including erosion and sediment deposition. Understanding the generation and propagation of tsunamis is

Fig. 11.16 Tsunamis: gravity ocean waves and their propagation

crucial for effective preparedness and response. Early warning systems, which monitor seismic activity and sea-level changes, play a vital role in providing advance notice to coastal communities. Education and public awareness campaigns are essential for ensuring that individuals are aware of tsunami risks and know how to respond in the event of an impending tsunami. In summary, tsunamis are complex and formidable natural phenomena with the potential for severe impact on coastal areas. They are generated by a variety of atmospheric and geological disturbances and can cause significant damage when they reach the shore. Effective monitoring, preparedness, and response are critical to mitigating the risks associated with tsunamis and protecting communities from their devastating effects.

Figure 11.16 illustrates the nature of tsunami waves, including their generation by underwater disturbances and their propagation across the ocean. It highlights the destructive potential of these waves upon reaching coastal areas.

11.3 Notable Historical Natural Disasters: Impact and Lessons

Natural disasters such as earthquakes, flash floods, forest fires, heat waves, and others are among the most impactful phenomena shaping the globe, often leaving profound consequences on human life, infrastructure, and ecosystems. Throughout history, the world has been shaped by numerous natural disasters that have caused profound devastation and loss of life. The severity and impact of these events can vary significantly depending on factors such as location, population density, and the level of preparedness. The information presented here is compiled from credible

Internet-based evidence and reflects the most significant events documented to date. However, it is important to recognize that numerous other notable natural hazards have also occurred throughout history, many of which have equally substantial consequences. Readers seeking more comprehensive or specific information on natural disasters are encouraged to explore additional reputable web portals and specialized resources. These platforms offer detailed insights and updated data that can help deepen understanding of the complex and diverse nature of Earth's dynamic processes. Below are listed the most catastrophic natural disasters in history, each representing a different type of disaster and illustrating the diverse ways in which such events can impact human societies and the environment:

1. The 1931 China Floods: Yangtze River Flood
2. Southern Chile Earthquake and Tsunami, May 22, 1960
3. The Bhola Cyclone, November 11, 1970
4. The Great Hurricane, October 10–16, 1780
5. Mount Tambora Eruption (1815)
6. The Peshtigo Fire: The Deadliest Fire in Human History, October 8, 1871
7. The 2003 European Heatwave
8. The Machchu-2 dam failure during a flash flood in 1979
9. Daulatpur–Saturia tornado in Bangladesh on April 26, 1989
10. The deadliest hailstorm in Moradabad, April 30, 1888
11. A single deadliest lightning strike in eastern Rhodesia (now Zimbabwe), December 23, 1975

11.3.1 The 1931 China Floods: Yangtze River Flood

The **1931 China Floods**, particularly along the **Yangtze River**, are regarded as one of the deadliest natural disasters in recorded history (Fig. 11.17). The floods were triggered by a combination of excessive rainfall, snowmelt, and the overflow of the Yangtze, Yellow, and Huai Rivers. Occurring between June and November 1931, the floods affected vast areas of central and eastern China, submerging over 400,000 square kilometers of land and displacing millions.

The death toll from the floods is estimated between one million and four million people, with most fatalities resulting from drowning, disease, and malnutrition. The destruction of farmland, especially rice fields, led to widespread famine and a collapse in agriculture, further exacerbating the crisis. The floodwaters inundated major cities like Nanjing and Wuhan, causing extensive damage to infrastructure and displacing more than 50 million people, making it one of the largest population displacements in history.

The catastrophic floods not only caused immense human suffering but also had significant social and political repercussions. The disaster underscored the vulnerability of densely populated riverine areas and highlighted the urgent need for improved flood management and disaster preparedness. The aftermath of the floods

Fig. 11.17 The devastating 1931 China floods on the Yangtze River, which caused over a hundred thousand deaths and threatened millions with starvation due to the destruction of rice fields. The floodwaters submerged vast agricultural areas, wiping out the main food source for millions. Emergency aid, including food distribution by the International Red Cross, was provided to help the starving population. Attribution: Bundesarchiv, Bild 102-12231/CC-BY-SA 3.0

played a role in shifting China's political landscape, particularly influencing the Nationalist government's policies. The 1931 floods remain one of the most devastating natural disasters of the twentieth century, with lasting impacts on China's infrastructure, agriculture, and population.

11.3.2 Southern Chile Earthquake and Tsunami, May 22, 1960

The largest recorded earthquake in history is the 1960 Valdivia Earthquake, which struck Southern Chile on May 22, 1960. Measuring 9.5 on the moment magnitude scale, it remains the most powerful earthquake ever documented (Fig. 11.18). The epicenter was located near Lumaco, around 570 km south of Santiago. This seismic event caused widespread destruction and a tragic loss of life. The earthquake triggered a devastating tsunami that affected Chile's coastline and reached distant regions such as Japan, the Philippines, New Zealand, and Hawaii. In some areas, the waves reached heights of up to 25 m, significantly exacerbating the destruction. The disaster resulted in over 1000 fatalities, thousands of injuries, and the displacement of around 2 million people in Chile. This earthquake occurred at the boundary between the Nazca Plate and the South American Plate, where the Nazca Plate

Fig. 11.18 The 1960 Valdivia Earthquake. The largest recorded earthquake in history, the 1960 Valdivia Earthquake in Chile, struck on May 22, 1960, with a magnitude of 9.5. It caused widespread destruction, triggered a devastating tsunami with waves up to 25 m high, and affected distant locations like Japan and Hawaii. The event resulted in over 1000 fatalities, thousands of injuries, and the displacement of around 2 million people. (Credit: Pierre St. Amand—NGDC Natural Hazards Slides with Captions Header, Public Domain, https://commons.wikimedia.org/w/index.php?curid=630576)

subducts beneath the South American Plate. Numerous significant aftershocks, some exceeding a magnitude of 7.0, followed the main event. The 1960 Valdivia Earthquake remains a critical case study in the field of plate tectonics and seismic activity, given its unparalleled magnitude and the extensive impact it caused.

11.3.3 The Bhola Cyclone, November 11, 1970

The 1970 Bhola cyclone is considered one of the deadliest tropical cyclones in recorded history. It struck on November 11, 1970, impacting East Pakistan (now Bangladesh) and parts of West Bengal in India. The cyclone was a Category 4 storm on the Saffir–Simpson scale, with winds reaching up to 115 mph (185 km/h).

However, its greatest devastation came from the storm surge and flooding that followed, severely affecting the low-lying regions of East Pakistan. Figure 11.19 shows a satellite image in black and white of the Bhola cyclone that occurred on November 11, 1970.

The death toll from the cyclone is estimated to be between 300,000 and 500,000 people, making it the deadliest tropical cyclone ever recorded. The majority of casualties were caused by the storm surge, which inundated coastal villages and towns (Fig. 11.20). The massive flooding left millions of people homeless and caused extensive damage to infrastructure, including roads, bridges, and buildings, complicating relief efforts.

The cyclone also led to a severe agricultural crisis, with crops destroyed and farmland washed away. The aftermath of the disaster, coupled with the delayed response from the government, intensified political unrest in East Pakistan. This

Fig. 11.19 Image of the 1970 Bhola cyclone taken on November 11, 1970. (Credit: From E.C. Barrett's Climatology from Satellites (London: Methuen, 1974), ISBN 0416659403))

Fig. 11.20 Aerial view of the devastation on Manpura Island, East Pakistan, following the 1970 Bhola cyclone. The tidal wave, which struck on November 12, 1970, left nearly 16,500 people dead and destroyed much of the island's infrastructure, including rice crops and fishing boats. Survivors, like local councilor Mr. Choudhary Kamal, faced dire conditions with limited food and clean water. (Courtesy: Photograph: Harry Koundakjian/AP)

discontent eventually played a role in the Bangladesh Liberation War and the country's subsequent independence in 1971.

The 1970 Bhola cyclone prompted significant changes in the way tropical cyclones were forecasted and in the development of early warning systems. It also highlighted the need for better disaster preparedness in vulnerable coastal areas.

11.3.4 The Great Hurricane, October 10–16, 1780

The Great Hurricane of 1780, also known as Hurricane San Calixto, is considered the deadliest hurricane in recorded history, with a staggering death toll that likely surpassed 22,000 people. This catastrophic event took place in the Caribbean, affecting several islands including Barbados, Saint Lucia, and Martinique, when the region had little to no warning systems or knowledge of impending storms.

The storm's path and intensity have been described as extraordinary. The hurricane's winds are believed to have reached well over 150 mph, placing it at the top of the Saffir–Simpson scale—Category 5, which is the strongest category on the modern scale. The hurricane's impact was compounded by a massive storm surge, causing extensive flooding and widespread destruction, wiping out towns, plantations, and infrastructure across the islands it struck. Thousands of people lost their homes and lives, as the high winds, torrential rains, and flooding overwhelmed both local populations and ships in the area (Fig. 11.21).

The hurricane also devastated the natural environment, completely flattening large swathes of forests and destroying crops, which led to a long-lasting economic

Fig. 11.21 Representation of the tragic loss of His Majesty's Ship Stirling Castle, dismasted during the Great Hurricane of 1780 (Hurricane San Calixto), which devastated the Caribbean on October 6. The storm, one of the deadliest in recorded history, caused widespread destruction, dismasting ships like the Stirling Castle and wrecking vessels near Hispaniola's Silver Keys. The hurricane claimed thousands of lives and left a lasting impact on the region. (Credit: By Lieutenant William Elliott; Valentine Green—http://collections.rmg.co.uk/collections/objects/140679, Public Domain, https://commons.wikimedia.org/w/index.php?curid=63134107)

impact. Many ships that had been docked in the Caribbean, including British warships and merchant vessels, were destroyed, causing significant maritime losses. The scale of the destruction and loss of life was so severe that it left a lasting mark on the historical records of the region.

In addition to the immediate physical toll, the Great Hurricane of 1780 had significant social and political implications. At the time, the Caribbean islands were under colonial rule, and the devastation of so many islands disrupted trade and colonial activities. The hurricane was part of a series of storms that plagued the region during that period, but none was as deadly or destructive.

The storm remains an important historical event, highlighting the immense power of nature and the vulnerability of coastal populations, especially during an era when storms were not tracked, and forecasting was nonexistent. The aftermath of the hurricane led to changes in how people understood tropical storms, though it would take many more decades before the science of meteorology advanced enough to provide effective early warnings or reliable forecasts.

11.3.5 Mount Tambora Eruption, April 10, 1815

The most powerful volcanic eruption in recorded history is the eruption of Mount Tambora in 1815. Located on the island of Sumbawa in present-day Indonesia, the eruption began on April 5, 1815, and culminated in a colossal explosion on April 10 (Fig. 11.22). This eruption is classified as a VEI 7 event, the highest possible level on the Volcanic Explosivity Index. The eruption caused the collapse of the volcano's

Fig. 11.22 Eruption of Mount Tambora in Indonesia. (Photo: Ron Wood from the cover of "A Year Without Summer")

summit, creating a caldera and releasing an immense volume of volcanic ash, gases, and aerosols into the stratosphere. This massive injection of particles led to the scattering of sunlight and a significant reduction in solar radiation reaching the Earth's surface.

The global climatic effects were profound, resulting in 1816 being dubbed "The Year Without Summer" or "Eighteen Hundred and Froze to Death." The year saw drastic global cooling, with unusual and extreme weather events such as widespread crop failures, summer frost, and even snowfall in northern hemisphere regions during June and July. The eruption's impact was felt worldwide, disrupting weather patterns and agricultural productivity across North America, Europe, and Asia.

The eruption of Mount Tambora had devastating effects on global temperatures and agriculture, leading to widespread food shortages and famine in various regions. The local population of Sumbawa suffered immensely, with tens of thousands of people perishing from the eruption's direct impacts, including pyroclastic flows and tsunamis. The resulting crop failures and ensuing famine triggered social unrest and migrations.

This catastrophic event was pivotal in advancing the early understanding of volcanic eruptions and their global consequences. It significantly contributed to the development of volcanology and highlighted the crucial link between volcanic activity and climate change. The 1815 eruption of Mount Tambora remains a landmark event in Earth's history, noted for its profound and enduring effects on both climate and human societies.

11.3.6 The Peshtigo Fire: The Deadliest Fire in Human History, October 8, 1871

The Peshtigo Fire, which occurred on October 8, 1871, is the deadliest wildfire in recorded history. It ravaged northeastern Wisconsin, claiming the lives of an estimated 1500–2500 people, though the exact death toll remains uncertain due to the destruction of many communities and a lack of records. The fire coincided with the Great Chicago Fire, which drew national attention away from the Peshtigo tragedy despite its far greater loss of life.

The fire was fueled by a combination of human activity and natural conditions. The Peshtigo region was a hub for logging, and the practice of slash-and-burn clearing left large amounts of dry, flammable debris. Weeks of drought had dried the landscape, creating tinderbox conditions. On the evening of October 8, strong winds fanned a small, controlled fire into an unstoppable inferno. The firestorm generated its weather, with flames reportedly reaching temperatures of up to 2000 °F (1093 °C). The intense heat caused trees to explode and superheated the air, killing many victims instantly.

The firestorm advanced so rapidly that residents had little time to react. Entire towns, including Peshtigo, were reduced to ashes within hours. Survivors described

Fig. 11.23 The Peshtigo Fire, Wisconsin, October 8, 1871. This wood engraving from Harper's Weekly depicts the deadliest wildfire in history, which claimed up to 2500 lives and destroyed over 1.2 million acres. (Credit: By Unknown author—Grainger Collection, Public Domain, https://commons.wikimedia.org/w/index.php?curid=73318156)

harrowing scenes of people fleeing toward rivers and wells, often succumbing to burns or suffocation as the fire consumed the oxygen around them (Fig. 11.23). Many who sought refuge in the Peshtigo River or other water sources survived but endured severe burns and trauma. The devastation was immense, with more than 1.2 million acres of forest, farmland, and settlements destroyed. Thousands of buildings were leveled, and entire families were lost. The Peshtigo Fire remains a stark reminder of the destructive power of wildfires, the role of human activity in exacerbating natural hazards, and the importance of fire safety and forest management practices.

11.3.7 The 2003 European Heatwave

Heat waves have become some of the most catastrophic and deadly natural disasters in history, with their frequency and intensity increasing due to rising global temperatures and climate change. These extreme weather events not only cause thousands of fatalities but also lead to widespread destruction of crops, ecosystems, and infrastructure. Some of the most deadly and devastating heat waves globally include:

The summer of 2003 witnessed an extraordinary heat wave in Europe, which became one of the deadliest in modern history. The heat wave affected much of Western and Southern Europe, with the most severe conditions observed in France (Fig. 11.24). Temperatures soared to 40 °C (104 °F) and higher, with some regions

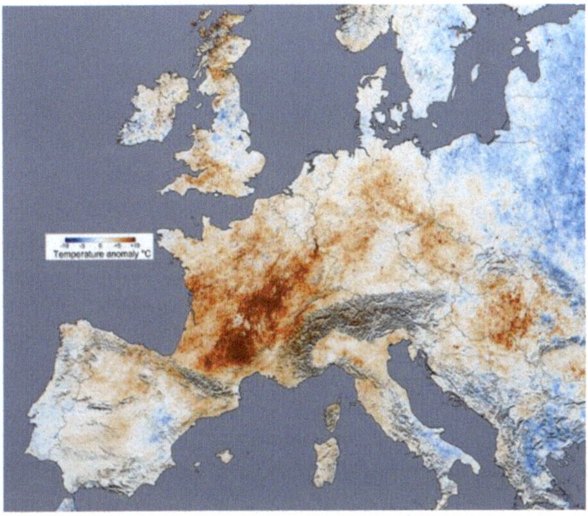

Fig. 11.24 Temperature variations during the 2003 European heatwave relative to July 2001. This image, derived from MODIS land surface temperature data on NASA's Terra satellite, shows the significant increase in daytime temperatures across Europe in July 2003. The differences are calculated by subtracting the average of cloud-free temperatures from 2000 to 2002 and 2004 from those measured in 2003 (July 20–August 20). (Credit: By "Image courtesy Reto Stockli and Robert Simmon, based upon data provided by the MODIS Land Science Team." http://earthobservatory.nasa.gov/IOTD/view.php?id=3714 (image), Public Domain, https://commons.wikimedia.org/w/index.php?curid=450988)

exceeding 45 °C (113 °F). The heat wave led to the deaths of more than 70,000 people across Europe, the majority of whom succumbed to heatstroke, dehydration, and other heat-related illnesses. France was particularly impacted, where the elderly population, lacking proper cooling facilities and vulnerable to the intense heat, saw a sharp rise in mortality rates.

In addition to the tragic loss of life, the heat wave caused widespread environmental damage. Prolonged high temperatures led to wildfires, the destruction of crops, and significant droughts, which further worsened the economic situation. France's rice and grape harvests were particularly affected, resulting in a reduced food supply and long-term economic challenges. The event highlighted the vulnerabilities of aging populations and underscored the importance of preparing for future extreme heat events, particularly with climate change.

11.3.8 The Machchu-2 Dam Failure During a Flash Flood in 1979

The Machchu-2 dam failure in 1979 is considered one of the worst and most catastrophic flash floods in history. Located near the town of Morbi in Gujarat, India, the dam collapsed after heavy rainfall in the upstream catchment area, resulting in a sudden release of water that flooded the surrounding region. The disaster struck in the early hours of August 11, 1979, when the dam's structural integrity was compromised by excessive water pressure.

The floodwaters rapidly surged through the town of Morbi and nearby villages, completely overwhelming the infrastructure. The town, which was home to around 100,000 people, saw most of its population swept away in the torrent. The Machchu-2 dam's failure not only affected Morbi but also inundated numerous surrounding villages, many of which were also washed away by the floodwaters.

The estimated death toll from the disaster varies widely, with figures ranging from 1800 to 25,000 people, making it one of the deadliest flash floods in recorded history. Most of the casualties were due to drowning, as the town was submerged in a matter of hours. In addition to the human toll, thousands of people were left injured and displaced, while vital resources such as food, water, and medical supplies were scarce.

The flood's impact extended beyond human suffering. It caused severe damage to infrastructure, including homes, roads, bridges, and agricultural land. The town's economy, which was heavily reliant on local industries and farming, was devastated. The rice and cotton crops in the affected regions were destroyed, and the damage to livestock further exacerbated the situation.

Moreover, the dam failure highlighted the dangers of poorly managed water resources and the need for improved infrastructure and disaster preparedness. Investigations into the cause of the dam's collapse revealed that a combination of factors, including inadequate dam maintenance, poor construction standards, and the overwhelming amount of rainfall, contributed to the disaster.

The Machchu-2 dam disaster is a poignant reminder of the potential dangers of large-scale infrastructure failures and the importance of proper engineering, maintenance, and disaster management to prevent such catastrophic events.

11.3.9 Daulatpur–Saturia Tornado in Bangladesh on April 26, 1989

The Daulatpur–Saturia tornado in Bangladesh, which struck on April 26, 1989, is one of the deadliest tornadoes in history. It affected the central regions of Bangladesh, particularly the towns of Daulatpur and Saturia, located in the Manikganj District, and is considered the most catastrophic tornado to ever hit the country (Fig. 11.25).

The tornado was an extremely powerful storm that caused widespread devastation, with wind speeds reaching over 200 mph (320 km/h). It cut a path through the area, destroying thousands of homes and buildings, and leaving behind a trail of destruction. The tornado's intensity and duration contributed to the heavy damage, which was worsened by the dense population and lack of sufficient infrastructure to cope with such extreme weather.

The death toll from this tornado is estimated to be between 1300 and 1500 people, with many more injured. The tornado also displaced over 50,000 people and

Fig. 11.25 The Daulatpur–Saturia Tornado (Bangladesh, April 26, 1989). The deadliest tornado in history, killing over 1300 people and causing extensive destruction in the Dhaka region. (Credit: https://www.britannica.com/event/Saturia-Manikganj-Sadar-tornado)

caused significant damage to crops, particularly rice and other agricultural products that are vital to the local economy. The destruction of homes, public buildings, and infrastructure left the affected areas in dire need of aid and support.

In addition to the physical devastation, the tornado caused a public health crisis, with many people suffering from injuries and a lack of basic medical care. Rescue operations were complicated by the widespread damage to roads and communication systems. The Daulatpur–Saturia tornado is a stark reminder of the vulnerability of densely populated regions to severe weather events, particularly in the case of tornadoes, which can cause immense damage and loss of life in a short amount of time. The disaster led to increased awareness and efforts to improve weather forecasting, early warning systems, and disaster preparedness in Bangladesh.

11.3.10 The Deadliest Hailstorm in Moradabad, April 30, 1888

The deadliest hailstorm in recorded history occurred on April 30, 1888, in Moradabad, a region in northern India. This devastating storm unleashed incredibly large hailstones, some reaching up to 13 cm (5 in.) in diameter. The storm struck with tremendous force, causing widespread destruction across the area (Fig. 11.26). Over 246 people tragically lost their lives, with most fatalities resulting from blunt force trauma caused by the massive hailstones. In addition to the human toll, the storm inflicted significant damage to buildings, crops, and livestock, further compounding the disaster. The hailstones were so large and heavy that they destroyed

Fig. 11.26 The 1888 Moradabad hailstorm in India, which killed 230 people and 1600 farm animals, is recognized as one of the deadliest. Hailstones up to 2 ft high, accompanied by squally winds, caused widespread destruction. (Credit: https://www.skymetweather.com/)

roofs, smashed windows, and caused serious injuries to those caught outside, especially those working in the fields or traveling.

The storm also had a major impact on the local agricultural economy, as it devastated crops just before the harvest season. The loss of both human life and livelihoods made this event one of the most catastrophic natural disasters in the history of India. Despite the lack of modern meteorological data, reports from the time suggest the storm lasted for several hours and affected not only Moradabad but also nearby regions. The 1888 Moradabad hailstorm remains a stark reminder of the immense power of nature, and it is still regarded as one of the deadliest and most destructive hailstorms ever recorded.

11.3.11 A Single Deadliest Lightning Strike in Eastern Rhodesia (Now Zimbabwe), December 23, 1975

The lightning strike that killed 21 people on December 23, 1975, in the Manica Tribal Trust Lands in eastern Rhodesia (now Zimbabwe), remains one of the deadliest lightning strikes in history. The incident occurred when a group of people sought shelter in a traditional hut during a thunderstorm. As the storm intensified, a powerful lightning bolt struck the hut directly, causing an immediate and devastating impact. Tragically, all 21 individuals inside the hut, including men, women, and children, were killed instantly. This lightning strike was particularly catastrophic due to the density of the strike and the vulnerability of the shelter. The hut, typically made of wood and thatch, offered little protection against the powerful forces of nature. The news of the incident spread quickly and was reported globally.

The primary sources documenting this event include a Reuters News Service release on December 24, 1975, and a listing in the Guinness Book of Records, which highlighted the number of casualties. It is regarded as one of the deadliest lightning events in recorded history, and its occurrence in a rural, remote area added to the tragedy, as it took emergency response teams some time to reach the site. This incident underscores the extreme danger posed by lightning strikes, especially in areas where shelters provide minimal protection and in communities that may lack access to lightning safety information.

References

Bavari, S., et al. (2019). The role of preparedness in mitigating biological hazards. *Nature Reviews Microbiology*.

Blöschl, G., et al. (2019). *Changing climate and flood risk in Europe*. Cambridge University Press.

Dai, A., et al. (2021). Drought under global warming: A review of recent advances in drought research. *Journal of Climate*.

Davis, K., & Hayes, B. (2023). *Advancements in meteorology: Early warning systems and climate resilience*. Cambridge University Press.

Elliott, J. R., & Oppenheimer, C. (2019). *Volcanic hazards: A global assessment of volcanic risk*. Wiley-Blackwell.

Gollust, S. E., et al. (2022). Public health communication in times of crisis: Lessons from the COVID-19 pandemic. *Health Affairs*.

Gonzalez, A., Perez, R., & Martinez, S. (2021). *Natural hazards and disaster management*. Wiley.

Heymann, D. L., et al. (2015). *Controlling infectious disease outbreaks: A global perspective*. Springer.

Intergovernmental Panel on Climate Change. (2021). *Climate change and health: Impacts and adaptation*. Cambridge University Press.

IPCC. (2021). *Climate change 2021: The physical science basis*. Cambridge University Press.

Jones, L., & McCabe, M. (2021). *Climate change and meteorological hazards: An overview*. Routledge.

Jones, K. E., et al. (2021). Global trends in infectious disease emergence. *National Academy of Sciences of the United States of America*.

Jongman, B., et al. (2015). *Reducing the risk of flood disasters: Trends and challenges*. World Bank.

Khakzad, N. (2020). *Vulnerability assessment of process vessels in the event of hurricanes*. IntechOpen.

Kundzewicz, Z. W., Mata, L. J., Wyżyński, P., & Zhai, P. (2020). *Flood risk management: From science to practice*. Springer.

López, J. C., & Azevedo, P. (2022). *Early warning systems for water-related disasters: Best practices and guidelines*. Springer.

Mackenzie, J. S., et al. (2019). *Emerging infectious diseases: A global perspective*. Wiley.

Mastrorillo, M., et al. (2016). The impact of drought on agricultural and economic outcomes. *Environmental Research Letters*.

Mokhtari, M., Faridi, P., Masoodi, M., & Ahmadi, S. M. (2022). *A global view of natural hazards related disasters*. IntechOpen.

Nedaei, M. (2021). *New insights into major seismic events*. IntechOpen.

Patterson, C. M., et al. (2017). *Biological hazards: Assessment and response*. Cambridge University Press.

Patton, J., Monfort, M., & Decker, R. (2022). *Geotechnical and geophysical monitoring for natural hazards*. Springer.

Richardson Varas, R. (2020). *Risk reduction*. IntechOpen.

Schwartz, S. Y., Leith, K., & Nunn, J. (2021). *Geological hazards: A global perspective*. Cambridge University Press.

Sharma, N., Saini, R., Sreehari, K., Pathaikara, A., Punde, P., & Attada, R. (2020). *Hydrological extremes in Western Himalayas*. IntechOpen.

Smith, K. F., et al. (2014). Globalization of human infectious disease: A review. *Journal of Globalization and Health.*

Smith, J., Roberts, A., & Clark, P. (2020). *Understanding weather disasters: A comprehensive guide*. Academic Press.

Smith, J., Johnson, L., & Brown, T. (2021). *Understanding natural hazards: A comprehensive guide*. Springer.

Smith V. K., Carbone J. C., Pope J. C., Hallstrom D. G., Darden M. E. (2022) Adjusting to natural disasters. The economics of environmental risk. Edward Elgar Publishing, pp 297–314. https://doi.org/10.4337/9781035301614.00030.

Spiridonov, V., & Ćurić, M. (2011). An introduction to meteorology (pp. 241). ISBN 978-608-65175-0-2 COBISS.MK-ID 83607306.

Spiridonov, V., Ćurić, M., Zafirovski, O. (2013). Weather and human health - Skopje : V. Spiridonov, 2013, pp. 345, ISBN - 978-608-65175-2-6, COBISS.MK-ID - 95152138.

Taylor, R., Johnson, D., & Yang, H. (2019). *Drought and agriculture: Impacts and adaptation strategies*. Springer.

Wang, X., & Liu, Y. (2022). *Climate change and natural disasters: Adaptation strategies*. IntechOpen.

Williams, T., & Burch, S. (2022). *Extreme weather and human impact: Case studies and solutions*. Wiley-Blackwell.

WMO. (2020). *World Meteorological Organization: Flood and drought management*. WMO Publications.

World Health Organization. (2020). *Biological hazards: Risk and response*. WHO Publications.

Chapter 12
The Planet's Natural Environment: Pathways to Sustainable Living

12.1 Definition of Environment

The term "environment" encompasses the surroundings or conditions in which a person, animal, or plant lives or operates. It includes both living (biotic) and non-living (abiotic) elements, such as air, water, soil, flora, and fauna, along with the interactions between them (Fig. 12.1). The environment plays a crucial role in sustaining life by providing essential resources and conditions necessary for health, development, and well-being (Odum & Barrett, 2005). Depending on the context, "environment" can also refer to specific circumstances, such as a workplace or a

© The Author(s), under exclusive license to Springer Nature Switzerland AG 2025

V. Spiridonov et al., *Atmospheric Perspectives*,
https://doi.org/10.1007/978-3-031-86757-6_12

Fig. 12.1 The natural environment encompasses everything that surrounds us

computer system. Beyond the natural world, it includes social, cultural, and built components, reflecting the diverse settings in which life occurs (Smith & Smith, 2015).

12.1.1 General Overview of Environment

Ecology is the scientific study of the relationships between living organisms and their environments. It encompasses the interactions between organisms and the influence of the surrounding environment on both individual species and entire ecosystems (Chapin et al., 2011). Ecology aims to understand the distribution and abundance of organisms, the dynamics of ecosystems, and the factors that influence these patterns over time. This field of study examines these relationships at various levels, including individual organisms, populations, communities, ecosystems, and the biosphere (Molles, 2019).

Environmental factors significantly impact ecosystem functions, such as productivity and nutrient cycling. Understanding these impacts is crucial for controlling the structure and functioning of ecosystems, thereby aiding ecologists in addressing environmental challenges (Odum & Barrett, 2005). Ecology's foundation was built on scientific research into how environmental factors affect organisms and ecosystem components. Historically, natural history explored these questions, but often without rigorous scientific methodology. Modern ecology, however, formulates hypotheses about the natural world's structure and functions, which are then tested through experiments and precise measurements in field or laboratory settings (Smith & Smith, 2015).

Ecologists also utilize non-experimental approaches to study the influence of environmental factors. This approach involves taking precise measurements of a specific factor's local intensity and observing plant growth at the same sites (Chapin et al., 2011). If the study reveals a predictable variation in the factor's intensity, it is considered an environmental gradient. Data collected from the field are then analyzed to determine mathematical relationships between factor intensity and plant responses, often using correlation analysis to measure the strength of these relationships (Molles, 2019). For example, if nitrate concentration significantly affects plant productivity, a strong correlation between the two factors would be observed.

The study of environmental influences in ecology has a rich history. Justus von Liebig (1803–1873), a pioneering researcher, discovered that nitrogen is essential for plant growth. His "law of the minimum" posits that plant growth is limited not by the total amount of resources available but by the scarcest resource (Odum & Barrett, 2005). This concept, known as the "principle of limiting factors," is fundamental to research on environmental factors that limit ecosystem productivity. For instance, in a freshwater lake, phosphate can be a limiting factor for phytoplankton growth. Managing these limiting factors, such as adding phosphorus to lakes, is essential for controlling the effects of eutrophication (Smith & Smith, 2015).

In summary, biological and ecological attributes are profoundly influenced by environmental factors, especially those that act as stressors or limit development, potentially leading to degradation.

12.1.2 Ecological Categories

Ecology can be divided into several categories, each focusing on different aspects of ecological interactions (Fig. 12.2). The main ecological categories are:

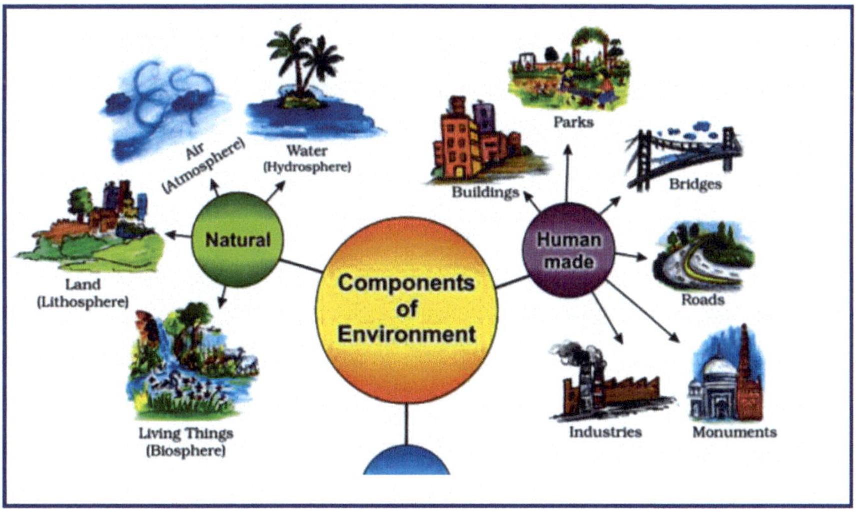

Fig. 12.2 Environmental components: natural (biosphere, atmosphere, lithosphere, hydrosphere) and anthropogenic (parks, buildings, industries, monuments, bridges, streets)

Fig. 12.3 Global ecology: examining the interactions between the globe's ecosystems, atmosphere, land, and oceans, focusing on how these relationships influence global patterns, climate systems, and environmental health

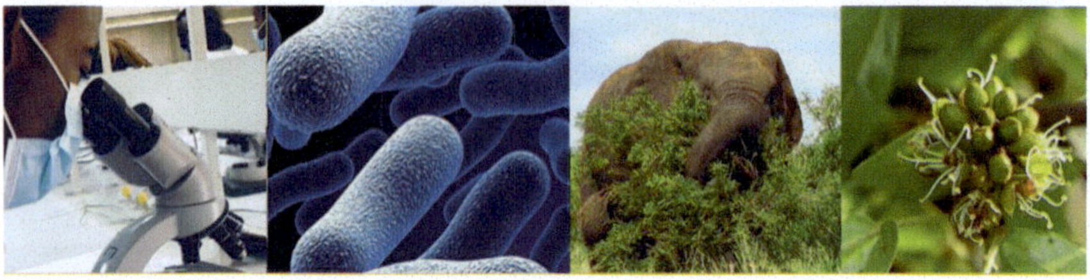

Fig. 12.4 Geographical ecology: exploring the transfer of energy, materials, organisms, and ecosystem products across different regions, with a focus on how human activities influence landscapes and ecological functions

Global Ecology: This category focuses on how the globe's ecosystems, land, atmosphere, and oceans interact (Fig. 12.3). It helps in understanding global relationships and their impact on the entire globe.

Geographical Ecology: Geographical ecology examines the transfer of energy, materials, organisms, and other ecosystem products across different regions (Fig. 12.4). Landscape ecology, a subfield of geographical ecology, emphasizes the impact of human activities on the characteristics and functions of landscapes.

Biological Ecology: Biological ecology studies how organisms respond to environmental stressors through behavior, morphology, physiology, and other adaptations. It explores the relationships between specific species and both biotic and abiotic components of their environment. Ecologists in this field investigate how organisms interact with living and non-living elements in their surroundings (Fig. 12.5).

Fig. 12.5 A vibrant blue starfish (*Linckia laevigata*) resting on hard coral structures, primarily Acropora and Porites. The scene also features colorful Anthiinae fish and delicate crinoids, highlighting the rich biodiversity of the coral reef ecosystem. Location: Lighthouse, Ribbon Reefs. (Credit: By Copyright © 2004 Richard Ling, CC BY-SA 3.0, https://commons.wikimedia.org/w/index.php?curid=78275)

12.2 Environmental Stressors

Environmental stressors are complex regimes of exposure intensities that impact organisms and ecosystem processes in various ways. These stressors, which vary in intensity, influence the performance of individuals and shape the characteristics of populations or communities (Tansley, 1935; Fitter & Hay, 2002). Stressors can limit productivity and reproductive fitness, leading to reduced growth and evolutionary consequences for populations and species (Grime, 2001).

On a broader scale, environmental stressors affect ecosystems, populations, and communities. The effects of stressors vary depending on the intensity and duration of exposure. Some stressors, termed "disorders," involve short but intense exposures, while others are chronic, leading to stable but long-term influences (Larcher, 2003). The study of environmental stressors began with Justus von Liebig, who explored factors limiting biological processes. Early studies centered on plant physiology, while later developments included ecologists and toxicologists examining chemical pollutants. By the twentieth century, system-level research expanded, thanks to contributions from ecologists like Eugene P. Odum and Howard T. Odum (Odum, 1983).

Environmental factors such as temperature, moisture, nutrients, and space can serve as stressors (Begon et al., 2006). These elements can limit species' development by impeding growth rates, leading to challenges during intense exposure.

Stressors are generally categorized as chronic or acute. Chronic stressors include climate and soil conditions that gradually affect ecosystems. Soil properties, for instance, significantly influence plant productivity and the overall functioning of ecological communities (Grime, 2001). In ecosystems like deserts or tundras, chronic stressors are particularly impactful, limiting species diversity and growth under high-intensity conditions (Whittaker & Levin, 1977).

Acute stressors, also called disturbances, represent short-term events that can severely damage dominant organisms in ecosystems. Examples include wildfires, hurricanes, floods, and biological outbreaks such as insect infestations or diseases (Holling, 1973). These disturbances often trigger ecological succession, a process of recovery and reorganization in the community (Pickett & White, 1985).

In contrast, localized disturbances may be minor, such as the fall of a large tree in a forest. This creates a gap that smaller plants quickly fill by utilizing the available resources. These disturbance regimes contribute to the complex structure of mature ecosystems, particularly in long-established forests (Connell, 1978).

12.2.1 Classification of Environmental Stressors

Environmental stressors manifest in various forms and affect different components of the environment, including air, water, soil, and living organisms. They challenge the health, functioning, and sustainability of ecosystems (Barrett & Odum, 2000). The manifestation of environmental stress is a complex process influenced by the intensity and interaction of environmental factors. Stressors can be categorized based on their occurrence as adverse events or persistent exposures. The basic classification of environmental stressors includes climatic, chemical, biological, and physical types (Chapin et al., 2002).

(a) **Climatic Stressors**

Climatic stressors are related to extremes in air temperature, solar radiation, wind, precipitation, and their combinations. These factors act as stressors when their conditions are either insufficient or excessive compared to the needs and comfort zones of organisms or ecosystem processes (Jones et al., 2012). For example, droughts, heat waves, and cold snaps can drastically alter the ecological balance, affecting species' survival and ecosystem productivity (IPCC, 2014).

(b) **Chemical Stressors**

Chemical stressors occur when the availability of certain substances is too low to meet biological needs or too high, causing toxicity or physiological damage (Carpenter et al., 1998). For instance, a forest fire represents a disturbance involving the burning of a significant portion of an ecosystem's biomass. This affects organisms not only through heat and physical damage but also through the release of toxic substances into the environment, such as carbon monoxide and particulate matter (Reid et al., 2016).

(c) **Biological Stressors**

Biological stressors involve interactions among organisms that negatively affect health, functionality, or ecosystem dynamics. These stressors include competition, predation, diseases, and parasites, which can alter population dynamics and ecosystem stability (Sutherland et al., 2013). Changes in environmental conditions, such as temperature, humidity, and food quality, can exacerbate the effects of biological stressors. Additionally, they may indirectly amplify physical or chemical stressors by increasing competition for nutrients, moisture, or space.

A critical aspect of biological stressors is their role in the spread of invasive species or pathogens, which can disrupt ecosystems and reduce biodiversity (Simberloff, 2013). These disruptions highlight the importance of understanding and mitigating biological stressors to maintain ecosystem balance and resilience.

(d) **Physical Stressors**

Physical stressors refer to non-biological factors that disrupt ecosystems and organisms by altering the physical environment. These stressors include exposure to intense kinetic energy, such as volcanic eruptions, seismic sea waves, and anthropogenic explosions, as well as gradual processes like soil erosion, habitat destruction, and pollution (Turner et al., 2001). These stressors can damage ecosystems by fragmenting habitats, limiting species migration and genetic flow, and degrading air and water quality. For instance, pollution introduces toxic substances that negatively impact both terrestrial and aquatic organisms (Foley et al., 2005).

The impact of physical stressors varies based on their intensity, duration, and interactions with other environmental factors. Effective management and mitigation of these stressors are essential to preserving ecosystem health and functionality (MEA, 2005).

12.2.2 Global Drivers of Ecological and Biodiversity Change

Global challenges to Earth's ecosystems and biodiversity refer to factors or conditions that disrupt the balance of natural systems on a large scale. While often driven by human activities, these challenges can also stem from natural processes, posing significant threats to the stability of the planet's climate and the vitality of its ecosystems.

Climate change is one of the most pressing global stressors, primarily resulting from human activities such as burning fossil fuels, deforestation, and industrial processes. These actions release greenhouse gases like carbon dioxide and methane into the atmosphere, causing global warming. Consequences include rising sea levels, extreme weather events, altered precipitation patterns, and disruptions to ecosystems, threatening both natural and human systems (Fig. 12.6).

Biodiversity, which encompasses the variety of all life forms on Earth, is essential for ecosystem stability and the provision of vital services such as food security,

Fig. 12.6 Rain, cold, heat, and wind—the same meteorological factors that support crop growth—can also become major stressors when they occur in excess. (Credit: Pinelopi Williams (Alltech))

air and water purification, and climate regulation (Fig. 12.7). The loss of biodiversity is another critical stressor, with far-reaching implications for ecosystem health. However, habitat destruction, pollution, overexploitation of resources, and climate change are leading to rapid declines in species diversity, weakening ecosystems and their ability to support life.

Land degradation, driven by activities such as unsustainable agriculture, deforestation, urbanization, and industrial expansion, leads to declining soil quality and the loss of arable land. This process undermines food production, reduces soil fertility, and disrupts ecosystem health, contributing to desertification and diminishing the resilience of ecosystems. Additionally, degraded land increases vulnerability to natural disasters, posing significant challenges to sustainable development and environmental stability (Fig. 12.8).

Ocean acidification, caused by the absorption of excess atmospheric carbon dioxide, is a critical threat to marine ecosystems. This process lowers ocean pH levels, compromising the ability of marine organisms like corals and mollusks to form calcium carbonate structures. The cascading effects of this phenomenon disrupt the marine food web, reducing biodiversity and threatening the livelihoods of communities dependent on these ecosystems (Fig. 12.9).

Pollution, encompassing air, water, and soil contamination, is a widespread environmental challenge with profound impacts on ecosystems and human health. Air pollution arises from industrial emissions, vehicle exhaust, and the burning of fossil fuels, while water pollution stems from agricultural runoff, industrial discharges, and untreated waste. Soil contamination results from improper waste disposal, pesticide use, and chemical spills. These pollutants degrade natural resources, disrupt ecosystems, reduce biodiversity, and pose significant health risks to both humans

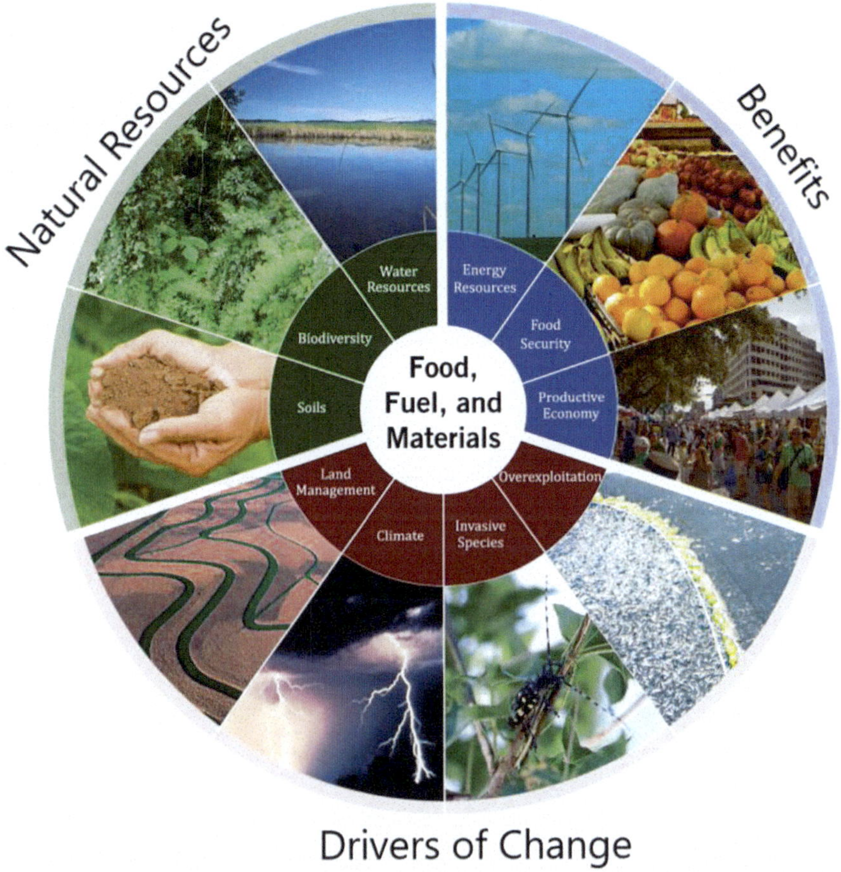

Fig. 12.7 Natural resources provided by biodiversity, including benefits and beneficiaries, and drivers of change. (Source: U.S. Environmental Protection Agency (EPA))

Fig. 12.8 Land degradation as a threat to food security and a driving factor for migration. (Credit: Alessandro Balduzzi, Food and Migration)

Fig. 12.9 Ocean acidification weakens the skeletons of coral reefs. (Source: ESA)

Fig. 12.10 Increased pollution of the environment and its impact on the natural world. (Source: Fairgaze)

and wildlife (Fig. 12.10). Addressing pollution is essential to safeguarding environmental integrity and promoting sustainable living.

Deforestation, the large-scale removal of forests for purposes such as agriculture, logging, or urban development, has severe consequences for both the climate and biodiversity. Forests serve as vital carbon sinks, and when they are destroyed, the stored carbon is released into the atmosphere, exacerbating global warming. Additionally, deforestation disrupts ecosystems and reduces habitats for numerous species. The loss of forests not only contributes to climate change but also leads to the decline of biodiversity, alters local and global weather patterns, and weakens the resilience of ecosystems (Fig. 12.11).

Fig. 12.11 Trees cut for processing, awaiting transportation and sale. (Credit: Photograph by Esemelwe. National Geographic)

Fig. 12.12 The world has undergone significant development since the era of industrialization in the twentieth century. One of the main phenomena resulting from industrialization is urbanization. (Source: Difference Between)

The overexploitation of natural resources, including overfishing, excessive logging, and wildlife harvesting, depletes populations and destabilizes ecosystems. Unsustainable resource use not only threatens species survival but also compromises ecosystem services essential for human well-being.

Urbanization and industrialization have both dramatically reshaped the globe, driving economic growth while imposing significant environmental challenges. Urbanization refers to the growth of cities and urban areas, which is driven by factors such as migration for work, modernization, and sociological changes. As people move from rural areas to cities, urban expansion accelerates, often leading to habitat loss, increased pollution, and altered local climates, such as the heat island effect. This process is a global phenomenon that transforms landscapes, social structures, and economies (Fig. 12.12).

Industrialization, however, represents a shift from agriculture to manufacturing, innovation, and other economic activities. It is closely tied to urbanization, as the growth of industries leads to the expansion of cities. While industrialization drives economic development, it also contributes to resource depletion, increased energy demand, and waste generation, placing immense pressure on natural systems. Industrial activities, such as emissions from factories and the production of goods, contribute significantly to air and water pollution and environmental degradation. The rapid growth of urban areas and the expansion of industrial activities bring significant environmental impacts. Habitat loss occurs as land is converted for development, and the demand for energy and resources increases, leading to higher pollution levels. Urbanization also disrupts local climate patterns, exacerbating issues like heat islands, where cities experience higher temperatures due to increased energy use and the concentration of buildings.

To mitigate these environmental stresses, coordinated international efforts are essential. Sustainable development practices, conservation initiatives, and the adoption of environmentally friendly technologies must be prioritized. Global frameworks, such as the United Nations Framework Convention on Climate Change (UNFCCC) and the Convention on Biological Diversity (CBD), play a key role in fostering collaboration and promoting sustainable practices. By protecting biodiversity, combating climate change, and managing resources wisely, we can work toward ensuring the health and stability of the planet for future generations.

12.2.3 Responses to Environmental Stressors

When an ecosystem is disturbed, changes occur in species composition, spatial distribution of biomass, and functional properties such as productivity and nutrient cycling. The process of ecological succession can help an ecosystem recover, potentially leading to the return of a mature ecosystem if the process is sustained long enough.

Chronic stressors, such as chemical and climatic influences, act continuously on ecosystems. Increasing the intensity of these stressors can reduce productivity and cause acute effects on organisms, including tissue damage and death. Species respond differently to increased stress intensity, which may result in decreased abundance or local extinctions. Species that are more tolerant of the altered conditions may replace those less able to cope, and natural selection may increase tolerance across populations in response to prolonged exposure to chronic stressors.

As environmental stress intensifies over time, ecosystems are expected to change structure and function. If the stressor regime becomes stable at higher intensities, ecosystems may experience long-term shifts toward simpler structures and functions, with reduced biodiversity, fewer top predators, and lower productivity. These changes can result from natural stressors, such as forest fires or glacial events, or anthropogenic factors, such as industrial pollution or resource exploitation. Such shifts are often interpreted as degradation of the ecosystem.

Determining whether environmental changes are 'good' or 'bad' is an important aspect of applied ecology. Changes caused by human activities, such as industrial processes, are frequently considered harmful. Conversely, reductions in stress intensity over time can lead to different ecological responses compared to those resulting from increased stress. Concepts such as environmental quality, ecological integrity, and ecosystem health are used to assess the impact of environmental changes and the overall health of ecosystems.

Ecosystems that are persistently exposed to intense natural stressors, such as deserts or tundra, tend to develop stable structures and functions adapted to these conditions. These ecosystems generally have low species richness, simple structures, and limited productivity, with slow rates of decomposition and nutrient cycling.

In contrast, ecosystems subjected to intense anthropogenic stressors, such as pollution or disturbances from industrial activities, may experience decreased ecological integrity. This degradation can manifest as reduced complexity, loss of large species or top predators, and increased presence of non-native species. Anthropogenic stressors often lead to significant damage, impacting the ecological integrity and quality of the environment.

12.3 Global Environmental Monitoring Programs

Environmental monitoring and research programs are essential for understanding and managing changes in ecological integrity. They play a crucial role in the early detection of environmental stressors and in providing data and knowledge that support informed decision-making (Lovett et al., 2007). These programs focus on tracking changes in environmental stressors and processes by measuring various indicators, such as chemical residues in organisms, shifts in species abundance, and other environmental metrics. Monitoring helps identify and analyze contamination issues and alterations in ecosystems, allowing for timely responses to emerging problems (Pereira et al., 2013).

In addition to monitoring, research programs aim to deeply understand the causes and consequences of environmental change. They explore complex ecosystem processes, species–environment interactions, and the long-term effects of stressors. This research is vital for uncovering the underlying mechanisms of change and developing effective management strategies (Vitousek et al., 1997). Environmental indicators are often used to simplify measurements, representing broader aspects of environmental quality and ecological integrity. Examples include chemical residues in the tissues of top predators, changes in species abundance, or multimetric indices such as biotic integrity scores. These indicators provide an overview of environmental conditions and support management and decision-making processes.

Both monitoring and research programs are crucial for managing environmental changes over time. They enable the tracking of ecosystems, and health, and the identification of biological and chemical changes caused by various stressors.

Additionally, these programs play a key role in education and public awareness, emphasizing the importance of sustainability and environmental protection.

12.4 Global Environmental Pollution

Global air, water, and land pollution encompass the widespread contamination and degradation of the Earth's atmosphere, water bodies, and soil. These forms of pollution are primarily caused by human activities that release harmful substances into the environment, leading to adverse effects on ecosystems, human health, and the planet's overall well-being. The sources of pollution include industrial activities, transportation, agriculture, and improper waste disposal.

Major pollutants consist of particulate matter, heavy metals, pesticides, plastics, and other harmful substances. Particulate matter, for instance, is linked to respiratory and cardiovascular diseases, while heavy metals like lead and mercury can accumulate in the food chain, posing serious health risks to humans and wildlife. Pesticides, commonly used in agriculture, have been associated with declining pollinator populations and loss of biodiversity, further destabilizing ecosystems (Hladik et al., 2014; Goulson, 2013).

The consequences of global pollution are extensive, including health risks, ecosystem disruption, climate change, and threats to biodiversity. Water pollution can lead to the eutrophication of lakes and rivers, resulting in harmful algal blooms that produce toxins affecting both aquatic life and human water supplies. Land pollution, often characterized by the accumulation of plastics and hazardous waste, threatens soil health and food security, as toxic substances can leach into the ground and contaminate crops.

Moreover, global pollution is not confined to local regions; it often transcends borders, highlighting the need for international cooperation. Transboundary pollution, such as acid rain and ocean plastic debris, necessitates collective action and shared responsibility among nations. Addressing these challenges requires sustainable practices, technological innovations, and awareness initiatives to mitigate pollution's impact on the environment and human health. For instance, adopting renewable energy sources can significantly reduce air pollution, while effective waste management strategies can minimize land and water pollution.

Innovative solutions, such as bioremediation and green chemistry, offer promising approaches to combat pollution at its source, restoring contaminated environments and promoting ecological health. Additionally, public awareness campaigns can empower communities to engage in responsible consumption and waste reduction practices, fostering a culture of sustainability. Ultimately, tackling global environmental pollution necessitates a multifaceted approach, integrating science, policy, and community engagement to create a healthier and more resilient planet for future generations.

12.5 Global Air Quality

Our planet is experiencing significant warming, with increasing emissions resulting from the use of non-clean fuels and technologies, leading to dangerously polluted air. According to the World Health Organization (WHO), nearly nine out of ten people breathe polluted air, which impacts around seven million people each year. The health effects of air pollution are severe, with nearly a third of deaths from stroke, lung cancer, and heart disease linked to air pollution and climate change (Ayres et al., 2006; Pallavi and Srivastava, 2020; Harrison & Hester, 2019).

The primary driver of climate change is the burning of fossil fuels, a major contributor to air pollution. Mitigating climate change can also improve air quality. The Intergovernmental Panel on Climate Change (IPCC), established by the World Meteorological Organization (WMO), highlights in its latest reports that to limit global warming to 1.5 °C, coal-based electricity must be phased out by 2050.

12.5.1 Atmospheric Pollution

Outdoor air pollution encompasses all chemical and biological components that alter the natural characteristics of the atmosphere, adversely affecting both human health and the environment. Air pollution results from various pollutants interacting with climatic forces, radiation, and other factors. The main types of air pollution include:

- Inert Dust: Particulate matter that does not chemically react but can cause physical harm to health and materials.
- Local Pollutants: Pollutants that originate from specific sources such as factories or vehicles, impacting localized areas.
- Toxic Substances: Harmful chemicals in the air, including heavy metals and volatile organic compounds (VOCs), that pose health risks.
- Photochemical Smog: A type of smog formed when sunlight reacts with pollutants like nitrogen oxides and VOCs, common in urban areas.
- Acid Rain: Rainfall with a lower pH due to the presence of sulfur dioxide (SO_2) and nitrogen oxides (NO_x) in the atmosphere.
- Reduction of Visibility: Decreased clarity of the atmosphere due to particulate matter and pollutants.
- Warming Caused by Greenhouse Effects: The trapping of heat in the atmosphere by greenhouse gases leads to global warming.
- Thinning of the Ozone Layer: The depletion of the ozone layer due to chlorofluorocarbons (CFCs) and other ozone-depleting substances.
- Climate Forcing Due to the Anthropogenic Factor: Climate changes caused by human activities, such as greenhouse gas emissions and deforestation.
- Atmospheric Aerosols: Tiny particles or droplets suspended in the air that can influence climate and air quality.

- Radiation: The presence of radioactive substances in the atmosphere that can have harmful effects on health and the environment.

Air pollution involves three main components:

1. Emitters: Sources of pollutants, including traffic, fuel combustion in industry and households, and various industrial processes.
2. Atmosphere: The medium through which pollutants are transported, spread, and undergo physical and chemical transformations.
3. Receptors: Entities affected by air pollution, including humans, animals, plants, and materials.

Understanding the interaction between these components is crucial for managing and mitigating air pollution.

12.5.2 Classification of Pollutants

In terms of chemical composition, air pollution results from the presence and activity of chemical components such as sulfur, nitrogen, carbon, halogens, toxic substances, and radioactive materials. Based on the phase state, atmospheric pollutants are divided into gaseous, liquid, and solid. According to the way they reach the atmosphere, they are divided into:

1. Primary pollutants that are directly emitted from existing sources
2. Secondary pollutants that are formed by chemical reactions between primary pollutants under standard atmospheric conditions

Primary pollutants include particulate matter (PM), sulfur dioxide, nitrogen oxides, volatile organic compounds, carbon monoxide, lead, and others. Secondary pollutants consist of sulfuric acid, air pollution in urban and industrial areas, and photochemical smog, which is a harmful mixture of gases and particles. According to the spatial distribution and their effects, pollutants are classified as local, regional, or global. The Environmental Protection Agency (EPA) defines six major pollutants for which there are six ambient air standards to protect human health. According to the EPA, the following measurable pollutants are defined:

1. Ozone (O_3)
2. Sulfur dioxide (SO_2)
3. Carbon monoxide (CO)
4. Nitrogen oxides (NO_x)
5. Lead (Pb)
6. Particulate matter (PM)

12.5.3 Good and Bad (Harmful) Ozone

Ozone (O_3) is a colorless gas composed of three oxygen atoms. Useful ozone (O_3) that occurs in the atmosphere is most abundant in the stratosphere. Ultraviolet (UV) radiation plays a key role in destroying ozone in the Earth's stratosphere. The ozone layer, located in the stratosphere, absorbs and protects the Earth's surface from most of the sun's harmful UV radiation. Here is a more detailed explanation of the role of UV radiation in ozone depletion:

Ozone (O_3) molecules in the stratosphere absorb incoming UV-B and UV-C radiation from the sun. When ozone molecules absorb UV radiation, the energy from the radiation causes the ozone molecule to break up into individual oxygen molecules:

$$O_3 + UV - \frac{B}{C} \rightarrow O_2 + O \tag{12.1}$$

In this reaction, one molecule of oxygen (O_2) and a free oxygen atom (O) are generated. The free oxygen atom produced during ozone photolysis is highly reactive and can subsequently react with another ozone molecule (O_3).

$$O + O_3 \rightarrow 2O_2 + O \tag{12.2}$$

The process produces one molecule of oxygen (O_2) and a free oxygen atom (O). The highly reactive oxygen atom can then react with another ozone molecule:

$$O + O_2 \rightarrow O_3 \tag{12.3}$$

This reaction regenerates one molecule of ozone (O_3). The depletion of the ozone layer (ozonosphere) due to human activities or the use of certain substances poses serious health risks and environmental hazards. Many environmental agencies worldwide are working to reduce the production of ozone-depleting substances to protect the ozone layer. This depletion is linked to increased health risks and environmental damage.

Ground-level ozone, often referred to as "harmful ozone," forms because of vehicle exhaust, power plants, industrial processes, and other sources that react chemically with sunlight. It is produced from nitrogen oxides (NO_x) and volatile organic compounds (VOC_s). Smog, a significant source of air pollution in large cities, is primarily caused by vehicle emissions. Ozone is often used as an indicator of smog conditions due to its relation to smog and its ease of measurement compared to other smog components.

Tropospheric ozone is a toxic gas that degrades air quality, with its formation driven by ozone precursors such as nitrogen oxides, carbon monoxide, methane, and other hydrocarbons.

Human activities such as biomass burning, inefficient coal combustion, waste treatment, industrial processes, and traffic produce ozone precursors. The complex

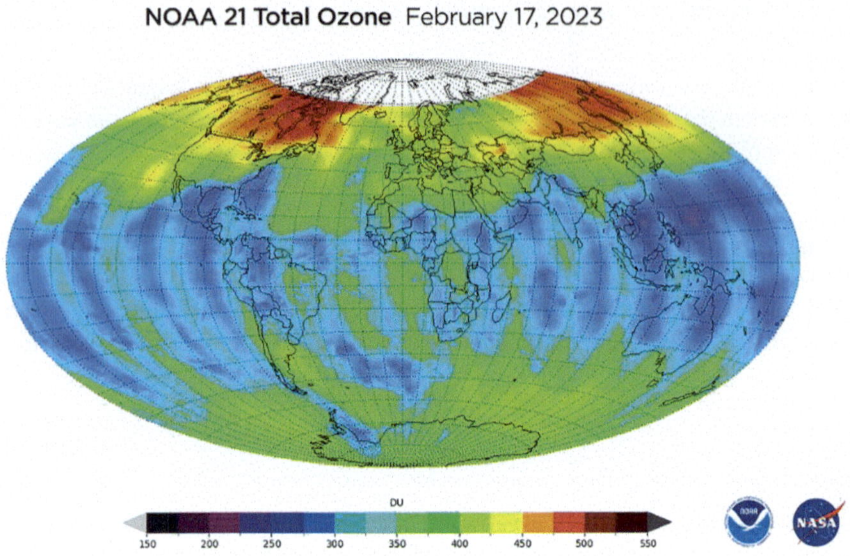

Fig. 12.13 Global distribution of ozone in Dobson Units (DU) for February 17, 2023. This global image was obtained by the Ozone Mapping and Profiling System (OMPS) on the NOAA-21 satellite for the first time. (Source: NASA)

chemistry of ozone complicates the quantification of the impact of poor local air quality. Uncertainties in pollutant emission inventories needed for air quality predictions can be as high as 50%. Additionally, the amount of ozone entering the troposphere from the stratosphere remains largely unexplained. The global concentration of ozone (O_3) in Dobson Units (DU) measured in the troposphere is illustrated in Fig. 12.13.

12.5.4 *The Main Components of Sulfur*

Sulfur dioxide (SO_2), a colorless gas with a sharp odor, is a major atmospheric pollutant. It originates from both natural and anthropogenic sources. Natural sources include marine plankton, seawater, bacteria, plants, and volcanic eruptions. Human activities contribute significantly through industrial processes, such as burning fossil fuels containing sulfur, mining, coal extraction, and oil refining.

Sulfur is present in fuels at varying concentrations. During the oxidation process, SO_2 in the atmosphere interacts with water in clouds, leading to the formation of sulfate aerosol particles. These sulfates, along with acid rain, are then deposited on the Earth's surface, contributing to environmental acidification.

They influence the radiation balance in the atmosphere and can lead to global cooling. The global spatial distribution of SO_2 emissions is illustrated in Fig. 12.14.

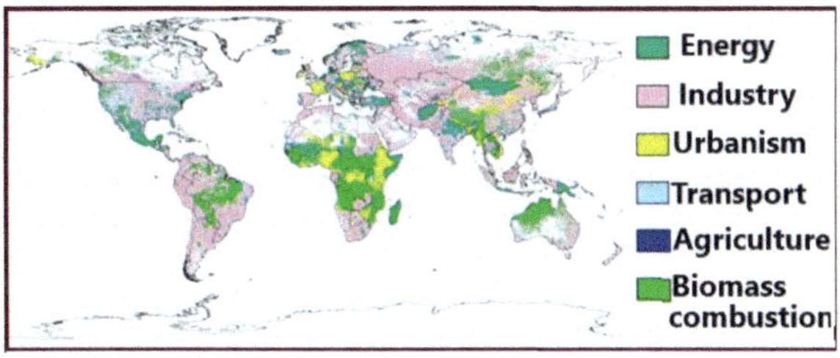

Fig. 12.14 Spatial distribution of SO_2 emissions from the sector with the predominant contribution to the network

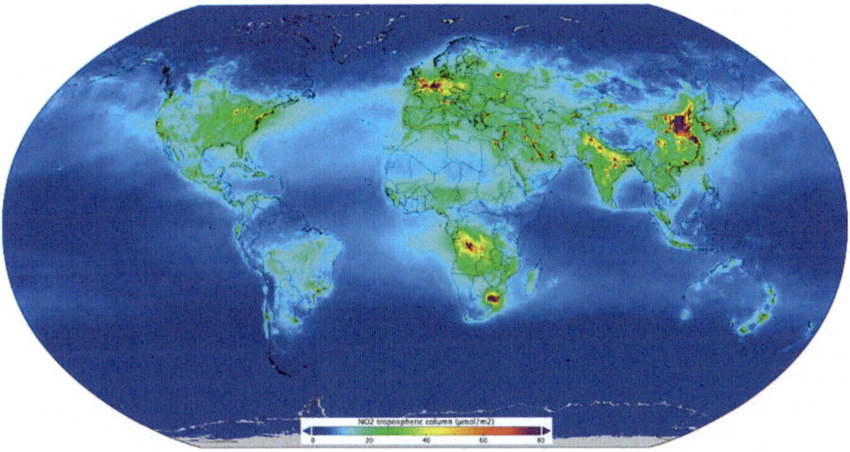

Fig. 12.15. Global distribution of NO_2 based on annual averages of tropospheric column NO_2 amounts, measured using high-resolution observations of multiple trace gases from Sentinel-5P/TROPOMI

12.5.5 Main Components of Nitrogen

Nitrogen (N_2) is the most abundant gas in the atmosphere, comprising approximately 78%. NO_x refers to a mixture of nitrogen oxides, including nitrogen monoxide (NO) and nitrogen dioxide (NO_2), which are unstable in the atmosphere. Figure 12.15 presents a global view of nitrogen dioxide based on satellite estimates for 2019. Nitrogen oxides (NO_x) are primarily formed at high temperatures during the combustion of fuels, particularly from motor vehicles. Anthropogenic sources of nitrogen include motor vehicles and biomass burning, while natural sources include bacterial activity, electrical discharges, and biomass burning. Nitrogen oxides are key contributors to the formation of harmful ground-level ozone.

12.5.6 *Main Components: Carbon Monoxide*

Carbon monoxide (CO) is a colorless, odorless, yet flammable gas that is a significant pollutant in urban areas, produced when fuel is incompletely burned. Anthropogenic sources include gasoline-powered motor vehicles, cigarette smoke, and biomass burning.

Natural sources of carbon monoxide include biomass burning. CO is also produced through the atmospheric oxidation of methane and other hydrocarbons (Fig. 12.16). In the atmosphere, carbon monoxide is oxidized to CO_2, a key greenhouse gas. Primary sources of CO_2 include the burning of fossil fuels, deforestation, and cement production.

12.5.7 *Particulate Matter (PM)*

Particulate matter, including PM_{10} and $PM_{2.5}$, consists of solid or liquid particles suspended in the air with very small dimensions. These particles are a mix of various organic and inorganic substances, such as sulfates, nitrates, ammonia, some crystals, sea salt, metal oxides, hydrogen ions, and water.

PM particles can be in either a solid or liquid state, and therefore, they appear in various forms, including soot, dust, ash, pollen, and smoke. Anthropogenic sources of particulate matter include biomass burning, industrial processes, and agricultural activities, while natural sources include sea salt, sand from desert storms, biomass combustion, and volcanic ash.

Air quality in urban areas is significantly reduced due to high concentrations of PM particles, which increasingly impact human health. PM_{10}, or coarse particles, are those with a diameter of less than 10 μm, such as dust or pollen. These particles

Fig. 12.16 Carbon emissions during biomass burning

mainly contain iron, aluminum, and silicates and are primarily generated by burning solid fuels and from vehicle and industrial exhaust gases. According to air quality directives, the maximum allowed concentration of PM_{10} particles over 24 h is 50 μg/m³; exceeding this threshold indicates poorer air quality, which over time can negatively affect health.

$PM_{2.5}$ or fine particles are particularly hazardous because they are smaller than 2.5 μm and can penetrate the body more easily and deeply. These particles are primarily produced by burning fossil fuels, such as car exhaust, wood, or coal heating. They mainly consist of sulfates, nitrates, ammonia, organic and elemental carbon, and heavy metals like lead, manganese, copper, cadmium, zinc, nickel, vanadium, and chromium. Due to their small size, $PM_{2.5}$ particles can easily enter the lungs, causing health issues when their concentration exceeds the allowed limit of 25 μg/m³.

They can cause eye irritation and pose significant health risks, with numerous studies showing a strong correlation between $PM_{2.5}$ exposure and premature mortality from cardiovascular and lung diseases. These particles can also exacerbate chronic conditions such as asthma, bronchitis, and other respiratory illnesses.

A global air quality map, depicting the mean concentration of $PM_{2.5}$ (particles with a diameter less than 2.5 μm) at a spatial resolution of $0.1° \times 0.1°$ for 2004–2008, is presented in Fig. 12.17. This map combines aerosol optical depth data retrieved from MODIS and MISR satellite instruments for the period 2001–2006.

Based on the measurements, the average concentrations of $PM_{2.5}$, considering the global population distribution, were most significantly influenced by the organic mass of particles, followed by secondary inorganic aerosol and mineral dust (Fig. 12.18). These estimates provide valuable information on global population exposure to the chemical components and sources of $PM_{2.5}$.

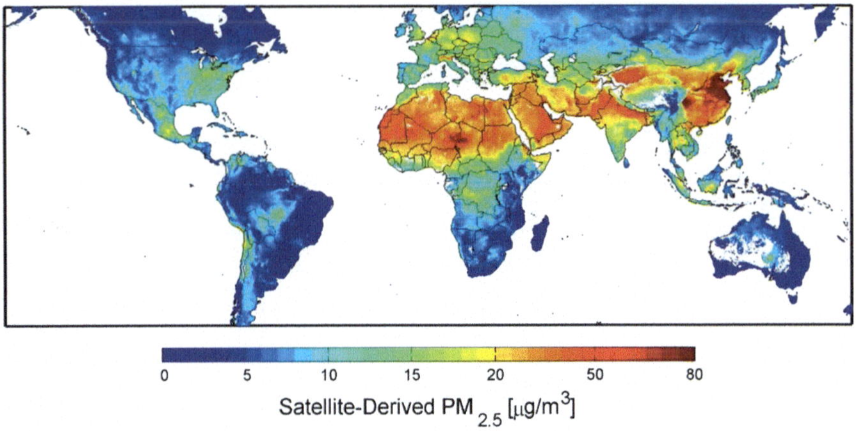

Fig. 12.17 Global distribution map of $PM_{2.5}$ (fine particles, 2.5 μm or less in diameter) from 2001 to 2006. (Source: NASA)

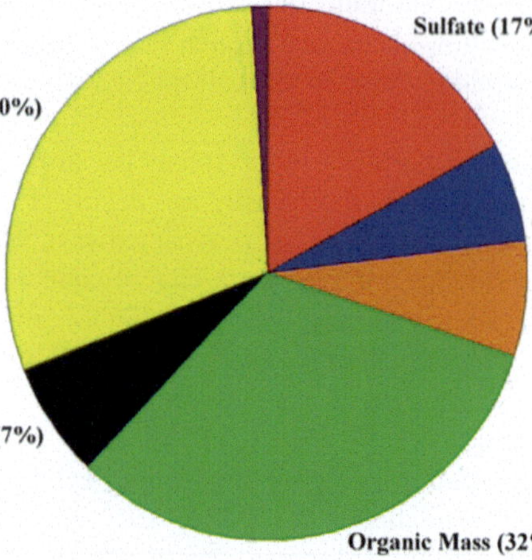

Fig. 12.18 Global percentage estimate of $PM_{2.5}$ concentrations, accounting for global population distribution. (Credit: Environ. Sci. Technol. 2014, 48, 22, 13060–13068)

12.6 Atmospheric Aerosols

Atmospheric aerosols are tiny solid or liquid particles suspended in the Earth's atmosphere, originating from natural sources, such as volcanic eruptions, sea spray, wildfires, and human activities, including industrial emissions, transportation, and agricultural practices. These particles play a critical role in various atmospheric processes and significantly affect climate, air quality, and human health.

The effects of aerosols in the atmosphere are multifaceted. High concentrations of aerosols can pose serious respiratory health hazards, particularly in urban environments where traffic and industrial activities are prevalent. Research has shown that exposure to particulate matter is associated with increased rates of respiratory and cardiovascular diseases (WHO, 2021). Additionally, aerosols scatter and absorb visible radiation, limiting visibility and affecting weather patterns. They influence the Earth's climate directly by scattering and absorbing radiation and indirectly by serving as nuclei for cloud formation, which can alter precipitation patterns (Haywood & Boucher, 2000).

NASA's advancements in satellite remote sensing technology have significantly enhanced our understanding of aerosol distributions and their impact on climate and air quality. The agency's Aerosol Robotic Network (AERONET) provides continuous global measurements of aerosol properties, helping scientists track changes in aerosol concentrations and types over time. Recent findings have highlighted the role of aerosols in modulating cloud properties, which can lead to more intense and prolonged weather events. For example, aerosols can influence the size and lifetime of cloud droplets, ultimately affecting precipitation efficiency and storm development (Liu et al., 2019).

Moreover, NASA's MODIS (Moderate Resolution Imaging Spectroradiometer) has been instrumental in monitoring aerosol optical depth (AOD), allowing researchers to assess the impact of aerosols on climate models and refine predictions related to climate change. Recent studies have indicated that changes in aerosol concentrations due to human activity have contributed to a cooling effect on the planet, partly counteracting the warming from greenhouse gases (Zhang et al., 2019). However, the diminishing role of aerosols, particularly in regions with stringent air quality regulations, raises concerns about future climate scenarios.

Aerosols also provide sites for surface and condensed-phase chemistry to occur in the atmosphere, leading to the formation of secondary pollutants, such as ozone and particulate matter. The interplay between aerosols and atmospheric chemistry underscores the importance of understanding aerosol sources and behavior to devise effective air quality management strategies.

Addressing the challenges posed by atmospheric aerosols requires integrated approaches combining satellite observations, ground-based measurements, and advanced modeling techniques to develop comprehensive strategies for improving air quality and mitigating climate change. Continuous monitoring and innovative research are essential for understanding the evolving role of aerosols in our atmosphere and their implications for public health and environmental sustainability.

12.6.1 Sources and Sinks of Aerosols

Atmospheric aerosols consist of a wide variety of condensed particles, typically spherical and ranging from 0.1 to 1.0 μm in size. These particles are significant air pollutants (Fig. 12.19). Aerosol concentration is highest near the Earth's surface and at emission sources and decreases with altitude.

Natural sources of aerosols include:

- Volcanic eruptions
- Forest fires
- Dust storms
- Sea spray
- Biogenic emissions (e.g., particles released from vegetation)

Anthropogenic sources involve:

- Industrial processes
- Emissions from vehicles
- Combustion of fossil fuels
- Agricultural activities
- Other human-related activities

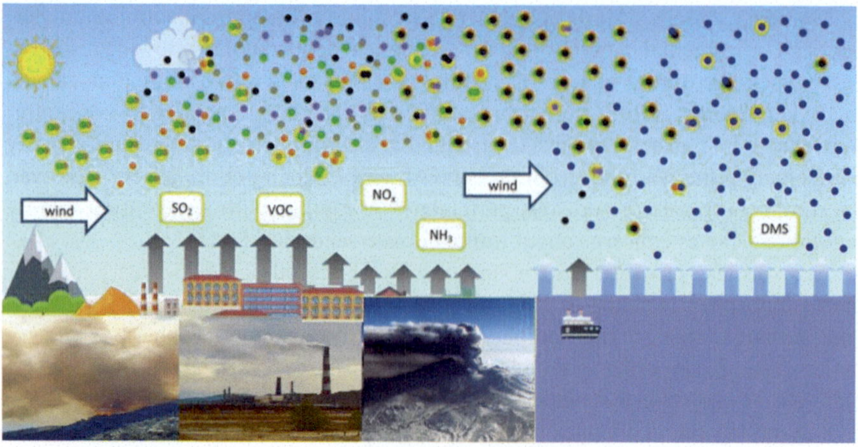

Fig. 12.19 A variety of aerosol particles originate from different sources, illustrating their diverse sizes, types, and internal structures as they are transported through the atmosphere. (Credit: https://eos.org/editors-vox/)

Primary sources of atmospheric aerosols include:

- Condensation and sublimation of vapors, and cloud formation during natural and anthropogenic combustion
- Reactions between atmospheric tracers (e.g., hydroxyl radicals (OH), reactive nitrogen species, hydrocarbons, ozone, sulfur compounds) facilitated by heat, radiation, or moisture
- Mechanical destruction and dispersion of materials from the Earth's surface, such as sea spray over oceans or mineral dust over continents
- Coagulation of nuclei, leading to the formation of larger particles with mixed compositions

H_2SO_4 has a low vapor pressure over H_2SO_4–H_2O solutions, leading to its condensation under various atmospheric conditions to form aqueous sulfate particles. The composition of these sulfate particles can be further modified by the condensation of other low vapor pressure gases, such as NH_3, HNO_3, and organic compounds. Organic carbon constitutes a significant portion of fine aerosols, primarily through the condensation of large hydrocarbons from both biogenic and anthropogenic sources.

Another crucial component of fine aerosols is soot, which forms from the condensation of gases during combustion processes. Soot, in this context, refers to both elemental carbon and black organic aggregates.

Additionally, the mechanical action of wind on the Earth's surface releases sea salt, soil dust, and vegetation debris into the atmosphere. These aerosols are predominantly coarse particles with radii ranging from 1 to 10 μm. Particles smaller than 1 μm are challenging to generate mechanically due to their large surface-to-volume ratio and high surface tension per unit volume. Conversely, particles larger than 10 μm are not easily lifted by wind and have a short atmospheric residence time due to their high sedimentation velocities.

12.6.2 *Aerosol Processes*

Several processes are involved in the production, growth, and removal of atmospheric aerosol particles (Fig. 12.20). Gas molecules typically range in size from 10^{-4} to 10^{-3} μm. Nucleation, or the clustering of gas molecules, produces ultrafine aerosols in the 10^{-3} to 10^{-2} μm size range. These ultrafine aerosols then grow into the fine aerosol size range of 0.01–1 μm through the processes of gas condensation and coagulation (collisions between particles during their random motions).

Aerosol particles formed from the condensation of gases generally accumulate in the 0.01–1 μm size range, known as the accumulation mode, in contrast to the ultrafine mode or coarse mode. These particles are too small to settle out of the atmosphere at a significant rate and are primarily removed through scavenging by cloud droplets followed by precipitation or by direct scavenging by raindrops.

Coarse particles carried by wind are also removed from the atmosphere by precipitation. Additionally, these particles settle considerably; for instance, a particle with a radius of 10 μm has a sedimentation rate of 1.2 cm/s at sea level, providing an alternative removal route.

12.6.3 *Vertical Distribution of Aerosols*

Most of the atmospheric aerosol mass is concentrated in the lower troposphere, where aerosols typically remain for a relatively short period, approximately 1–2 weeks, before settling. Aerosol concentrations in the upper troposphere are

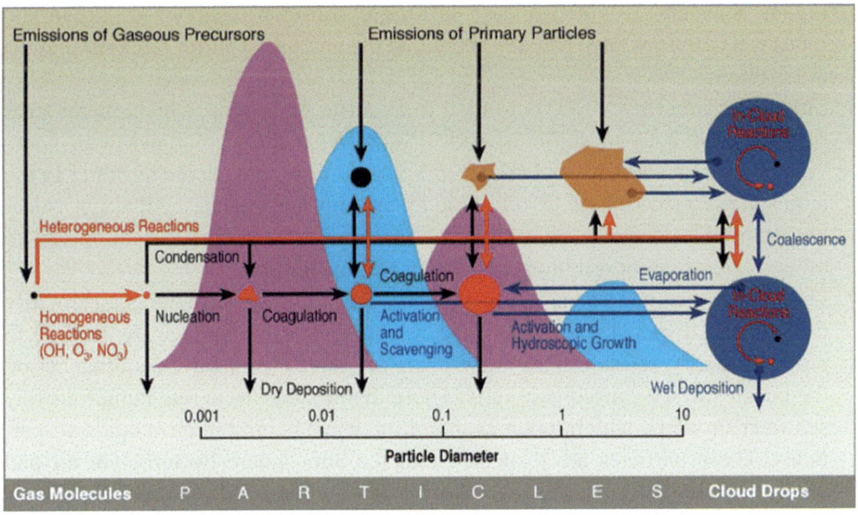

Fig. 12.20 Production, growth, and removal of atmospheric aerosols, illustrating the processes from nucleation of gas molecules

generally 1–2 orders of magnitude lower than those in the lower troposphere. In contrast, the stratosphere contains a pervasive H_2SO_4–H_2O aerosol layer located 15–25 km above sea level, which plays a significant role in the chemistry of stratospheric ozone. This layer forms from the oxidation of carbonyl sulfide (COS), a biogenic gas with a long atmospheric lifetime that allows it to reach the stratosphere.

The layer can be episodically enhanced by the oxidation of SO_2 from large volcanic eruptions, such as the 1991 eruption of Mount Pinatubo. Although the contribution of H_2SO_4 from carbonyl sulfide (COS) oxidation in the stratosphere is less than 0.1% of the tropospheric source, stratospheric aerosols have a much longer atmospheric lifetime due to the absence of precipitation.

12.6.4 *Origin of Aerosols*

Aerosol particles originate from three primary sources:

1. **Cosmic:** Cosmic dust enters the Earth's atmosphere directly from space or results from the combustion of smaller celestial bodies, such as meteorites. Although this dust contributes only a small fraction of the total atmospheric aerosol load, approximately a thousand tons of cosmic dust are deposited into the atmosphere daily.
2. **Continental:** The continents are a major source of both hygroscopic and non-hygroscopic particles that are carried into the atmosphere by winds, particularly in arid and desert regions. For instance, it is estimated that around 3.5 million tons of desert dust are transported from North Africa to Europe annually. Another significant source of hygroscopic particles is the combustion and decay of organic substances, which release sulfur and nitrogen compounds. Urban and industrial activities have recently increased the amount of sulfur released into the atmosphere.
3. **Sulfur dioxide** (SO_2): SO_2 is formed from sulfur emissions, which, under solar radiation, oxidizes to sulfur trioxide (SO_3). When SO_3 combines with water vapor, it forms sulfuric acid (H_2SO_4). Volcanic eruptions also contribute to the atmospheric aerosol load by releasing hygroscopic substances. Although volcanic eruptions generally contribute less to atmospheric aerosols compared to other sources, major eruptions can have significant impacts.
4. **Oceanic:** Oceanic areas are crucial sources of atmospheric aerosols, primarily through the dispersion of tiny droplets from the ocean surface. According to Aitken's theory, sea salt in the atmosphere originates from wave disturbances on the ocean surface caused by winds. These disturbances generate numerous tiny seawater droplets, which, upon evaporation, leave behind microscopic salt particles. These particles are dispersed into the atmosphere by turbulent air and winds, traveling distances of several hundred kilometers. Measurements indicate that the smallest sea salt particles have diameters around 0.2 μm. The average concentration of sea salt particles in the atmosphere near wave-breaking areas is

approximately a hundred particles per cubic centimeter, with a formation rate of about a hundred particles per square centimeter per second.

12.6.5 Dimensions and Concentrations of Atmospheric Aerosols

Typical total aerosol number concentration values are (10^3 cm^{-3}) over oceans, (10^4 cm^{-3}) over rural areas, and land (10^5 cm^{-3}) in air-polluted areas. The concentration of aerosols changes in space and time. Concentrations are usually highest near the Earth and sources.

The concentration of aerosols decreases with increasing altitude. Small ions play almost no role in atmospheric condensation because of the very high saturation necessary for condensation. Basic particles are only able to stay in the air for a limited time. Aerosols have a wide range of concentrations varying from 10^{-6} to 10^7 cm^{-3}. Aerosol concentrations decrease very rapidly with increasing aerosol size. Total aerosol concentrations are on average highest in urban polluted areas and lowest in clean marine air. Based on size, atmospheric aerosols are classified as Aitken, large, and giant aerosols:

Aitken aerosols, also known as nucleation mode particles, typically have a radius $r \leq 0.1$ μm (or diameter $d \leq 0.1$ μm). These are small fine particles that often form through nucleation processes. Aitken aerosols play a role in cloud formation and can influence visibility and atmospheric chemistry. Due to their small size, they can remain suspended in the atmosphere for extended periods.

Large aerosols, with a radius of $0.1 \leq r \leq 1.0$ μm (accumulation mode) or $0.1 \leq r \leq 2.5$ μm, are relatively larger than Aitken aerosols. These particles settle more quickly due to their size and mass, influencing air quality and deposition processes. Their larger size means they can have a more significant impact on atmospheric visibility and contribute to environmental and health-related issues.

Giant aerosols are even larger particles with a radius $r > 1.0$ μm (coarse mode) $d > 2.5$ μm including coarse particles. They settle quickly due to their size and contribute to surface-level contamination. These particles are often associated with industrial emissions, dust, or other natural sources. Liquid particles are always spherical (minimize surface tension), while solid particles can have any shape (Fig. 12.21).

This size-based classification of aerosol particles aids in understanding their behavior, distribution, and impact within the atmosphere. Each size range possesses distinct characteristics that affect their role in atmospheric processes, air quality, and potential impacts on climate and health.

Atmospheric visibility refers to the ability of the human eye to distinguish an object from its background. Aerosols limit visibility primarily through the scattering of solar radiation (Fig. 12.22). Without aerosols, the visual range would be

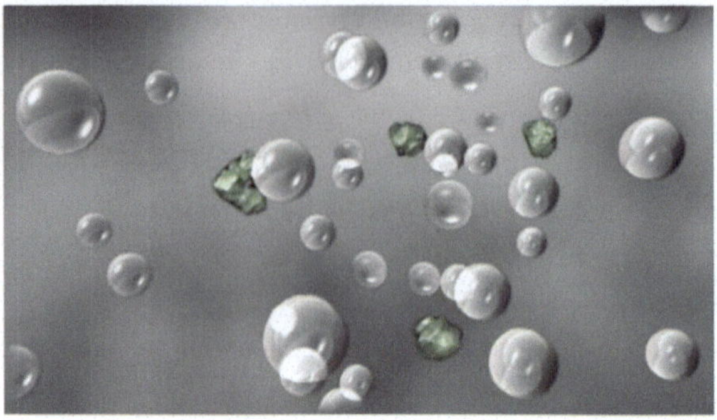

Fig. 12.21 Shape of aerosol particles, illustrating the various forms and structures that can be found in atmospheric aerosols

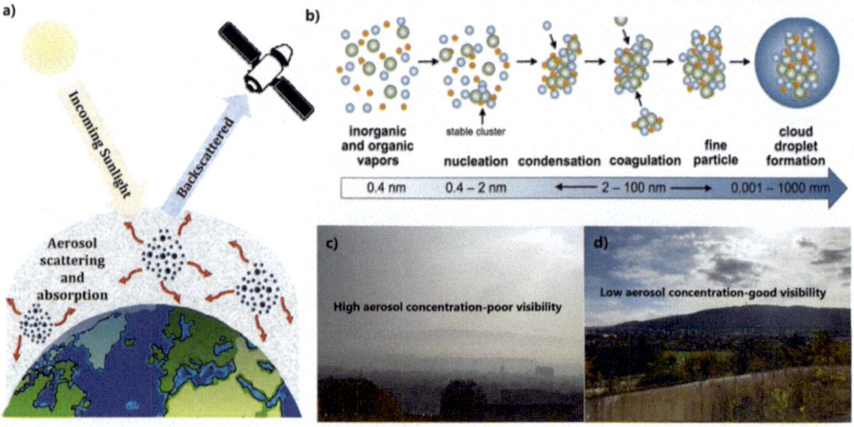

Fig. 12.22 The impact of aerosols on visibility. (**a**) Incoming solar radiation is scattered and absorbed by aerosols, with a portion being backscattered. (**b**) Aerosols of varying sizes and shapes participate in this scattering process. (**c**) High aerosol concentrations lead to reduced visibility by diminishing the contrast between an object and its background. (**d**) Low aerosol concentrations result in better visibility. Aerosol scattering of solar radiation in the field of view reduces object contrast, while scattering from aerosols outside the field of view further diminishes visibility

approximately 300 km, constrained only by scattering from air molecules. In urban environments, anthropogenic aerosols typically reduce visibility by an order of magnitude compared to unpolluted conditions. The reduction in visibility is most pronounced at high relative humidity, as aerosols absorb water and swell, increasing their cross-sectional area and enhancing scattering. This phenomenon, known as fog, significantly diminishes visibility.

12.7 Factors and Conditions Affecting Air Pollution

Atmospheric conditions play a crucial role in determining air quality. Factors such as thermal conditions, atmospheric stability, turbulence, wind flow, and other parameters influence the transport, dispersion, and redistribution of air pollutants. They also affect physical and chemical processes in clouds, the deposition of pollutants, and their qualitative and quantitative characteristics. Air pollution episodes can result not only from increased emissions but also from changes in atmospheric conditions and physical processes.

Two key atmospheric conditions affecting the transport and distribution of pollutants are:

Atmospheric Stratification: The vertical temperature gradient, which defines the stability of the air. This stratification affects how pollutants are trapped or dispersed in the atmosphere.

Wind Characteristics: Wind plays a significant role in the transport and distribution of pollutants. Prevailing wind conditions can either mitigate or exacerbate air quality. Higher wind speeds generally enhance the dispersion of pollutants, leading to lower concentrations. Conversely, low wind speeds can lead to the accumulation of pollutants around the emission source, potentially causing elevated concentrations and extreme air pollution episodes under certain atmospheric conditions (Fig. 12.23).

Fig. 12.23 Prevailing wind and dispersion of pollution. The figure illustrates how wind direction and speed influence the transport and dispersion of pollutants away from their source

12.8 Acid Rain

Acid rain refers to precipitation, such as rain, snow, or fog, that has become more acidic than normal due to chemical alterations. This increased acidity primarily results from atmospheric pollutants, particularly sulfur dioxide (SO_2) and nitrogen oxides (NO_x) . These pollutants undergo chemical reactions with water vapor to form sulfuric acid (H_2SO_4) and nitric acid (HNO_3) The formation of these acids in the atmosphere is facilitated by a complex series of reactions involving sunlight, water vapor, and other trace gases. Clouds play a crucial role in the transport, distribution, chemical transformations, washing, and deposition of these pollutants. The chemistry of clouds, including the processes of acid formation and precipitation, is essential for understanding atmospheric composition, air quality, and the broader impacts on global climate (Fig. 12.24).

Acid rain has far-reaching environmental impacts. The reduction in pH due to acid rain adversely affects natural ecosystems in several ways:

- Water Bodies: Acid rain leads to the acidification of lakes, rivers, and streams, which can harm aquatic life by lowering the pH of the water. This process can result in the loss of fish species and disrupt aquatic ecosystems. Acidification also reduces the availability of essential nutrients and increases the solubility of toxic metals like aluminum.
- Soil: The increased acidity can leach important nutrients from the soil, such as calcium and magnesium, which are vital for plant growth. This nutrient depletion can lead to decreased agricultural productivity and damage to forests. Acid rain also affects soil pH, which can alter microbial activity and reduce soil fertility.

Fig. 12.24 Global simulation of sulfur dioxide and sulfate aerosols. Visualizations by William Patman, NASA. Originally posted on December 10, 2014, and updated on November 14, 2023, at 12:28 pm

- Vegetation: Acid rain can damage vegetation by affecting the leaves, stems, and roots of plants. Acids can weaken plant tissues, making them more susceptible to disease, pests, and environmental stress. Forests can suffer from reduced growth and increased mortality rates, impacting biodiversity and ecosystem stability.
- Built Environment: Acid rain can accelerate the decay of buildings and infrastructure. The acids react with materials such as limestone, marble, and concrete, leading to erosion and structural damage. Historic monuments and buildings are particularly vulnerable to this form of degradation.

In addition to environmental effects, acid rain poses health risks by contaminating drinking water sources. The increased acidity can lead to the mobilization of harmful substances, such as heavy metals, which can have detrimental effects on human health if consumed. Acid rain also affects agricultural productivity, potentially leading to reduced crop yields and impacting food security.

Efforts to mitigate the impacts of acid rain include reducing emissions of sulfur dioxide and nitrogen oxides through regulatory measures and adopting cleaner technologies. Monitoring and research continue to play a crucial role in understanding and addressing the complex interactions between acid rain and environmental systems.

12.9 Air Pollution in Urban Areas

Air pollution in urban areas such as megacities (Fig. 12.25) has become a significant social, health, environmental, and economic issue. As our planet warms and emissions rise due to the use of non-clean fuels and technologies, the air we breathe is

Fig. 12.25 Air pollution in urban areas and megacities. High concentrations of particulate matter and other pollutants reduce air quality, posing health risks and impacting visibility. (Credit: NOAA)

increasingly polluted. On average, an adult inhales about 15,000–20,000 L of air daily. According to the World Health Organization (WHO), nearly nine out of ten people now breathe polluted air, contributing to approximately seven million deaths annually. An unpolluted natural environment is essential for a high quality of life and health.

The WHO and other global agencies report that over 80% of urban residents are exposed to poor air quality, with pollutant levels exceeding established thresholds and guidelines. Worldwide, exposure to particulate matter (PM) is a major public health issue, linked to approximately seven million premature deaths each year. Urban populations are particularly vulnerable, facing high risks of respiratory diseases and other chronic health issues.

Air pollution is influenced by various factors, including uncontrolled emissions, fossil fuel use for heating, waste treatment, traffic, and urbanization. Atmospheric and weather conditions also play a crucial role in determining air quality. For instance, stable anticyclonic weather with high pressure and minimal airflow can lead to increased relative humidity, supersaturation, and fog formation. In such stagnant conditions, particularly in basin-like urban areas, pollution accumulates, and air quality deteriorates significantly. Conversely, when the atmosphere destabilizes—due to cold air intrusion, frontal systems, or other disturbances—winds can disperse pollutants, reduce their concentrations, and improve air quality.

The adverse effects of air pollution are linked to ischemic heart diseases, strokes, and lung cancer, which are the leading causes of death in urban areas. However, not all residents experience health problems, as short-term pollution episodes can vary based on active emissions.

The level of total human exposure to air pollution varies based on the amount of time spent in polluted environments, the infiltration of pollutants into indoor spaces, and the pollution levels in homes and offices. Given that most individuals spend a significant portion of their lives indoors, air pollution becomes a major health concern, particularly for children.

The health effects of air pollutants are categorized into acute (short-term) and chronic (long-term) impacts. Acute effects may include sudden irritation of the eyes and throat, while episodes of high pollution can lead to severe outcomes such as respiratory or cardiovascular failure. Chronic effects, however, develop over extended exposure periods and can include respiratory diseases and cardiovascular conditions.

Certain pollutants, such as sulfur dioxide, have established exposure thresholds. Below these thresholds, health effects may be less pronounced but still present. It is crucial to assess and manage risks, especially for vulnerable groups, including infants, young children, the elderly, pregnant women, and individuals with pre-existing health conditions. Identifying and protecting these high-risk groups is essential, as they are often the first to experience adverse health effects from increased pollution levels.

12.9.1 Ambient Air Quality Forecast System

Urban air quality has emerged as a critical concern for local authorities in large cities, particularly due to deteriorating air quality, which is often exacerbated during the winter season in temperate regions. In these areas, pollution levels can reach alarming levels, significantly impacting public health. Addressing the problem of air pollution is complex, as it involves a range of contributing factors including urbanization, specific topographical features, emission sources, environmental conditions, and meteorological variables.

To manage and mitigate these challenges, an ambient air quality forecast system is essential. Such systems integrate various data sources and analytical tools to predict pollution levels and provide actionable insights. These forecasts help in anticipating periods of high pollution, enabling timely interventions to protect public health. The system typically includes monitoring networks that track pollutant concentrations, meteorological data to understand dispersion patterns, and predictive models to estimate future air quality scenarios.

Effective air quality forecasting not only helps in issuing health advisories and managing pollution sources but also supports long-term planning and policymaking aimed at improving urban air quality. By utilizing advanced forecasting techniques, cities can better prepare for and respond to pollution events, ultimately reducing the adverse health impacts associated with poor air quality.

For this purpose, systems are employed for global, regional, and local air quality modeling, leveraging advanced atmospheric chemical models. These systems incorporate a wide range of data, including emissions from anthropogenic sources, biogenic sources, fossil fuels, and mobile sources.

The modeling frameworks are designed to capture complex interactions and transport processes of pollutants, providing detailed forecasts of air quality across various spatial and temporal scales. The prototype of a modern PM-alarm system, illustrated in Fig. 12.26, exemplifies such a system. This prototype is engineered to

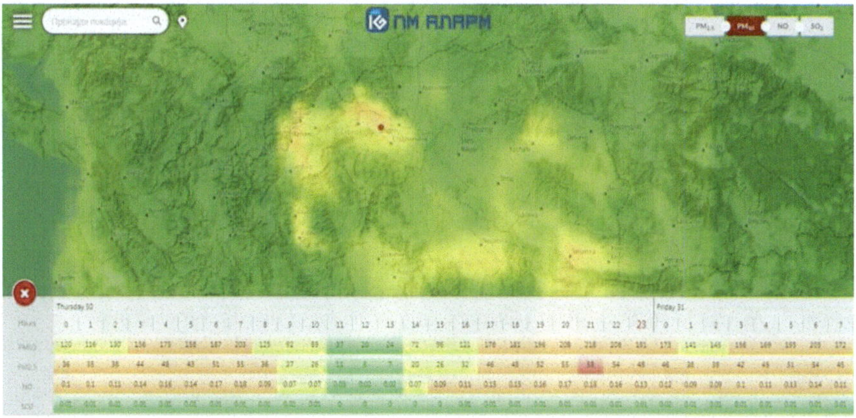

Fig. 12.26 Prototype of the PM-alarm system for detailed forecasting and notification of air pollution levels

deliver high-resolution forecasts and alerts for air pollution over regional domains. It includes a user-friendly website and a mobile application that offers real-time assessments of pollutant concentrations (e.g., PM_{10}, $PM_{2.5}$, SO_2, NO) specific to individual locations. The system features an alarm mechanism that alerts users to potential extreme air pollution episodes up to 5 days in advance, enabling timely public health interventions and personal precautions.

12.9.2 Air Quality Indices

Advancements in air quality monitoring and the growing public awareness of the detrimental effects of air pollution have spurred the development of health-oriented tools to manage air quality. Air Quality Indices (AQIs) are crucial in this regard, offering a means to gauge air quality and its impact on health. These indices are typically classified into two main types:

1. Air Quality Index (AQI): This index is calculated by comparing the concentrations of specific air pollutants to established air quality standards. It identifies the pollutant with the highest concentration relative to its standard, and the AQI is determined based on this pollutant. The AQI provides a snapshot of overall air quality and helps individuals understand the potential health impacts of current pollution levels.
2. Air Quality Health Index (AQHI): This index focuses on the relationship between short-term exposure to air pollutants and adverse health effects. It is represented by numerical values ranging from 1 to 10+, which indicate the level of health risk associated with local air quality. The AQHI offers both a current diagnostic value of air quality and a forecast of maximum air quality levels several days in advance. It also includes health advisories to guide individuals in mitigating adverse effects, especially for sensitive groups such as children, the elderly, and those with pre-existing health conditions.

These indices are essential for informing the public about air quality conditions and helping individuals make informed decisions to protect their health in response to varying pollution levels.

References

Ayres, J., Maynard, R., and Richards, R (2006) Air Pollution and Health. Air Pollution Reviews: Volume 3. pp. 264. World Scientific. https://doi.org/10.1142/p165.

Barrett, G. W., & Odum, E. P. (2000). *Fundamentals of ecology* (5th ed.). Brooks/Cole.

Begon, M., Townsend, C. R., & Harper, J. L. (2006). *Ecology: From individuals to ecosystems* (4th ed.). Blackwell Publishing.

Carpenter, S. R., Caraco, N. F., Correll, D. L., Howarth, R. W., Sharpley, A. N., & Smith, V. H. (1998). Nonpoint pollution of surface waters with phosphorus and nitrogen. *Ecological Applications, 8*(3), 559–568.

Chapin, F. S., III, Matson, P. A., & Mooney, H. A. (2002). *Principles of terrestrial ecosystem ecology*. Springer.

Chapin, F. S., Matson, P. A., & Vitousek, P. (2011). *Principles of terrestrial ecosystem ecology* (2nd ed.). Springer.

Connell, J. H. (1978). Diversity in tropical rainforests and coral reefs. *Science, 199*(4335), 1302–1310.

Fitter, A., & Hay, R. K. M. (2002). *Environmental physiology of plants* (3rd ed.). Academic Press.

Foley, J. A., DeFries, R., Asner, G. P., et al. (2005). Global consequences of land use. *Science, 309*(5734), 570–574.

Goulson, D. (2013). An overview of the environmental risks posed by neonicotinoid insecticides. *Journal of Applied Ecology, 50*(4), 977–987.

Grime, J. P. (2001). *Plant strategies, vegetation processes, and ecosystem properties* (2nd ed.). John Wiley & Sons.

Harrison, R. M. and Hester R. E. (2019) *Indoor Air Pollution*. Royal Society of Chemistry; 1st edition.

Haywood, J. M., & Boucher, O. (2000). Estimates of the direct and indirect radiative forcing due to tropospheric aerosols: A review. *Reviews of Geophysics, 38*(4), 513–543.

Hladik, M. L., Kolpin, D. W., Kuivila, K. M. (2014) Widespread occurrence of neonicotinoid insecticides in streams in a high corn and soybean producing region, USA. *Environ Pollut, 193,* 189–196.

Holling, C. S. (1973). Resilience and stability of ecological systems. *Annual Review of Ecology and Systematics, 4*, 1–23.

IPCC. (2014). *Climate change 2014: Impacts, adaptation, and vulnerability*. Cambridge University Press.

Jones, P. D., Harpham, C., & Briffa, K. R. (2012). Temperature and precipitation. In A. Henderson-Sellers & P. J. Robinson (Eds.), *Contemporary climatology* (2nd ed.). Longman.

Larcher, W. (2003). *Physiological plant ecology* (4th ed.). Springer-Verlag.

Liu, Y., Zhang, H., Chen, H., & Li, Z. (2019). The influence of aerosols on cloud properties and precipitation: A review. *Atmospheric Chemistry and Physics, 19*(3), 1921–1943.

Lovett, G. M., Burns, D. A., Driscoll, C. T., Jenkins, J. C., Mitchell, M. J., Rustad, L., et al. (2007). Who needs environmental monitoring? *Frontiers in Ecology and the Environment, 5*(5), 253–260. https://doi.org/10.1890/1540-9295(2007)5[253:WNEM]2.0.CO;2

Millennium Ecosystem Assessment (MEA). (2005). *Ecosystems and human well-being: Synthesis*. Island Press.

Molles, M. C. (2019). *Ecology: Concepts and applications* (8th ed.). McGraw-Hill Education.

Odum, H. T. (1983). *Systems ecology: An introduction*. Wiley.

Odum, E. P., & Barrett, G. W. (2005). *Fundamentals of ecology* (5th ed.). Brooks Cole.

Pallavi, S., and Srivastava, A. (2020) *Air Pollution and Environmental Health*. Springer. ISBN: 978-981-15-3481-2. https://doi.org/10.1007/978-981-15-3481-2.

Pereira, H. M., Ferrier, S., Walters, M., Geller, G. N., Jongman, R. H., Scholes, R. J., et al. (2013). Essential biodiversity variables. *Science, 339*(6117), 277–278. https://doi.org/10.1126/science.1229931

Pickett, S. T. A., & White, P. S. (1985). *The ecology of natural disturbance and patch dynamics*. Academic Press.

Reid, C. E., Brauer, M., Johnston, F. H., Jerrett, M., Balmes, J. R., & Elliott, C. T. (2016). Critical review of health impacts of wildfire smoke exposure. *Environmental Health Perspectives, 124*(9), 1334–1343.

Simberloff, D. (2013). *Invasive species: What everyone needs to know*. Oxford University Press.

Smith, T. M., & Smith, R. L. (2015). *Elements of ecology* (9th ed.). Pearson.

Sutherland, W. J., Dicks, L. V., Ockendon, N., & Smith, R. K. (Eds.). (2013). *What works in conservation*. Cambridge University Press.

Tansley, A. G. (1935). The use and abuse of vegetational concepts and terms. *Ecology, 16*(3), 284–307.

Turner, B. L., Matson, P. A., McCarthy, J. J., et al. (2001). *Climate change science: An analysis of some key questions*. National Research Council/National Academies Press.

Vitousek, P. M., Mooney, H. A., Lubchenco, J., & Melillo, J. M. (1997). Human domination of Earth's ecosystems. *Science, 277*(5325), 494–499. https://doi.org/10.1126/science.277.5325.494

Whittaker, R. H., & Levin, S. A. (1977). *Niche: Theory and application*. Dowden, Hutchinson & Ross.

World Health Organization (WHO). (2021). *Air quality and health*. WHO.

Zhang, Q., et al. (2019). The role of aerosol-cloud interactions in climate variability and change: A review. *Earth-Science Reviews, 188*, 9–31.

Chapter 13
The Planet Under Pressure: Challenges in the Twenty-First Century

13.1 The Climate System: Structure and Dynamics

The climate system is a complex interplay of various components that govern the long-term behavior of the Earth's atmosphere, hydrosphere, cryosphere, lithosphere, and biosphere (Fig. 13.1). While weather refers to short-term atmospheric conditions experienced daily, climate encompasses these patterns over extended periods—decades, centuries, and even millennia (National Oceanic and Atmospheric Administration NOAA, 2022). Modern climate science goes beyond mere averages of temperature and precipitation; it delves into the variability, extremes, and

© The Author(s), under exclusive license to Springer Nature Switzerland AG 2025
V. Spiridonov et al., *Atmospheric Perspectives*,
https://doi.org/10.1007/978-3-031-86757-6_13

Fig. 13.1 The climate system and its integral components: atmosphere, hydrosphere, cryosphere, geosphere, and biosphere. (Source: Wikipedia)

evolving trends that shape regional and global climates. Climate defines not only the expected conditions in a region but also its response to natural and anthropogenic factors (IPCC, 2021; Hartmann, 2016).

Understanding climate requires an integrated approach that considers both the fundamental drivers of climate patterns and the feedback between them. The climate system's complexity lies in the interactions between its different elements, each influencing one another over various timescales. Central to this system are the following key factors:

1. Solar Radiation: Solar energy is the primary driver of the climate system, influencing temperature, atmospheric circulation, and ocean currents. Variations in solar activity and the Earth's orbital characteristics (Milankovitch cycles) can lead to long-term climate shifts, such as ice ages (Berger et al., 2010).
2. Atmospheric Composition: The concentration of greenhouse gases (CO_2, methane, nitrous oxide, and water vapor) determines the Earth's ability to trap heat. Human activities have significantly altered this composition, leading to enhanced greenhouse effects and global warming (Hartmann, 2016).
3. Ocean Circulation: Oceans store and distribute heat across the globe through currents like the Gulf Stream and deep ocean circulation. Changes in oceanic processes, such as the Atlantic Meridional Overturning Circulation (AMOC), can drastically influence regional climates and weather patterns.
4. Cryosphere: Glaciers, ice caps, and sea ice reflect solar radiation and regulate the Earth's energy balance. As the cryosphere melts due to warming, it not only reduces albedo (reflectivity) but also contributes to rising sea levels.
5. Land Surface: Topography, vegetation cover, and land use significantly impact local and global climates. Mountain ranges affect wind patterns and precipita-

tion distribution, while deforestation alters the carbon cycle and hydrological processes.

6. Anthropogenic Influences: Human activities—industrialization, deforestation, and urbanization—have become dominant forces reshaping the climate system. These activities contribute to the increasing concentration of greenhouse gases and aerosols, leading to climate change.

13.1.1 Defining the Climate in the Modern Era

The modern definition of climate encompasses both the statistical description of weather patterns and the occurrence of extreme events over an extended period, typically 30–100 years (NRC, 2002; Geiger and Pohl, 1998). This definition captures the inherent variability in climate, as regions can experience long-term cycles of wet and dry periods, heat waves, cold snaps, and changes in wind patterns. Importantly, the boundary between weather and climate is not rigid; they exist on a continuum of timescales. Certain atmospheric oscillations, such as the North Atlantic Oscillation (NAO) and the Madden–Julian Oscillation (MJO), further blur this distinction, as they manifest over weeks to months, affecting both short-term weather and long-term climate variability. One of the most influential large-scale phenomena affecting global climate is the El Niño-Southern Oscillation (ENSO), which involves periodic changes in sea surface temperatures in the equatorial Pacific. ENSO events, occurring every 2–7 years, significantly alter atmospheric circulation patterns and global precipitation regimes. Its counterpart, La Niña, also has wide-reaching impacts on weather systems. These periodic fluctuations illustrate the crucial role of the ocean–atmosphere interface in regulating the Earth's climate (Cai et al., 2020).

13.1.2 Climate System Feedback and Climate Change

The feedback mechanisms within the climate system are key to understanding its sensitivity to perturbations. For instance, the ice-albedo feedback—where melting ice reduces the Earth's reflectivity and accelerates warming—is a major amplifier of polar climate change. Similarly, the carbon cycle feedback, where warming causes the release of stored carbon in forests and permafrost, enhances greenhouse gas concentrations and accelerates global warming (IPCC, 2021). In the twenty-first century, human-induced climate change has emerged as the most pressing environmental challenge. The Intergovernmental Panel on Climate Change (IPCC) has documented that the ongoing increase in global temperatures, sea-level rise, and the frequency of extreme weather events are clear indicators of a rapidly changing climate system. Understanding these processes, their interactions, and the projected impacts is critical for formulating strategies for mitigation and adaptation (IPCC, 2023a, b, c).

13.1.3 Assessing Climate Variability and Predictability

The predictability of the climate system hinges on our ability to model its many components accurately. Climate models simulate the behavior of the atmosphere, oceans, and land surfaces, offering valuable insights into how the climate may evolve under different scenarios of greenhouse gas emissions. These models are essential tools for forecasting regional climate impacts, such as changes in monsoon systems, shifts in agricultural zones, or increased risk of droughts and floods. However, there are uncertainties in climate projections due to the inherent variability of the system and the influence of unforeseen factors, such as volcanic eruptions or abrupt changes in ocean circulation. Despite these uncertainties, climate models remain critical for informing policy decisions aimed at reducing greenhouse gas emissions and preparing for climate-related challenges (Hartmann, 2016). In conclusion, the climate system is a highly dynamic and interconnected network of processes that govern the Earth's long-term atmospheric behavior. Its complexity demands a multidisciplinary approach to understanding, predicting, and managing the impacts of ongoing climate change. As human activities continue to alter the balance of natural systems, addressing the challenges posed by a changing climate is essential for the future sustainability of the globe.

13.2 Climate Types and Sensitivity to Climate Change

The earliest known climate classification, attributed to the ancient Greeks, divided each hemisphere into hot, temperate, and frigid zones. Modern climate classification systems, such as the **Köppen-Geiger classification**, refine these early attempts by considering temperature, precipitation, and seasonal patterns to define climate zones (Köppen, 1936; Peel et al., 2007). The Köppen system identifies five primary climate groups:

A: Humid tropical climate
B: Dry climate
C: Temperate mid-latitude climate with wet conditions and mild winters
D: Continental mid-latitude climate with wet conditions and cold winters
E: Polar climate

Four of these groups (A, C, D, and E) are primarily defined by air temperature, while the dry climate (B) is distinguished by precipitation as the determining factor.

13.2.1 Climate Sensitivity to Disturbed Climate System

Each of these climate types exhibits varying sensitivity to changes in the global climate system. The impacts of climate change are distributed unevenly across regions, with some climate zones undergoing more profound transformations than others due to shifts in atmospheric circulation, temperature anomalies, and precipitation patterns. Below is an examination of these climate types in the context of a changing climate:

A: Humid Tropical Climate. Located along the equator, this climate type is characterized by high temperatures and year-round precipitation. As global temperatures rise, tropical regions are likely to experience increased frequency and intensity of extreme weather events, such as tropical cyclones, heavy rainfall, and floods. Changes in the Intertropical Convergence Zone (ITCZ) could lead to shifts in rainfall distribution, affecting the dense ecosystems that thrive in tropical rainforests. Although the average temperatures may rise only slightly, the impacts on precipitation and seasonal variability can disrupt the balance of these ecosystems.

B: Dry Climate. Covering about 30% of Earth's land surface, dry climates are highly sensitive to changes in precipitation patterns. These regions are already defined by an annual water deficit, where evaporation exceeds rainfall. With climate change exacerbating drought conditions, arid regions may face more severe water scarcity, intensifying desertification and reducing agricultural productivity. The Köppen formula, which considers temperature and precipitation, becomes increasingly relevant as slight shifts in these variables can push dry regions toward even more extreme aridity.

C: Temperate Mid-Latitude Climate. Mild winters and moderate precipitation define this climate type. As global temperatures rise, temperate zones may experience warmer winters, potentially disrupting ecosystems that rely on cold-season processes. Changes in precipitation patterns, such as more frequent extreme rainfall or prolonged droughts, may also occur. Regions with a temperate climate are vulnerable to shifting seasonal cycles, which could affect agriculture, water resources, and biodiversity.

D: Continental Mid-Latitude Climate. Continental climates, characterized by large annual temperature variations, are particularly sensitive to changes in global circulation patterns. These regions may experience both warmer winters and hotter summers, leading to more extreme temperature fluctuations. Increased variability in snowpack, earlier snowmelt, and more frequent heat waves are expected in continental zones, affecting water resources and increasing the likelihood of severe weather events like floods and droughts.

E: Polar Climate. Polar regions, where the average temperature of the warmest month is less than 10 °C, are experiencing some of the most dramatic effects of climate change. The phenomenon of Arctic amplification—the enhanced warming in polar regions compared to the rest of the globe—leads to faster ice melt and significant changes in temperature patterns. These changes disrupt ecosystems that depend on permafrost and ice cover while also contributing to global sea-level rise. Polar climates are classified as humid due to low evaporation rates, but they receive minimal precipitation, and the loss of ice and snow cover further exacerbates warming in these areas.

- Tundra Climate (ET): The permafrost in tundra regions is particularly vulnerable to thawing, which releases greenhouse gases like methane, further accelerating global warming.
- Ice Cap Climate (EF): Permanent ice cover is rapidly diminishing, with implications for global sea levels and climate patterns.

Mountain climates are highly sensitive to altitude-related changes in temperature and precipitation. The effects of climate change are amplified in mountainous regions, where glaciers are melting, and snowpacks are decreasing. These shifts affect water resources, ecosystems, and local weather patterns. Enhanced orographic precipitation may result in more intense localized storms, while higher temperatures could lead to biodiversity shifts as species migrate to higher altitudes in search of cooler conditions.

13.2.2 Linking Climate Types to a Changing System

As the global climate system becomes increasingly disturbed, the sensitivity of different climate types to temperature and precipitation changes plays a crucial role in determining regional impacts. Dry regions face heightened risks of desertification, while tropical zones may see shifts in the ITCZ, altering precipitation patterns. Continental and temperate zones may experience more intense seasonal extremes, and polar regions are on the front lines of rapid ice loss and warming. Mountain climates, with their unique microclimates, are already undergoing significant changes, particularly in terms of snow and glacier melting. The evolving nature of the Earth's climate system underscores the need for an updated understanding of how climate types interact with the broader patterns of change. By analyzing the sensitivity of these climate zones, we can better anticipate the regional consequences of global warming and develop strategies for adaptation and resilience.

13.2.3 Climatic Factors and Controllers

The Köppen climate classification provides a framework for understanding global climate zones by analyzing recurring meteorological patterns. These patterns are influenced by several interconnected factors that shape atmospheric processes and climate variability. Latitude is a primary determinant of climate, as it influences the amount of solar energy a region receives. Areas closer to the equator experience more direct sunlight and higher temperatures, while higher latitudes receive less solar energy, resulting in cooler climates. The distribution of land and water significantly affects regional climates. Large bodies of water, such as oceans and seas, have moderate temperature fluctuations, creating maritime climates with mild variations in temperature. In contrast, regions distant from these water bodies, known as continental climates, experience greater extremes in temperature due to the absence of this moderating effect. Geographical location, combined with prevailing wind directions, also plays a critical role in shaping climate. Coastal areas, particularly those on the windward side, are influenced by moist air from nearby water bodies, resulting in milder and more stable temperatures. Mountains, or orography, act as natural barriers to air masses, especially moist maritime air. As these air masses are forced to rise over mountain ranges, they cool and condense, leading to orographic precipitation on the windward side. The leeward side, however, often experiences dry conditions due to the rain shadow effect.

Ocean currents further contribute to climate variability by transferring heat across different regions. Warm currents, such as the Gulf Stream, elevate air temperatures along coastal areas, often making them warmer than other regions at similar latitudes. Finally, the global distribution of high- and low-pressure systems, along with associated wind patterns, has a profound influence on precipitation and temperature. These systems drive the movement of air masses and govern the distribution of weather phenomena, shaping the diverse climates observed across the globe. These factors work in tandem to shape the complex and dynamic climate system, producing diverse regional climates across the globe.

13.3 Climate in Crisis: Understanding the Pressures of the Twenty-First Century

The Earth's climate system is a highly intricate and interconnected network of processes that regulate temperature, weather patterns, and the distribution of life across the globe. It comprises several key components: the atmosphere, hydrosphere (including oceans and freshwater bodies), cryosphere (frozen water systems), lithosphere (Earth's solid crust), and biosphere (all living organisms). These components continuously interact, driven by energy from the sun, which creates the conditions necessary for life as we know it.

At the heart of the climate system is the energy balance, where the amount of solar energy absorbed by the Earth is countered by the energy radiated back into space. This balance is crucial for maintaining a stable climate. However, human activities, particularly the burning of fossil fuels and deforestation, have disrupted this equilibrium by increasing the concentration of greenhouse gases, such as carbon dioxide, methane, and nitrous oxide, in the atmosphere. These gases trap heat, resulting in a warming effect known as the greenhouse effect, which is now being amplified far beyond natural levels (IPCC, 2021).

The consequences of this disruption are profound. The warming of the atmosphere and oceans is altering weather patterns, shifting the distribution of precipitation, and increasing the frequency and intensity of extreme weather events, such as hurricanes, droughts, and heat waves. The melting of polar ice and glaciers contributes to rising sea levels, threatening coastal communities and ecosystems.

Furthermore, the warming climate is impacting ocean circulation patterns, which play a critical role in regulating global temperatures and climate stability (NOAA, 2023).

As the climate system comes under increasing strain, the impacts on the biosphere become more evident. Species are being forced to migrate, ecosystems are destabilized, and biodiversity is under threat. The cascading effects of these changes are not isolated; they ripple through economies, societies, and the environment, creating complex and far-reaching challenges. Understanding the Earth's climate system is essential for grasping the full scope of these challenges. It is not just a scientific endeavor but a fundamental requirement for informed decision-making and effective action (Houghton et al., 2001).

As we continue to exert pressure on this delicate system, understanding its workings, vulnerabilities, and resilience becomes ever more critical. The choices we make today will determine how well the climate system can withstand these pressures and what kind of planet we leave for future generations. Grasping climate change requires a thorough understanding of the fundamental characteristics of the climate system and its dynamics. At a planetary scale, the global climate is primarily driven by the energy received from the Sun. Changes in this energy balance directly influence the global climate. The global climate system consists of several key physical components: the atmosphere, oceans, cryosphere (ice caps), biosphere (living organisms), and geosphere (land, sediments, and rocks). Each component plays a role in transferring heat across the Earth's surface and interacts with others across various spatial and temporal scales. The atmosphere does not function in isolation; it exchanges energy with other parts of the climate system, notably the oceans. For instance, ocean currents transport heat from equatorial regions to colder polar areas. Additionally, moisture transfer plays a critical role: water evaporating from the ocean surface absorbs heat, which is then released during condensation to form clouds and precipitation. The oceans, which store significantly more heat than the atmosphere, have about 30 times the heat content above 200 m depth compared to the atmosphere. This substantial heat capacity means that energy flow between the ocean and the atmosphere can profoundly impact climate change. Theoretical frameworks for climate change highlight that the global climate must be viewed as an interconnected system involving the atmosphere, oceans, ice, and land. Changes in this complex system can create forcing agents—both internal and external—that drive climate change. External forcing mechanisms originate outside the climate system, such as changes in solar radiation or volcanic eruptions, while internal forcing mechanisms arise from interactions within the climate system itself.

13.3.1 Forcing Mechanisms and Earth's Climate Evolution

Climate change involves complex processes that redistribute energy within the Earth's climate system. These processes are shaped by two primary mechanisms: non-radiative and radiative forcing, both of which influence the planet's climate over a wide range of timescales.

Non-radiative forcing refers to mechanisms that affect the climate system without directly influencing the atmospheric energy budget, which is the balance between incoming solar radiation and outgoing Earth radiation. These mechanisms generally operate over extensive geological timescales (10^7 to 10^9 years) and include factors such as the position and size of mountain ranges and the state of ocean basins. Such features impact climate through their effects on heat distribution and atmospheric circulation. In contrast, radiative forcing directly alters the Earth's energy budget. This category includes changes in Earth's orbit around the Sun, variations in solar radiation, volcanic eruptions, and shifts in atmospheric composition.

Radiative forcing mechanisms are pivotal for understanding how energy input and output changes affect the global climate system. Climate change operates across various timescales, driven by both random and periodic forcing factors. Random forcing factors are stochastic mechanisms resulting from the complexity and chaotic behavior of the climate system. Influenced by internal feedback processes, these changes occur over timescales ranging from a few years to hundreds of millions of years, contributing to unpredictable aspects of climate variability. Conversely, periodic forcing factors offer more predictability and are linked to external or internal climate processes. For instance, variations in Earth's orbital parameters, as proposed by Kroll in the mid-nineteenth century, recur over a range of periodicities (200–500 million years) and are associated with the solar system and galactic movements. These mechanisms significantly influence Earth's climate by causing shifts in energy distribution and temperature patterns. Earth's climate history spans billions of years, marked by dramatic shifts driven by internal and external influences. Around 4.6–4 billion years ago, during Earth's early Hadean Eon, the planet was a molten mass with an unstable atmosphere dominated by volcanic outgassing. The subsequent cooling period led to the formation of oceans and the transformation of Earth's atmosphere. The Great Oxygenation Event, approximately 2.4 billion years ago, drastically altered atmospheric composition, introducing significant oxygen and driving climate evolution (Lyons et al., 2014). During the Proterozoic Eon, Earth experienced extreme glaciation episodes known as "Snowball Earth," where the entire globe may have been ice-covered. These frozen intervals alternated with warmer interglacial periods (see Fig. 13.2), highlighting the dynamic nature of Earth's climate system (Hoffman et al., 2017). The Phanerozoic Eon, starting 541 million years ago, witnessed the emergence of a more recognizable climate framework. The Carboniferous period featured lush forests and swamps, while the subsequent Permian period ended with global cooling and one of Earth's largest mass extinctions (Benton, 2019). The Mesozoic era, characterized by high temperatures and elevated (CO_2) levels, fostered tropical climates, whereas the Cenozoic era saw a gradual cooling trend, culminating in the ice ages of the Pleistocene epoch (Zachos et al., 2001).

Understanding today's global climate requires studying the complex interactions among the atmosphere, oceans, land surfaces, and biosphere, along with external influences like solar radiation and volcanic activity. These interactions create diverse climate zones ranging from polar extremes to tropical regions. Global climate reflects the cumulative behavior of Earth's climate system, encompassing energy

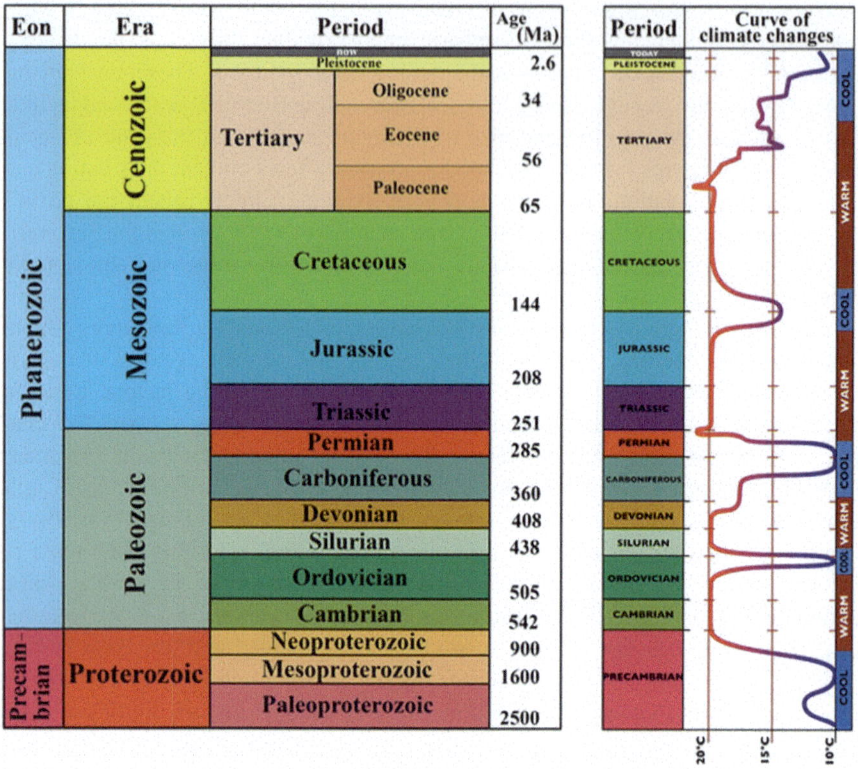

Fig. 13.2 A comprehensive timeline of key developments in climate science, illustrating the evolution of understanding regarding the role of carbon dioxide (CO_2) and other greenhouse gases in Earth's climate system. (Credit: Skeptical Science)

exchange, moisture distribution, and gas interactions. Examining Earth's deep climate history reveals how forcing mechanisms have shaped the planet over billions of years. Insights from these processes provide a framework for understanding modern climate change, including the risks posed by rising (CO_2) levels due to human activities. As we confront contemporary environmental challenges, understanding the scale and scope of climate dynamics becomes critical for mitigating impacts and fostering sustainable global systems.

13.4 Brief History of Climate Science

The understanding of carbon dioxide (CO_2) as a greenhouse gas, capable of trapping heat and warming the Earth, has roots that trace back nearly 200 years. This journey began with early research into how various gases impact Earth's climate, eventually leading to the identification of carbon dioxide's critical role in climate change. In

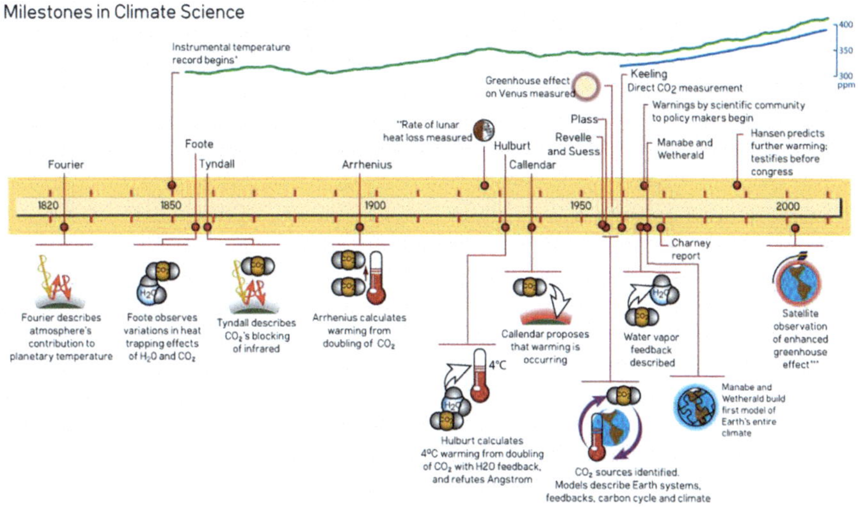

Fig. 13.3 Climate science—the complete timeline of the history of climate science, Credit: Skeptical Science

the early nineteenth century, scientists such as John Tyndall conducted experiments that demonstrated the heat-absorbing properties of gases like CO_2 and water vapor. Tyndall's work laid the foundation for understanding how these gases contribute to the greenhouse effect. By the late nineteenth century, the connection between CO_2 concentrations and global temperature was further elucidated by Svante Arrhenius, who proposed that increases in atmospheric CO_2 from fossil fuel combustion could lead to significant global warming (Arrhenius, 1896). Over the past two centuries, scientific discoveries have linked (CO_2) to historical climate shifts and underscored the dangers of rising CO_2 concentrations resulting from fossil fuel consumption. Research from ice core samples and climate models has illustrated how elevated CO_2 levels correspond with past warming events, shaping our contemporary understanding of climate change and its far-reaching consequences (see Fig. 13.3) (Lüthi et al., 2008; Marcott et al., 2013). Today, the scientific consensus recognizes that human activities, primarily the burning of fossil fuels and deforestation, are driving unprecedented increases in atmospheric CO_2 levels, resulting in climate change with potentially devastating impacts on ecosystems and human societies.

13.4.1 *The Beginnings: Fourier and the Greenhouse Effect*

The groundwork for the greenhouse effect was laid in the 1820s by French physicist Jean Baptiste Joseph Fourier (1768–1830). Fourier investigated heat behavior and calculated that, based on its distance from the Sun, the Earth should be significantly colder than it is. He hypothesized that an unknown factor was keeping the planet warm, suggesting that while sunlight penetrates the atmosphere to heat the Earth's surface, heat radiates back as infrared energy that struggles to escape. This led to the idea of the atmosphere acting as an insulating blanket, trapping heat. Despite the limitations of his era's technology, Fourier's work remarkably foreshadowed our understanding of atmospheric heat retention.

13.4.2 1850s to Early 1860s: Foote, Tyndall, and Heat-Trapping Gases

About 35 years after Fourier's theories, significant advancements in the study of heat-trapping gases emerged. American scientist Eunice Foote was among the first to identify carbon dioxide and water vapor as significant heat-trapping agents, although her contributions were largely unrecognized at the time. Around the same period, Irish physicist John Tyndall conducted experiments that confirmed that water vapor and carbon dioxide effectively trap heat. His research established a critical milestone in climate science by demonstrating how changes in atmospheric CO_2 could influence global temperatures, reinforcing Fourier's earlier theories.

13.4.3 Carbon Dioxide: Arrhenius Makes a Discovery

Building on the insights of Foote and Tyndall, Swedish scientist Svante Arrhenius (1859–1927) further explored CO_2's role in regulating Earth's temperature in the late nineteenth century. Arrhenius noted that while water vapor fluctuated in concentration, CO_2 remained in the atmosphere for much longer. His calculations predicted that increased CO_2 levels would lead to warming, which would subsequently raise water vapor levels, creating a positive feedback loop. He estimated that doubling atmospheric CO_2 could increase global temperatures by 5–6 °C, though this was perceived as a distant concern at the time.

13.4.4 Controversy: Arrhenius vs. Ångström

In the early twentieth century, Arrhenius's findings faced significant skepticism. Critics raised concerns regarding potential oversimplifications, neglect of cloudiness changes, and experiments conducted by Knut Ångström. Ångström's research measured infrared radiation absorption through a short tube filled with carbon dioxide, leading him to conclude that CO_2 quickly saturated, rendering it ineffective in radiation absorption. He argued that water vapor's presence would overshadow the effects of increasing CO_2 levels. However, Ångström's measurements were later proven flawed; his reported decrease in absorption was closer to 1% than 0.4%, a crucial factor in understanding planetary temperatures. Arrhenius countered that water vapor's limited presence in the upper atmosphere would diminish its impact, but his rebuttals were largely overlooked, causing the carbon dioxide greenhouse effect hypothesis to remain dormant for over two decades.

13.4.5 Renewed Interest: Hulburt and Callendar in the 1930s

A revival of interest in climate science emerged in the 1930s, spearheaded by American physicist E.O. Hulburt, who revisited the effects of doubling CO_2, estimating a warming of approximately 4 °C when factoring in water vapor. He challenged Ångström's conclusions, asserting that the escape of infrared radiation to space was the key issue rather than convective processes. Despite Hulburt's significant insights, his work went unnoticed due to publication in a lesser-read journal. Seven years later, English engineer Guy Callendar analyzed temperature records, identifying an early twentieth-century warming trend alongside a 10% increase in atmospheric CO_2. He posited that this rise might be responsible for the observed warming and predicted a potential transition to a permanently warmer climate. Callendar's ideas, however, faced skepticism as critics questioned historical CO_2 measurement accuracy and the ocean's capacity to absorb excess gas. He contended that while the ocean's upper layers might become saturated, limiting their CO_2 absorption, concerns over overlapping absorption bands of water vapor and CO_2 persisted. Critics noted Callendar's omission of convection—an important heat transfer mechanism—yet his calculations indicated a possible 2 °C rise from doubled CO_2 concentrations, which resonated amid the limited data of the time.

13.4.6 Climate Science Advances Post-1945

Following World War II, a renewed focus on climate research emerged, fueled by advancements during the Cold War. Understanding atmospheric processes became crucial for military applications, particularly for infrared radiation, which could guide missiles to heat sources like jet exhaust. Reassessments of Ångström's experiments revealed limitations due to outdated measuring equipment. Modern technologies identified distinct absorption lines for CO_2 and water vapor, confirming minimal overlap and highlighting the necessity to consider characteristics of different atmospheric layers. This validation acknowledged earlier theories proposed by Hulburt, Callendar, and Arrhenius. In the mid-1950s, scientists like Gilbert Plass utilized computers for accurate analyses of atmospheric layers. Plass's 1956 paper reinforced the understanding that increased CO_2 would lead to warming, estimating a 3–4 °C rise with doubled CO_2 levels. Despite this, the scientific community remained cautious, with critics questioning water vapor, cloudiness, and ocean absorption roles in climate dynamics.

13.4.7 Carbon Isotopes and Climate Research

The 1950s marked a transformative period in climate research, primarily driven by the fallout from nuclear tests introducing carbon-14 into the atmosphere. Unlike stable carbon-12, carbon-14 undergoes radioactive decay, enabling scientists to track its distribution and establish rapid mixing of long-lived gases in the atmosphere. Chemist Hans Suess, studying carbon isotopes, found that younger trees had elevated levels of carbon-12 and carbon-13 relative to carbon-14, indicating fossil fuel emissions. Suess and Roger Revelle's research emphasized the oceans' role in CO_2 absorption. They discovered that while oceans could initially absorb emitted CO_2, buffering chemicals limited this capacity. Revelle estimated a potential 40% increase in atmospheric CO_2 over centuries if current emission rates continued, underscoring humanity's significant geophysical experiment with the planet's climate. By the late 1950s, accurate atmospheric CO_2 monitoring became critical, with Charles David Keeling pioneering improved measurement techniques at Mauna Loa, Hawaii. Keeling's findings confirmed rising CO_2 levels, with seasonal fluctuations aligned with plant growth cycles. By 1965, a multidisciplinary approach emerged, integrating atmospheric scientists, biologists, and geochemists, culminating in warnings that increased atmospheric CO_2 could lead to significant climate changes by the year 2000.

13.4.8 Mid-1960s: Development of the First Climate Model

In 1967, **Syukuro Manabe** and **Richard Wetherald** developed the first comprehensive computer model simulating Earth's climate. This pioneering model revealed the crucial role of heat movement through convection in regulating temperatures, suggesting that doubling atmospheric CO_2 could lead to a temperature rise of about 2 °C. The model highlighted how convective updrafts transport substantial heat from the Earth's surface to the upper troposphere, enhancing heat transfer as surface temperatures rise. This insight addressed longstanding questions in climate science that had previously remained unexplored.

13.4.9 1970s: Doubts and Speculations

Despite the advancements in understanding climate dynamics, skepticism persisted within the scientific community. Climatologist Helmut Landsberg remarked in 1970 that a CO_2 increase might only yield a 2 °C rise over four centuries, which he deemed "hardly cataclysmic." Researchers like S. Ichtiaque Rasool and Stephen Schneider also investigated how pollution, including aerosols, could induce cooling episodes, raising speculation about a potential new ice age.

13.4.10 Continued Research in the 1970s and 1980s

Research progressed throughout the 1970s and 1980s, aligning with a renewed warming trend. Scientists identified other potent greenhouse gases and recognized limitations on CO_2 absorption by plants and oceans. By 1981, climatologists Tom Wigley and Phil Jones warned that CO_2's effects might remain undetectable until the turn of the century, by which time concentrations would likely be high enough to provoke significant climate changes.

13.4.11 Ice Cores and Ancient Air

The drilling of ice cores in Greenland and Antarctica became essential for reconstructing past climates. These cores trap tiny bubbles of prehistoric air, enabling scientists to analyze atmospheric composition over ice ages and interglacials. For instance, during the last ice age, CO_2 levels were recorded below 200 parts per million. The mid-1980s Vostok core, drilled to 2 km, provided 150,000 years of climate history, illustrating a consistent relationship between CO_2 fluctuations and temperature changes. Ice core data also indicated feedback mechanisms whereby warming releases greenhouse gases from oceans and permafrost, amplifying temperature increases. As CO_2 levels climbed to unprecedented heights, surpassing those seen over the last two million years, the greenhouse effect emerged as a significant concern. Although future temperature rises remained uncertain, models suggested a likely increase of 3 °C with doubled pre-industrial CO_2 concentrations. Research into long-term climate changes, driven by geological processes like plate tectonics, revealed that warmer periods in Earth's history corresponded with elevated CO_2.

13.5 Climate System Modeling

Climate models aim to simulate the behavior of the climate system by representing key physical, chemical, and biological processes that drive climate dynamics. These models provide a framework to understand past climates through comparison with empirical observations and enable predictions of future climate changes. Climate models operate at various spatial and temporal scales (Warner, 2011), ranging from regional to global. Regional climate studies often focus on localized phenomena, while global-scale models simulate the planet's overall climate system.

The core physical processes essential for climate modeling include

1. **Radiative Processes**—encompassing the transfer, absorption, reflection, and scattering of radiation within the climate system
2. **Dynamic Processes**—covering the horizontal and vertical transfer of energy through mechanisms such as advection, convection, turbulence, and diffusion
3. **Surface Processes**—involving interactions between the Earth's surface (land, ocean, ice) and the atmosphere, including albedo effects, emissivity, and energy exchange

Accurately representing these processes is crucial for capturing the behavior of the climate system. The fundamental laws and relationships governing the climate system are expressed through a set of equations, which can be empirical or derived from primitive equations of motion. Solving these equations typically involves finite difference methods, making it vital to carefully consider the spatial and temporal resolution of the model.

13.5.1 Simplifying the Climate System

All climate forecasting models significantly simplify the complex climate system due to both limited knowledge and computational constraints. Simplifications are applied in various ways, including spatial dimensionality, spatiotemporal resolution, and the parameterization of processes. The simplest models operate in zero dimensions, where the state of the system is represented by a single global average. As complexity increases, models expand to 1D, 2D, and 3D representations, each adding layers of spatial detail. However, even within these spatial dimensions, models often simplify further by reducing spatial resolution.

Temporal resolution in climate models also varies widely, ranging from minutes to years, depending on the model's purpose and the specific phenomena being studied. Maintaining a stable model requires careful alignment between spatial and temporal resolutions. In many cases, atmospheric processes occurring at scales smaller than the model's resolution cannot be explicitly resolved. To address this, these processes are parameterized, meaning they are represented by simplified or semi-empirical functions or parameters. For example, turbulent processes within thunderstorms are typically parameterized due to their subscale nature.

Parameterization not only addresses unresolved processes but also helps reduce computational demands. Some processes may be entirely omitted if their influence is negligible on the timescale of interest. For instance, deep ocean circulation can be excluded from models focused on changes over years to decades. While some models may accurately simulate radiative transfers, they may need to parameterize or simplify energy transport. Conversely, other models may offer detailed 3D representations but with less precise treatment of radiative transfer processes.

13.5.2 Climate Models

The most common division of climate models is the following:

- Energy balance models (EBM).
- (1D) radiative-convective models (RCMs).
- (2D) statistical-dynamic models (SDMs).
- (3D) general circulation models (GCMs). Energy balance models are the simplest (0D) models that simulate energy balance in the atmosphere.

The remaining 1D, 2D, and 3D models have a gradual increase in complexity depending on the extent to which they simulate individual processes and their temporal and spatial resolution. The simplest models allow little interaction between primary processes, radiation, dynamics, and surface processes, while the most complex models are fully interactive and coupled (Trenberth, 2010).

Energy balance models (EBMs) simulate two basic processes that regulate the state of the climate: (a) global radiative balance (between incoming solar radiation and outgoing Earth radiation) and (b) geographic transfer of energy (from the equator to the poles). Radiation-convective models: Radiative-convective models (RCMs) are either 1D or 2D, with height as a dimension that is currently irreversible. RCMs simulate in detail the transfer of energy through the thickness of the atmosphere, including (a) radiative transformations that occur when energy is absorbed, emitted, and scattered and (b) the role of convection, the transfer of energy by atmospheric vertical motion, in maintaining stability. Two-dimensional models (RCMs) can also simulate horizontally average energy transfers. Statistical-dynamic models: Statistical-dynamic models (SDMs) are essentially two-dimensional (2D) in form, usually with one horizontal and one vertical dimension, although two versions of horizontal dimensions have been developed. SDM models combine horizontal energy transfer modeled by energy balance schemes (EBM) with RCMs with a radiation–convection approach. The energy transfer from the equator to the pole is simulated in a sophisticated way, based on theoretical and empirical cell flow relationships between latitudes. Parameters such as wind direction and speed are modeled with statistical relationships to obtain a measure of energy diffusion, such as EBM, by applying laws of motion. General circulation models (GCMs): General circulation models (GCMs) represent the most sophisticated attempt to simulate the climate system.

The three-dimensional (3D) formulation of the model is based on the basic laws of physics: (a) conservation of mass; (b) conservation of momentum; and (c) conservation of energy. A series of primitive equations describing these laws are solved, summarizing the estimate of the wind field, which is expressed as a function of temperature. Processes such as cloud formation are also simulated. To calculate the basic atmospheric variables at each grid point, it is necessary to use a large amount of computer time, due to the large number of points, and therefore, global circulation models (GCMs) that have a higher resolution are used. 3D models can provide a reasonable representation of the planetary climate and, unlike simple models, can simulate processes at global and continental scales (such as the effects of mountain ranges on atmospheric circulation). However, many GCMs are unable to simulate synoptic (regional) weather phenomena such as tropical storms, which play an important role in the geographic transfer of energy and the amount of motion. The spatial resolution of GCMs is also limited in the vertical dimension. Consequently, many processes in the boundary layer must be parameterized. New generation models are GCMs with transient atmosphere–ocean coupling to accurately simulate the climate system (Bretherton et al., 1990; Gates et al., 1992). Such coupled models have created great difficulties for computing atmospheric and oceanic components at the corresponding different timescales. In summary, GCMs cannot reasonably be considered to simulate global and continental-scale climates but lack regional-scale reliability.

13.6 Climate Scale Analysis: From Local to Planetary Scale

Atmospheric processes exhibit a wave-like nature, encompassing a vast range of spatial scales, from millimeter-sized turbulent eddies to kilometer-spanning planetary waves, and temporal scales, from mere seconds to multiple years. While weather refers to the current atmospheric state, quantitatively expressed through the distribution of key meteorological parameters (e.g., temperature, moisture, pressure, precipitation) over a specific area and time, climate represents the average pattern of weather conditions over extended periods, typically over a standard 30-year period (normal climatic period) or even longer-term records. To grasp the complex concept of climate change, it's essential to first understand climate variability. A useful way to differentiate between weather, climate variability, and climate change is by considering their distinct timescales. As illustrated in Fig. 13.4, weather operates on shorter timescales (hours to days), climate variability spans months to decades, and climate change unfolds over centuries or longer. While the term "climate change" implies long-term shifts in climate patterns, "climate variability" refers to deviations or fluctuations within the broader climate context. Climate variability is often expressed through recurring patterns in temperature or other climate variables, quantified by various indices. These indices are valued for their simplicity and comprehensiveness, each representing the status and timing of a specific climate factor. Indices provide a condensed yet complete description of the atmosphere or oceans, offering insights into the factors influencing the global climate system.

Climate oscillations, or climate cycles, refer to any recurring patterns within global or regional climates. These fluctuations in air temperature, sea surface temperature, precipitation, or other parameters may be quasi-periodic, occurring over interannual, multiannual, centennial, millennial, or even longer periods. A prominent example is El Niño, a component of the Southern Oscillation that affects sea surface temperatures in the equatorial Pacific and the western coast of tropical South America, influencing climate patterns across the entire globe.

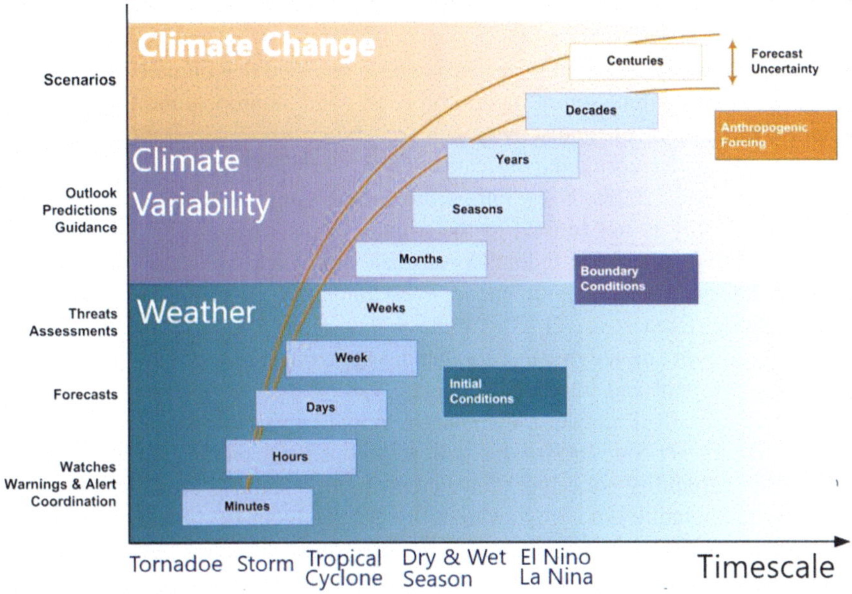

Fig. 13.4 Time and spatial scales of atmospheric processes. The framework illustrates the range of atmospheric processes, from millimeter-scale turbulence to kilometer-scale planetary waves. It also highlights the temporal spectrum (timescale), from seconds (turbulent processes) to centuries (climate changes), underscoring the varying scales at which weather and climate phenomena operate

13.7 Climate Variability

Understanding the distinctions between weather, climate variability, and climate change is crucial, as each operates on different temporal scales. Weather reflects atmospheric conditions over short periods, such as hours or days, while climate variability encompasses fluctuations that occur over months, years, or decades. Climate change, however, refers to long-term shifts in climate patterns, spanning centuries or even longer. Climate variability represents the natural variation around the long-term average climate, manifesting through different patterns of temperature, precipitation, and other climatic variables. These variations are often quantified using climate indices, which simplify and encapsulate complex atmospheric and oceanic processes. Each index typically represents a specific aspect of the climate system, offering a snapshot of the status and timing of key climate factors.

For example, the El Niño-Southern Oscillation (ENSO) is a well-known index that tracks large-scale fluctuations in sea surface temperatures (SSTs), precipitation, and atmospheric circulation across the tropical Pacific (NOAA, 2023). Beyond ENSO, other critical modes of climate variability include the North Atlantic Oscillation (NAO), Arctic Oscillation (AO), and the polar vortex. Recent studies highlight that these indices not only reflect atmospheric conditions but also interact with oceanic patterns, enhancing our understanding of climate impacts on different regions (Cohen et al., 2022).

Recent advancements in climate science have revealed that climate variability is influenced by a combination of natural processes and anthropogenic factors. Solar radiation variability, volcanic activity, and greenhouse gas concentrations modulate these patterns, underscoring the intricate interplay between human actions and natural systems (IPCC, 2023a, b, c). Furthermore, emerging research emphasizes that climate variability is not limited to the tropics; unusual sea surface temperature anomalies in the Atlantic or Indian Oceans can significantly alter the seasonal climate of adjacent continents, influencing everything from monsoon patterns to hurricane activity (Saha et al., 2023). Additionally, factors such as snow cover, soil moisture, and vegetation dynamics are increasingly recognized as vital components in shaping seasonal and interannual climate variability. For instance, changes in snow cover can affect albedo, influencing regional temperatures and precipitation patterns (Higgins & Shi, 2000). The feedback loops between land surface changes and climate variability are critical for understanding local and global climate systems. The predictability of climate variability is a key focus in climate science, as it holds the potential for improving seasonal forecasts. Long-term climate patterns, like ENSO, can have widespread impacts, influencing weather patterns, agricultural productivity, water resources, and even socioeconomic conditions globally. Understanding these patterns and their drivers is essential for developing robust climate adaptation and mitigation strategies. Advances in machine learning and data assimilation techniques are enhancing our ability to forecast climate variability, offering new tools for decision-makers in sectors, such as agriculture, water management, and disaster preparedness (Graham et al., 2023).

13.8 Global Climate Change

Climate change encompasses long-term alterations in average weather patterns that define Earth's climate, including shifts in temperature, precipitation, and other atmospheric conditions. In contrast to climate variability, which refers to short-term fluctuations, climate change denotes persistent, statistically significant shifts occurring over extended periods—often spanning decades to centuries. The primary driver of recent climate change is the rise in greenhouse gas (GHG) emissions, predominantly stemming from human activities, such as fossil fuel combustion, deforestation, and industrial processes (NASA, 2023; NOAA, 2023). These anthropogenic actions modify the atmosphere's composition, intensifying the greenhouse effect and driving global warming (IPCC, 2023a, b, c; Bennett & O'Neill, 2021).

The consequences of climate change are complex and far-reaching, impacting both abiotic and biotic processes worldwide. This intricate relationship manifests in various climate parameters, including rising global average temperatures, changing precipitation patterns, and an increase in the frequency and intensity of extreme weather events. For instance, the warming of the Arctic region has resulted in significant ice melting, contributing to rising sea levels and altering both oceanic and atmospheric circulation patterns (Met Office, 2023; IPCC, 2023a, b, c). Scientific advancements have markedly enhanced our understanding of climate change, its causes, and its potential impacts. This knowledge is essential for devising effective mitigation and adaptation strategies, enabling policymakers to frame climate change within the larger context of global challenges, such as food security, water resources, and economic stability (NOAA, 2023; Rosen, 2023). Although uncertainties persist—especially regarding the scale and timing of specific impacts—there is overwhelming evidence, supported by multiple research avenues, that the climate is changing and that these alterations are predominantly driven by human activities (NASA, 2023; IPCC, 2023a, b, c; Stern, 2021). The scientific consensus on climate change has endured rigorous scrutiny and debate, affirming the robustness of the underlying theories and models. As climate change continues to pose significant risks to human societies and natural systems, it emerges as one of the most urgent and interdisciplinary challenges of the twenty-first century. Addressing this global issue demands coordinated efforts across all sectors of society to reduce emissions, enhance resilience, and safeguard vulnerable ecosystems and communities from the escalating impacts of a changing climate (Met Office, 2023; IPCC, 2023a, b, c; Boulanger & Gassmann, 2023).

13.8.1 Overview of Past and Future Temperature Trends

The past and future trends in global mean temperature over the last 67 million years illustrate significant climate shifts, as depicted in Fig. 13.5. Oxygen isotope values from deep-sea benthic foraminifera, derived from sediment cores, serve as indicators of global temperature and ice volume, with temperature relative to the 1961–1990 global mean. Ice core records spanning the last 25,000 years depict the

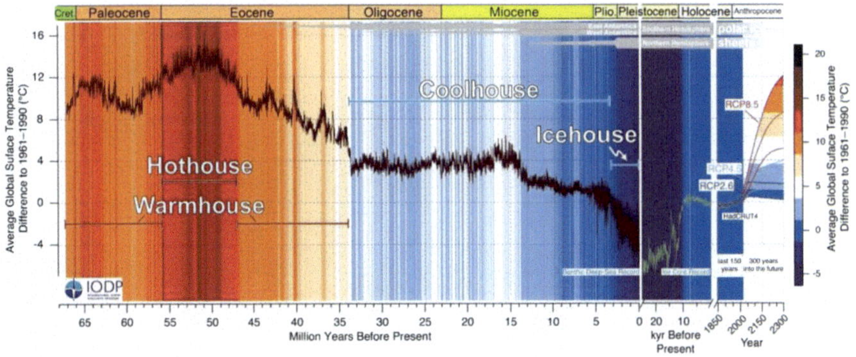

Fig. 13.5 Past and future trends in global mean temperature spanning the last 67 million years. (Credit: Westerhold et al., CENOGRID)

transition from the last glacial period to the current warmer era, known as the **Holocene**. Historical data from 1850 to the present reveal a marked temperature increase post-1950, signaling the onset of the **Anthropocene**. Future projections for global temperatures, based on three representative concentration pathways (RCP) scenarios, suggest that by 2100, the climate may resemble that of the Miocene Climate Optimum (approximately 16 million years ago), surpassing the threshold for the formation of continental ice sheets. If emissions remain constant beyond 2100 and are not stabilized before 2250, global climate by 2300 may enter a hot-house state akin to the early Eocene (around 50 million years ago), characterized by multiple global warming events and the absence of large ice sheets at the poles. This is a general overview of the past and future trends in climate change. However, more detailed information will be provided in the next section. The continuous record of the past 66 million years shown in Fig. 13.5 illustrates that natural climate variability due to changes in Earth's orbit around the sun is significantly smaller than the projected future warming driven by greenhouse gas emissions.

13.8.2 Anthropocene

We are currently living in a period often referred to as the Anthropocene (e.g. Steffen et al., 2007; Zalasiewicz, et al., 2010) a time when human activities have become the dominant force shaping the planet (Fig. 13.6). Our species has drastically influenced the climate, landscapes, ecosystems, and even the atmosphere, leading to widespread global warming and environmental transformation.

The term "Anthropocene" is derived from the Greek words for human ("anthropo") and new ("cene"). Its use has been a subject of debate, with its origins tracing back to the 1980s. It gained prominence in 2000 when atmospheric chemist Paul J. Crutzen and diatom expert Eugene F. Stoermer suggested that humanity's influence was so profound that it warranted the recognition of a new geological

Fig. 13.6 Earth's Planet enters New Era: The Anthropocene—"Epoch of Humans." This illustration depicts the profound impact of human activities on the planet, highlighting the intersection of nature and technology, the reshaping of landscapes, and the intricate web of environmental changes that define our current geological epoch

epoch (Crutzen & Stoermer, 2000). For thousands of years, Earth's climate remained relatively stable, but the effects of global warming are becoming most pronounced in polar regions like Svalbard. This remote Norwegian archipelago, located between the northern coast of Norway and the North Pole, has seen record-high temperatures, making it one of the fastest-warming places on Earth. During the summer of 2022, the extreme heat led to unprecedented ice melt, captured in stunning detail by the Copernicus Sentinel-2 mission. The satellite image shown in Fig. 13.7 provides a clear visual record of the accelerating changes in the Arctic landscape, further signaling that we may indeed be entering a new epoch—the Anthropocene.

The profound impact of *Homo sapiens* on Earth's systems and biodiversity is widely acknowledged, with many believing that our legacy will leave a lasting—and perhaps irreversible—mark on the globe. Our actions are reshaping ecosystems, altering the planet's fundamental processes, and threatening the stability of its natural systems. The Earth, which is approximately 4.5 billion years old, has seen modern humans for only about 200,000 years. Despite this relatively short span, our influence has been profound, particularly over the last few decades. This recent phase of intense human activity is often referred to as the "Great Acceleration." During this time, we have witnessed rapid increases in carbon emissions, global temperatures, ocean acidification, deforestation, biodiversity loss, and the extraction of natural resources, all contributing to significant planetary change.

However, there remains ongoing debate about whether these transformations constitute sufficient evidence to officially declare a new geological epoch, the Anthropocene. Researchers and scientists around the world continue to explore and discuss the implications of humanity's influence on Earth's history, with no consensus yet on the formalization of this epoch. To understand the geological context of

Fig. 13.7 Satellite photo of Svalbard made by Copernicus Sentinel-2 mission, August 2022. (Credit: European Space Agency (ESA))

the Anthropocene, it's important to look at Earth's deep history, which spans about 4.5 billion years. Scientists organize this immense timeline into eons, eras, epochs, and ages, each marked by significant changes in Earth's geology, climate, and life forms. Epochs, which can last for millions of years, are characterized by shifts in rock layers, including changes in mineral composition and the presence of distinct fossils. These variations serve as markers of major climatic transitions. For the past 11,500 years, Earth has been in the Holocene Epoch. This epoch began at the end of the last ice age, a time when the vast glaciers that once covered much of the planet retreated, giving way to the warmer and more stable climate conditions that have prevailed since.

The Holocene has been a period of remarkable transformation for both the planet and its inhabitants. It has witnessed the dramatic growth of the human population and the rise of complex civilizations. Over these 11,500 years, humanity has built cities, developed agriculture, and made extraordinary technological advancements. Since the end of the last ice age, Earth's climate has remained relatively warm and stable, conditions that have allowed human societies to thrive and expand.

Indicators of the Anthropocene

Earth's climate is rapidly warming, and scientists agree that human activity—through agriculture, urbanization, deforestation, and pollution—is the primary cause. These actions have triggered extraordinary changes to the planet. Geologists remain divided on whether humanity's impact will leave a lasting mark on Earth's geological record. Given the brief period that humans have existed, it is unclear if our influence will be visible in the fossil layers millions of years from now.

The International Commission on Stratigraphy (ICS), which oversees geological epochs, is debating the evidence for the Anthropocene. They are searching for a "geological benchmark," a significant marker in the rock record that could define this new epoch and differentiate it from the Holocene. Some believe the Anthropocene began during the eighteenth-century Industrial Revolution, when fossil fuel use expanded rapidly, driving industrial growth and environmental change. The resulting rise in carbon emissions reshaped Earth's climate. Other viewpoints suggest an earlier start, linked to the dawn of agriculture, while some argue it began in 1950 with nuclear weapons spreading radioactive elements globally. These materials could serve as the "Standard Reference Point" geologists seek to mark this new epoch. It's important to recognize that industrialized nations have contributed more to pollution and climate change than developing countries. The term "Anthropocene" is increasingly used across various cultural and scientific spheres, adding to the complexity of its meaning. Researchers, conservationists, poets, philosophers, politicians, and activists employ the term, often with differing interpretations. At its core, the Anthropocene refers to the period in which human activities have significantly influenced the Earth. Regardless of the debate over whether we have entered a new geological epoch, it is evident that we are part of a multifaceted global system, with clear signs of our impact becoming increasingly apparent. While many associate this human influence primarily with climate change, specifically, the warming of the atmosphere, air, and oceans due to fossil fuel consumption, the concept of the Anthropocene encompasses more than just climate issues. A crucial aspect of the Anthropocene is our growing awareness of the planet's current condition and the repercussions of our actions.

Throughout human history, our planet's global ecology has never faced such a critical state as it does today. However, we also find ourselves better equipped than ever with the knowledge and tools necessary to comprehend the challenges we face and the actions required to address them.

13.9 Basic Factors for Climate Change

Climatological studies reveal that Earth's climate has undergone significant variations throughout its history. Climate records and paleoclimatic research indicate periods where the planet experienced extensive warmth, with minimal or no ice at the poles, as well as periods where large portions of the temperate zones were covered by thick ice sheets. Understanding the primary drivers of these climate changes is essential, and they can be broadly categorized into three groups:

1. Astronomical factors
2. Geophysical factors
3. Biotic factors

Astronomical and geophysical factors are external agents of climate change because they originate outside the Earth's atmosphere. Astronomical factors are primarily related to the Sun and the dynamic relationships between celestial bodies, including Earth's distance from the Sun, orbital trajectories, relative positions, and axial tilt. These factors influence Earth's climate over extremely long timescales, often spanning tens of thousands to hundreds of thousands of years, making them less relevant when considering climate changes over a century.

13.9.1 Astronomical Theory of Climate Change

The astronomical theory of climate change, famously developed by Serbian mathematician, astronomer, and geophysicist Milutin Milankovitch, provides a crucial framework for understanding long-term climatic variations on Earth. Born on May 28, 1879, in Dalj, then part of the Austro-Hungarian Empire, Milankovitch made significant contributions to the understanding of Earth's climate through his theory of orbital variations and their correlation with climate change over geological timescales. His work is foundational in explaining how Earth's orbital changes act as climate-forcing mechanisms. Milankovitch identified three key types of orbital changes that influence Earth's climate (Fig. 13.8):

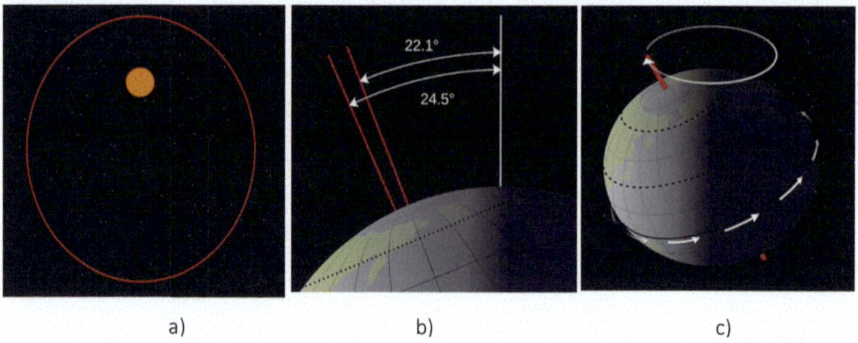

Fig. 13.8 The three primary cycles in Milankovitch's astronomical theory of climate change, illustrating the mechanisms that influence Earth's long-term climate patterns: (**a**) Eccentricity—the variation in the shape of Earth's orbit around the Sun, occurring over 100,000–400,000 years; (**b**) Inclination—the tilt of Earth's axis, affecting the distribution of solar radiation, with a cycle of approximately 41,000 years; and (**c**) Precession—the wobble in Earth's rotational axis, with a cycle of about 21,000 years. These cycles, known collectively as Milankovitch cycles, are pivotal in understanding past and future climate changes. (Credit: NASA/JPL-Caltech)

1. **Eccentricity:** This refers to the variation in Earth's orbital shape around the Sun, ranging from more circular to more elliptical, with cycles spanning 100,000–400,000 years. These changes affect the distribution of solar energy received by Earth, influencing long-term climate patterns.
2. **Obliquity (Axial Tilt):** This involves the variation in the angle of Earth's axial tilt relative to its orbital plane, occurring over a cycle of approximately 41,000 years. Changes in obliquity influence the contrast between seasons, with greater tilt leading to more extreme seasonal variations.
3. **Precession:** This describes the gradual shift in the orientation of Earth's rotational axis, with a cycle of about 21,000 years. Precession affects the timing of seasons and their intensity, further modulating climate patterns.

While Milankovic's theory provides a robust explanation for long-term climate shifts, solar variability remains a topic of ongoing research and debate as another potential external forcing factor. Despite numerous studies attempting to establish a statistical relationship between solar activity cycles and global climate fluctuations, compelling mechanisms that fully explain this connection remain elusive. In addition to these external factors, internal forcing mechanisms play a significant role in climate change. These mechanisms can be radiative or non-radiative and operate over various timescales, from decades to millions of years. For instance, tectonic activities, such as continental drift, mountain formation (orogeny), and vertical movements of the Earth's crust, contribute to long-term climate change by altering sea levels and land configurations. Volcanic activity, ocean circulation patterns, and changes in atmospheric composition also act as critical internal factors influencing Earth's climate on shorter timescales. The response of Earth's climate system to these forcing factors is complex and varies across different components of the system. The atmosphere, cryosphere (snow and ice), and biosphere (vegetation) can respond rapidly to climate forcings, often within scale frame from hours to days. In contrast, other components, such as the oceans and ice sheets, may take centuries or even millennia to fully respond to these changes. This diversity in response timescales further complicates the prediction and understanding of climate change, emphasizing the need for continued research and monitoring.

13.9.2 Climate System Dynamics: Timescales, Feedback, and Sensitivities

Surface oceans respond to climate changes over the years, while deep oceans and mountain glaciers have changed hundreds of years. Large ice sheets evolve over millennia, while parts of the geosphere react over periods spanning hundreds of millions of years. The climate system's response to these forcing events can be likened to a form of resonance, where the impact is greatest when the forcing period aligns closely with the natural response time of a particular climate system component. The global climate's stability hinges on the balance between incoming solar radiation and outgoing terrestrial longwave radiation. This balance maintains the Earth's climate in equilibrium, where the planet neither continues to warm nor cool excessively. However, when the climate system encounters radiative forcing, this equilibrium is disrupted, creating an imbalance between incoming and outgoing radiation. To restore this balance, the global climate adjusts by warming or cooling, depending on the nature of the initial forcing. Although the climate system strives for equilibrium, it is inherently dynamic and constantly in flux. The system continually adapts to disturbances induced by various forcings. This adaptability contributes to climate change as the effects of these forcings cascade through different components of the climate system. In some cases, the system may amplify the initial disturbance (positive feedback), while in others, it may dampen the disturbance (negative feedback). Climate sensitivity refers to the degree of climate system response to a given radiative forcing. It reflects the combined effects of various feedback mechanisms that either amplify or reduce the initial temperature change induced by the forcing. Understanding climate sensitivity is crucial for predicting the ultimate impact of any changes in radiation on the global climate.

To comprehend the current climate and forecast future climate changes, a combination of theoretical models and empirical observations is essential. Studying climate change involves constructing or rehabilitating time series of climate data that provide quantitative or qualitative measures of climate change over time. These climate statistics encompass a broad range of data types, including temperature, precipitation, wind, humidity, evaporation, pressure, and solar radiation, all of which are vital for understanding the complexities of climate change.

13.9.3 Solar Activity, Climate Change, and the Overriding Role of Human Influence

The solar cycle, lasting around 11 years, is characterized by fluctuations in solar phenomena such as sunspots and solar flares. These changes arise from shifts in the Sun's magnetic field and can slightly alter the amount of solar radiation reaching Earth. While the Sun is undeniably the planet's primary energy source, the overall impact of solar cycles on long-term climate patterns is minimal compared to the influence of human-induced factors. Although short-term climate variations are affected by solar activity, their influence pales in comparison to the effects of rising greenhouse gas concentrations due to human activities.

For instance, satellite measurements over the last four decades have shown that the Sun's energy output has varied by less than 0.1%. This small fluctuation is insufficient to drive the substantial warming trends seen today. The cumulative effect of human-generated greenhouse gases since the Industrial Revolution has caused over 270 times more warming than any solar variability observed in that same period. Thus, while solar cycles contribute to natural climate dynamics, their role is minor when considering the rapid climate changes driven by fossil fuel combustion and other human activities.

Are We Headed for Another "Grand Minimum"?

There has been speculation about whether we are entering a prolonged period of low solar activity, known as a "grand minimum," like the Maunder Minimum (1645–1715), which coincided with a period of cooler global temperatures, particularly in the Northern Hemisphere, referred to as the "Little Ice Age." During a grand minimum, solar magnetism weakens, sunspots become rare, and ultraviolet radiation decreases, potentially leading to cooler conditions on Earth. However, research suggests that even if a grand minimum were to occur, it would likely only reduce global temperatures by about 0.3 °C—a modest and temporary effect (see Fig. 13.9). Such a cooling effect would merely slow, but not reverse, the broader trend of human-caused global warming. The reduction in solar energy during a grand minimum would be counteracted by the continued rise in greenhouse gas concentrations, with just a few years of current carbon dioxide emissions offsetting any cooling effect. Therefore, even in the unlikely event of a significant solar decline, the long-term trajectory of global warming remains dominated by anthropogenic factors.

Recent Solar Cycle 25 Activity

Figure 13.10 illustrates the rise in sunspot numbers as a sign of increasing solar activity during Solar Cycle 25, which began in late 2019. By January 2023, the monthly sunspot count surged to 144, a peak not seen in the last 9 years and like the

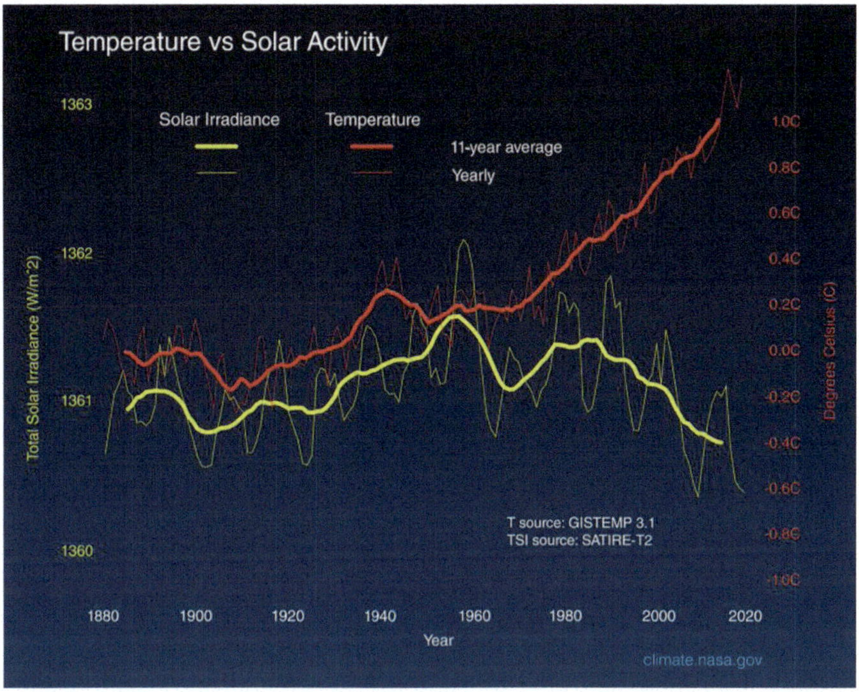

Fig. 13.9 Comparison of global surface temperature changes (red line) and solar energy received by Earth (yellow line) from 1880 to the present. The lighter/thinner lines represent annual values, while the thicker lines display 11-year averaged trends, which help reduce year-to-year variability and reveal long-term patterns. Solar energy levels have followed the Sun's natural 11-year cycle of minor fluctuations, with no net increase since the 1950s, whereas global temperatures have risen significantly during this period, indicating that the observed warming trend is unlikely to be driven by changes in solar activity. (Source: NASA/JPL-Caltech)

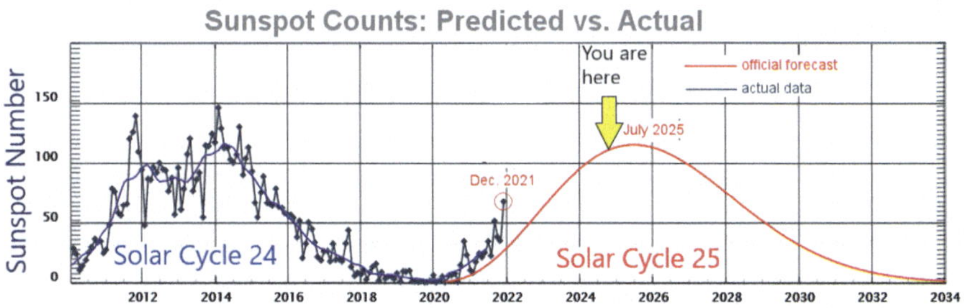

Fig. 13.10 The solar cycle. This figure illustrates the solar cycle, showcasing variations in solar activity over time, including sunspot numbers and solar radiation fluctuations. The figure also presents a preliminary forecast for Solar Cycle 25, which predicts the expected levels of solar activity based on observed patterns and predictive models. Solar cycles, typically spanning approximately 11 years, play a role in influencing space weather and have minor implications for Earth's climate. The forecast helps prepare for potential impacts on satellite operations, communications, and power grids due to heightened solar activity. (Credit: NOAA333)

146 sunspots recorded at the peak of Solar Cycle 24 in February 2014. Earlier predictions had suggested that Solar Cycle 25 would continue the subdued activity of Solar Cycle 24, which was one of the weakest cycles of the past century. These predictions implied that solar activity was in a long-term decline, potentially approaching a Maunder Minimum-like event.

However, recent data has shifted this perspective. Forecasts now indicate that Solar Cycle 25 could exceed initial expectations, with the solar maximum likely to occur between 2024 and 2025. This suggests that solar activity may not be in continuous decline, and the peak of Solar Cycle 25 might surpass that of Cycle 24. Even so, such increases in solar activity, while noteworthy, are unlikely to significantly impact the long-term trend of global climate change. In conclusion, while solar cycles play a role in natural climate variability, their overall impact is dwarfed by the effects of human activity. The overwhelming scientific consensus points to human-caused greenhouse gas emissions as the dominant force behind recent and ongoing global warming, with solar variability contributing only minor, short-term fluctuations.

As a constant signal of Solar Cycle 25's intensity, the number of sunspots has reached a peak not seen in the past 9 years. In January 2023, the monthly sunspot count surged to 144, closely mirroring the peak of Solar Cycle 24, which recorded 146 sunspots in February 2014. Initially, projections indicated that Solar Cycle 25 would mirror the subdued activity of Solar Cycle 24, one of the weakest cycles in the past century, following a prolonged period of low solar activity. This expectation suggested that the decline in solar cycle amplitude observed from cycles 21–24 had come to an end, with no imminent approach toward a Maunder-type minimum prolonged period of minimal solar activity observed in the seventeenth century. However, recent forecasts indicate that Solar Cycle 25 might exceed previous expectations and show significant activity. The solar maximum, anticipated to occur between 2024 and 2025, may see further strengthening, potentially surpassing the activity levels of Solar Cycle 24.

13.9.4 Geophysical Factors for Climate Change

Geophysical factors, or planetary factors, are crucial in understanding climate change as they relate to the Earth's tectonic activities and planetary phenomena. These include volcanic eruptions, tectonic movements, and changes in Earth's surface. Major volcanic eruptions and large meteor impacts can significantly affect the climate. When such events occur, they eject large quantities of aerosols into the stratosphere. These aerosols can remain in the upper atmosphere for extended periods, leading to a reduction in incoming solar radiation and thus cooling the Earth's surface. Additionally, volcanic eruptions release substantial amounts of carbon dioxide (CO_2) into the atmosphere, which can contribute to long-term warming. However, in recent times, there has been no significant planetary event that could explain the observed global warming. Therefore, while geophysical factors remain important, they are not currently considered primary drivers of the recent trend of global temperature rise.

13.9.5 Biotic Factors of Climate Change

Biotic factors influencing climate change can be divided into anthropogenic and non-anthropogenic categories. Non-anthropogenic biotic factors involve natural ecosystem processes such as primary production (photosynthesis) and the cycling of water and nutrients. These processes can impact climate, but their effects are typically less pronounced compared to human activities. Anthropogenic biotic factors are those resulting from human actions. Contemporary climate change is largely attributed to these human-induced factors. Key anthropogenic influences include industrialization, deforestation, urbanization, irrigation, and land-use changes. The combustion of fossil fuels, the expansion of nuclear power, and the emission of various aerosols have reached a global scale, contributing to significant changes in atmospheric composition. The release of greenhouse gases such as carbon dioxide (CO_2), methane (CH_4), nitrous oxide (N_2O), chlorofluorocarbons (CFCs), and water vapor has intensified the greenhouse effect, leading to global warming. The impact of these anthropogenic factors is particularly severe at the poles, where rapid warming is causing the accelerated melting of ice sheets. Conversely, in dry regions and around the Intertropical Convergence Zone (ITCZ), shifts in climate patterns are leading to extended and intensified droughts moving toward higher latitudes. These changes are expected to have profound global repercussions, affecting human lives, infrastructure, and the economy. The interplay of rapid globalization, technological advances, industrial activities, and uncontrolled greenhouse gas emissions is disrupting the Earth's energy balance, leading to alterations in atmospheric circulation and increased occurrences of extreme weather events such as droughts, floods, tornadoes, and severe storms.

13.10 Key Drivers of Climate Change

The primary factors influencing climate change include the concentrations of greenhouse gases and aerosols in the atmosphere, as well as changes to the Earth's surface. These elements significantly impact how solar energy is absorbed or reflected, thereby shaping global temperature patterns. Greenhouse gases, such as carbon dioxide (CO_2), methane (CH_4), and nitrous oxide (N_2O), have risen markedly since the Industrial Revolution due to human activities like burning fossil fuels, deforestation, agricultural practices, and alterations to the natural environment. The enhanced greenhouse effect caused by these gases traps more heat in the atmosphere, driving global warming. Aerosols, including particulate matter and pollutants, influence the climate in complex ways. Some aerosols reflect sunlight, causing a cooling effect, while others absorb heat, contributing to warming. The overall impact of aerosols depends on their composition and distribution. Surface characteristics, such as deforestation, urbanization, and land-use changes, also play a significant role. These modifications can alter the Earth's albedo (reflectivity), affecting how much solar energy is absorbed or reflected. For example, replacing forests with urban areas or agricultural land typically increases heat absorption due to lower albedo. Observational data underscores the impact of these drivers. Over the past century, global air temperatures have risen by about 0.74 °C, and sea levels have increased by roughly 17 cm, driven partly by melting glaciers and polar ice. Additional changes include rising Arctic temperatures, shifts in ocean salinity, changes in wind and precipitation patterns, increased drought intensity, and more frequent heat waves and tropical cyclones. Future projections suggest temperatures will continue to rise by approximately 0.2 °C per decade over the next 20 years if current trends persist. Continued greenhouse gas emissions are expected to cause further warming, which, over centuries, could lead to the complete melting of ice caps and a potential sea-level rise of about 7 m. Such warming would profoundly impact natural systems, leading to ecosystem loss, reduced biodiversity, and heightened risks of species extinction as ecosystems struggle to adapt. The drivers of climate change are interconnected and multifaceted. Rising greenhouse gas concentrations, combined with changes in aerosols and surface characteristics, are transforming the Earth's climate system. Addressing these challenges requires a comprehensive understanding of these drivers and the implementation of effective mitigation strategies.

13.10.1 Greenhouse Effect

The greenhouse effect is a crucial process for understanding global warming, driven by the presence of greenhouse gases in the Earth's atmosphere. Shortwave radiation from the Sun passes through the atmosphere and is absorbed by the Earth's surface, causing it to warm. This absorbed energy is then re-emitted as long-wave infrared radiation. While some of this infrared radiation escapes into space, a significant portion is trapped by greenhouse gases, such as carbon dioxide (CO_2), methane (CH_4), and nitrous oxide (N_2O). These gases absorb and re-radiate the heat back toward the Earth's surface, enhancing the warming effect and creating a positive feedback loop. The intensification of the greenhouse effect is largely due to increased concentrations of greenhouse gases, which have risen significantly since the Industrial Revolution. Earth's climate has undergone significant changes over its history. In the past 800,000 years alone, the planet has experienced 8 cycles of ice ages followed by warmer interglacial periods. The last major ice age ended approximately 11,700 years ago, ushering in the current climate conditions that have supported human civilization. Historically, these shifts in climate were largely driven by subtle variations in Earth's orbit, which altered the amount of solar energy reaching the planet. However, the ongoing warming trend is distinctly different. Beginning in the mid-nineteenth century, it has been driven predominantly by human activities, particularly the burning of fossil fuels, deforestation, and industrial processes. This has led to an accelerated rate of warming unseen in recent millennia. Figure 13.11 illustrates the long-term history of atmospheric CO_2 concentrations, showing fluctuations over millennia and the stark increase in recent times. The data, which combines atmospheric samples from ice cores with more recent direct measurements, provides compelling evidence of the significant rise in CO_2 levels since the onset of the Industrial Revolution.

The ice core samples offer insights into the Earth's atmospheric composition dating back hundreds of thousands of years, while modern measurements track the unprecedented surge in CO_2 concentrations, reflecting human influence on the planet's climate system. By 2017, human-induced warming had reached approximately 1 °C (1.8 °F) above pre-industrial levels. If the current trend continues, global temperatures are projected to rise to 1.5 °C (2.7 °F) by around 2040. The green portion of the accompanying diagram highlights the range of uncertainty regarding how much further temperatures could increase before stabilizing, assuming immediate and substantial reductions in carbon dioxide emissions, to reach net-zero emissions by 2055.

The evidence is clear: human actions have increased concentrations of greenhouse gases, such as carbon dioxide and methane, in the atmosphere, trapping more solar energy within the Earth's system. As a result, this excess energy has heated the atmosphere, oceans, and land, causing widespread and rapid changes across the globe, including in the cryosphere and biosphere. The primary contributor is CO_2, which accounts for approximately 75% of global greenhouse gas emissions. This increase is mainly due to burning fossil fuels like coal, oil, and natural gas. Other greenhouse gases, such as methane and nitrous oxide, also contribute to this warming effect, although in smaller amounts compared to CO_2. Methane is emitted from agriculture and landfills, while nitrous oxide is released from agricultural practices and industrial processes.

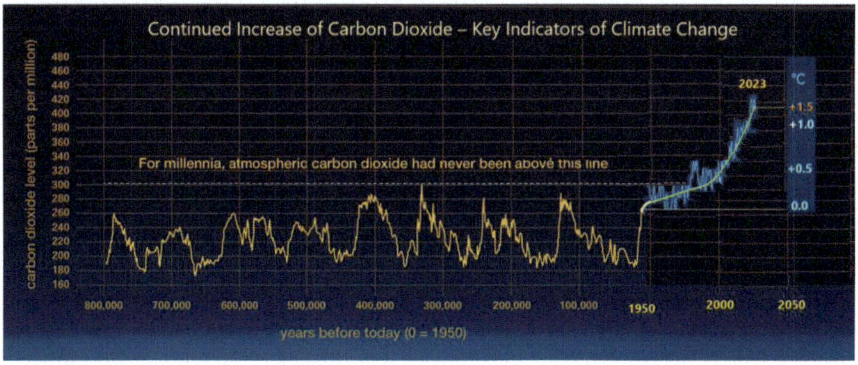

Fig. 13.11 Carbon dioxide concentrations through millennia to present: Credit: Data: Luthi, D., Et Al. 2008; Etheridge, D.M., Et Al. 2010; Vostok Ice Core Data/J.R. Petit Et Al.; NOAA Mauna Loa CO2 Record. Some Description Adapted from the Scripps CO_2 Program Website, "Keeling Curve Lessons"

The enhanced greenhouse effect leads to a rise in global temperatures, resulting in various climate changes such as more frequent and severe heat waves, altered precipitation patterns, and melting polar ice. These changes profoundly impact the environment, sea levels, and weather patterns. Reducing greenhouse gas emissions through measures such as adopting renewable energy sources and improving energy efficiency is essential to mitigating the effects of global warming and addressing climate change.

13.10.2 Global Warming

Global warming refers to the rise in average global temperatures of the air, Earth's surface, and oceans, observed since the mid-twentieth century. Climate models suggest that a 40% increase in atmospheric carbon dioxide could lead to a significant rise in average annual temperatures, particularly in lower latitudes, potentially by about 2 °C. This warming is expected to shift the subtropical belt poleward, exacerbating drought conditions and accelerating the melting of polar sea ice. The intensity and frequency of weather extremes, including storms and heat waves, are anticipated to increase as a result. Recent observations confirm some of these predicted impacts, such as shrinking ice cover and more frequent extreme weather events. However, precise future projections remain challenging due to uncertainties in climate models and our incomplete understanding of complex feedback mechanisms within the climate system. These feedback processes—such as changes in cloud cover, ice albedo, and greenhouse gas emissions from permafrost—can amplify or mitigate warming effects, adding layers of complexity to predictions.

The warming induced by human activities, particularly from fossil fuel emissions, is expected to persist for centuries to millennia, driving long-term changes in climate patterns. This prolonged impact will likely lead to further shifts in weather systems, sea levels, and ecosystems, underscoring the urgent need for mitigation strategies to address these ongoing changes and their potential consequences.

13.11 Global Trends

Climate change stands out as one of the most pressing challenges of the twenty-first century. Recent global data indicates a rise in average temperatures of approximately 1.1 °C since the pre-industrial era (1850–1900). The last 5-year average global temperature is the highest recorded to date. In Europe, temperatures have risen by nearly 2 °C since the latter half of the nineteenth century, highlighting a more pronounced regional trend. This warming contributes to widespread environmental and socioeconomic impacts, including more frequent and severe heat waves, shifting weather patterns, and accelerating ice melting. These trends underscore the urgency of addressing climate change through mitigation and adaptation strategies to manage its impacts and reduce future risks.

While natural variability does contribute to climate fluctuations, the overwhelming evidence indicates that human activities, particularly the emission of greenhouse gases, are primarily responsible for global warming. According to the latest analysis from NASA's Goddard Institute for Space Studies (GISS), Earth's average global temperature has increased by at least 1.1 °C since 1880, at a rate of approximately 0.15–0.20 °C per decade. As shown in Fig. 13.12, 2022 was the fifth warmest year on record, and the past 9 years have been the warmest since modern temperature records began in 1880. In October 2023, global warming reached an estimated 1.25 °C. If the current trend continues, the global temperature could rise by 1.5 °C above pre-industrial levels by May 2034 (Fig. 13.13).

The factors driving climate change encompass astronomical, geophysical, and biotic influences. Among these, biotic factors are particularly significant, stemming from both non-anthropogenic processes like photosynthesis and natural nutrient cycling, as well as anthropogenic activities, such as industrial emissions and deforestation. It has been evident for decades that human activities are a major driver of climate change, with the evidence for this influence being indisputable. Climate change is accelerating, leading to more frequent and severe climatic and weather extremes with profound socioeconomic consequences globally. This is especially critical for less developed and vulnerable regions, including those within our own country, which face greater risks from climate impacts. Recent years have not seen significant geophysical events such as earthquakes, volcanic eruptions, or tsunamis on the European continent that would drastically alter the climate. Therefore, uncontrolled human emissions remain the primary factor influencing climate

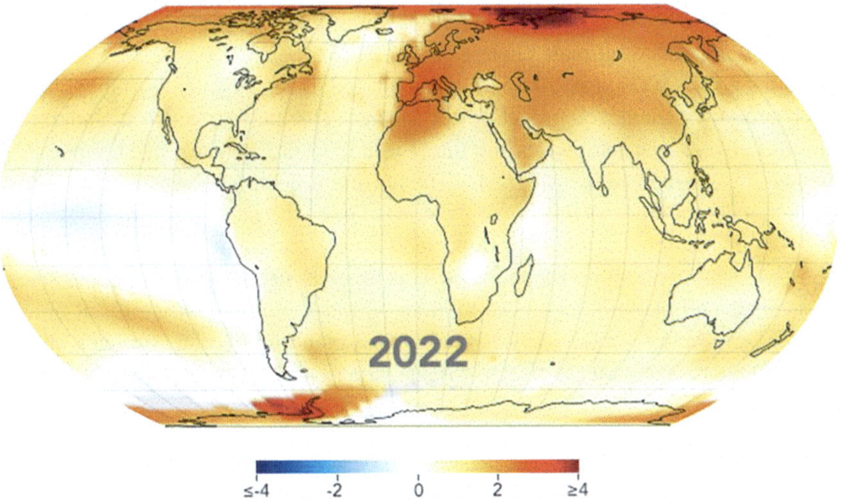

Fig. 13.12 Anomaly of the average global temperature (°C), compared to the average for the normal climate period 1951–1980. This figure illustrates deviations in global temperature from the long-term average, highlighting the significant warming trend since the mid-twentieth century. (Source: NASA)

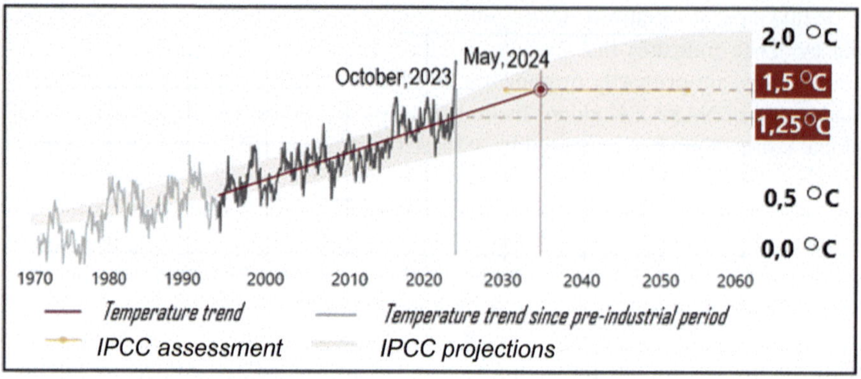

Fig. 13.13 Global trend in average surface temperature relative to the pre-industrial period. The graph illustrates the projected temperature increase, highlighting a potential rise to 1.5 °C above pre-industrial levels by May 2034 if current trends persist. (Source: IPCC)

change. Since 1950, the disruption of the climate system has been marked by unprecedented changes, with the current rate of global warming being the fastest observed in over 1300 years. According to the World Meteorological Organization, atmospheric concentrations of greenhouse gases have surged since 1750, largely due to human activities. In 2016, carbon dioxide levels surpassed 400 ppm (parts per million), methane reached over 1800 ppb (parts per billion), and nitrous oxide exceeded 320 ppb. These levels represent significant increases from pre-industrial times—over 40% for CO_2, 150% for methane, and 20% for nitrous oxide. The primary driver of the CO_2 increase is the burning of fossil fuels, with land-use changes contributing to a lesser extent. These figures highlight the urgent need to address the causes and impacts of climate change.

For the past 10,000 years, humanity has thrived under a relatively stable climate, which supported the development of our species. This period, which included fluctuations between warmer phases and the "Little Ice Age," saw average temperatures vary within a narrow range of just 1°. The current trajectory, however, represents a significant deviation from this historical norm. Given the increasing risks associated with climate phenomena and their impacts, it is essential to integrate these considerations into spatial planning frameworks. This involves updating and adapting spatial assessments and planning practices to align with the evolving dynamics of climate change. Developing a modern system of spatial planning that incorporates climate change impacts is crucial for effective management and adaptation strategies.

13.12 Climate Change in the Twenty-First Century

It has been evident for decades that the Earth's climate is changing, with human influence on the climate system now well established. The pace of climate change is accelerating, leading to increasingly severe climatic and weather extremes with significant socioeconomic and developmental impacts globally. This is particularly critical for less developed regions and communities, which are especially vulnerable to climate impacts.

Over the past century, the average global air temperature has risen by approximately 0.74 °C, and the global sea level has increased by 17 cm, partly due to snow and ice melting from polar regions and mountains.

Additional observed changes include shifts in Arctic temperatures, alterations in ice and ocean salinity, changes in wind patterns, increased frequency and intensity of droughts, precipitation variations, heat waves, and more frequent and severe tropical cyclones. Future projections indicate a temperature rise of about 0.2 °C per decade, with continued greenhouse gas emissions leading to further increases in global temperatures. If current trends persist, the melting of ice sheets could raise global sea levels by approximately 7 m, significantly altering ecosystems.

Simulations from global climate models suggest that a 40% increase in atmospheric CO_2 could raise average annual temperatures in lower latitudes by about 2.0 °C. The potential consequences of global warming include:

1. Displacement of the subtropical high-pressure belt toward higher latitudes
2. More frequent and intense heat waves and extreme heat events
3. Longer and more severe droughts
4. Sudden melting of polar ice seas
5. Increased intensity and frequency of weather events globally

The latest insights into these trends are detailed in the Sixth Climate Change Assessment Report by the Intergovernmental Panel on Climate Change (IPCC), part of the World Meteorological Organization (WMO). This report integrates contributions from three IPCC working groups and includes a comprehensive synthesis of global monitoring data, theoretical climate studies, model simulations, and future projections. According to the report, global warming has increased by about 1.0 °C relative to pre-industrial levels in the latter half of the twentieth century and is projected to reach approximately 1.5 °C between 2030 and 2052 if current trends continue. The Working Group I (Science Basis) report, the first part of the IPCC's Sixth Assessment Report (AR6), reflects recent advances in climate science and provides crucial information for climate negotiations and policymaking.

13.12.1 Current Status of Climate Change

Human influence on the climate system is now undeniable, with anthropogenic greenhouse gas emissions at unprecedented levels, impacting both human and natural systems globally. Carbon dioxide (CO_2) concentrations have risen to levels approximately 50% higher than those before the large-scale burning of fossil fuels that began with the Industrial Revolution. Recent measurements from Hawaii's Mauna Loa Observatory indicate that in February and March 2021, CO_2 levels in the atmosphere surpassed 417 parts per million (ppm). For comparison, pre-industrial levels were around 278 ppm (Fig. 13.14).

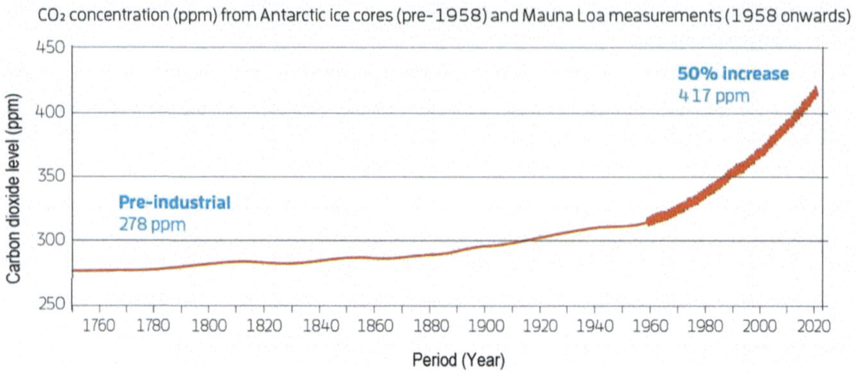

Fig. 13.14 Global atmospheric CO_2 concentrations from 1700 to 2021. Recent data shows a significant increase in CO_2 levels, reaching over 417 ppm in early 2021, compared to pre-industrial levels of about 278 ppm. (Source: Met Office)

(a) The warming of the climate system since 1950 has been unequivocal. Observations confirm that both the atmosphere and oceans are experiencing significant warming; snow and ice cover are diminishing; and sea levels are rising. Furthermore, anthropogenic greenhouse gas emissions have increased markedly, contributing to these changes.
(b) Climate change has had profound impacts on both natural and human systems. These impacts are evident regardless of the underlying causes and highlight the sensitivity of ecosystems and human societies to climatic shifts. The repercussions are observed across various sectors, including agriculture, water resources, and infrastructure.
(c) Extreme weather and climate phenomena have intensified since 1950. Notable changes include a reduction in extreme cold events, an increase in extreme heat, rising sea levels, and more frequent and intense precipitation events in many regions. These shifts are increasingly linked to human activities, highlighting the growing influence of anthropogenic factors on the climate.

The effects of climate change are not uniform and vary significantly across different regions. For example, land areas are warming at a rate higher than the global average, with Arctic regions experiencing twice the average warming rate. The latest reports forecast that at 1.5 °C of global warming, we will see increased frequency and severity of heat waves, extend hot seasons, and reduced cold seasons. If global warming reaches 2 °C, heat extremes will likely surpass critical thresholds, leading to more severe impacts. Furthermore, ongoing warming will exacerbate changes in precipitation patterns, snow and frost regimes, and coastal and oceanic conditions, reinforcing the need for urgent and effective climate action. Climate change is amplifying the water cycle, leading to more intense rainfall and flooding, as well as severe droughts in various regions. Climate change is intensifying the water cycle, resulting in more extreme weather patterns, including heavier rainfall, flooding, and severe droughts in various regions.

Rainfall patterns are shifting, with higher latitudes expected to see increased precipitation, while large portions of the subtropics are projected to face declines. Monsoon systems are also likely to undergo significant changes.

Coastal areas will continue to be impacted by rising sea levels throughout the twenty-first century. This will lead to more frequent and severe flooding in low-lying regions and increased coastal erosion.

In the Arctic and glacial regions, further warming is accelerating the thawing of permafrost, reducing seasonal snow cover, and causing the continued loss of glaciers, ice sheets, and summer Arctic Sea ice.

Oceans are undergoing profound changes, including warming, more frequent marine heat waves, acidification, and declining oxygen levels, all of which are closely tied to human activity. These changes are severely affecting marine ecosystems and the human communities that depend on them.

Urban areas are facing heightened climate-related challenges. Cities may experience amplified heat due to the "urban heat island" effect, more intense flooding from heavy rainfall, and rising sea levels threatening coastal urban centers.

The Sixth Assessment Report (IPCC, 2023a, b, c) provides a detailed regional analysis of climate change impacts, offering valuable insights for risk assessment, adaptation, and decision-making. It introduces a new framework to translate physical climate changes—such as heat, cold, rainfall, drought, and coastal flooding—into their societal and ecological implications. The report highlights significant scientific advances in understanding how climate change intensifies specific weather events, including extreme heat waves and heavy rainfall. It also underscores that human actions are crucial in shaping the future climate. The predominant driver of climate change is carbon dioxide (CO_2), along with other greenhouse gases and air pollutants.

Stabilizing the climate necessitates urgent and substantial reductions in greenhouse gas emissions, aiming for zero (CO_2) emissions. Limiting other greenhouse gases, particularly methane, can yield both health and climate benefits. The coming decades will present multiple climate hazards, and even a temporary exceedance of 1.5 °C global warming will lead to significant, potentially irreversible impacts, increasing risks to infrastructure and coastal communities.

13.12.2 Future Climate Projections

The IPCC Sixth Assessment Report highlights that limiting global warming to 1.5 °C or even 2 °C will be increasingly difficult without immediate and substantial reductions in greenhouse gas emissions. Since the pre-industrial period (1850–1900), human-induced greenhouse gas emissions have contributed to approximately 1.1 °C of warming. Enhanced observational data and advancements in climate science indicate that global temperatures will likely reach or exceed 1.5 °C within the next two decades, depending on emission pathways. Projections for the period 2070–2100 show continued temperature increases, with different emissions scenarios suggesting varying degrees of warming (Fig. 13.15). These projections reflect the long-term consequences of sustained greenhouse gas emissions and are critical for understanding potential climate impacts. Key elements of future climate projections include:

- Surface Temperature Rise: All emission scenarios predict substantial increases in global surface temperatures throughout the twenty-first century. As shown in Fig. 13.15a, projected temperature changes up to 2100 indicate that higher emissions scenarios lead to more significant warming. The global temperature rise will vary regionally, with some areas experiencing more extreme warming than others.
- Cumulative CO_2 Emissions: The extent of future warming directly correlates with cumulative CO_2 emissions. The higher the total emissions, the greater the surface temperature rise. Continued emissions at current rates will likely lead to warming well beyond 2 °C by the end of the century, contributing to more severe climate impacts, including extreme weather events and shifts in ecosystems.
- Snow and Ice Reduction: The reduction in snow and ice cover is a critical concern, contributing to sea-level rise due to melting glaciers and ice sheets. Projections indicate that by 2100, Arctic sea ice could decline significantly, affecting global weather patterns and marine ecosystems.
- Ocean Changes: Ocean warming and acidification will continue, disrupting marine ecosystems and further contributing to rising sea levels. Projections show

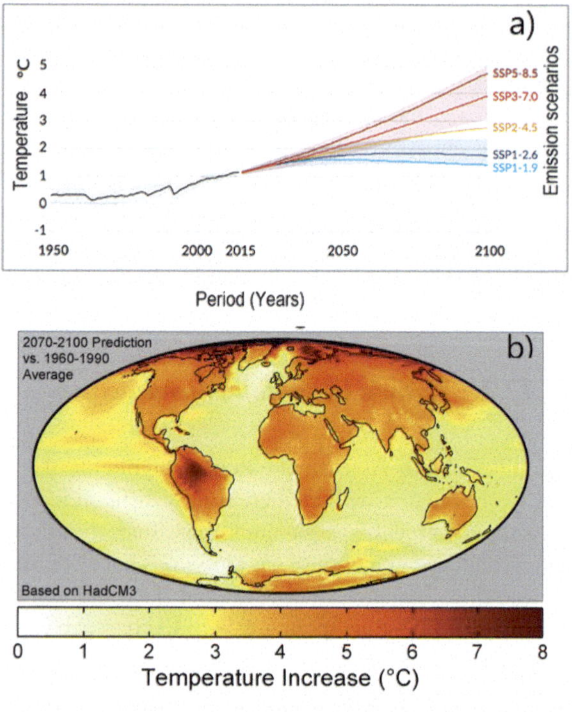

Fig. 13.15 (**a**) Change in the mean global surface temperature (at ground level) relative to the period 1850–1900, including projections for temperature changes up to the year 2100 based on various emission scenarios. (Credit: IPCC, 2023a, b, c). (**b**) Global warming 2070–2100 averaged surface global temperature prediction with HadCM3, Met Office climate coupled atmospheric-ocean model for climate prediction. (Credit: https://creativecommons.org/licenses/by-sa/3.0/)

significant changes in ocean circulation, with implications for weather patterns, marine life, and coastal regions.
- Extreme Weather Events: The frequency and intensity of extreme weather events such as heat waves, heavy rainfall, and storms are expected to rise significantly in many regions. Projections indicate that the likelihood of more intense and prolonged heat waves, flooding, and droughts will increase under all emissions scenarios. This trend poses major challenges to agriculture, infrastructure, and human health.

As global temperatures rise, the impacts of climate change will intensify, affecting ecosystems, human health, and economies worldwide. Climate change will exacerbate existing vulnerabilities, particularly in regions already facing economic or social challenges. Urgent action to reduce emissions and enhance adaptation efforts will be crucial in managing these risks and limiting the severity of future climate impacts.

13.13 Climate Change and Weather Extremes

Climate change has emerged as one of the most pressing challenges of our time, profoundly altering the frequency, intensity, and nature of weather extremes across the globe (Fig. 13.16). Rising global temperatures, driven primarily by human-induced greenhouse gas emissions, have led to significant shifts in climatic patterns. As a result, we are witnessing an increase in the occurrence of extreme weather events, including intense heat waves, heavy rainfall, and prolonged droughts. These phenomena have far-reaching implications, affecting ecosystems, human health, agriculture, and water resources. The interactions between climate change and hydrological cycles are particularly critical, as they influence the severity of water-related events, including floods and droughts, which can devastate communities and disrupt livelihoods.

Understanding the definition and characteristics of these extremes is essential for developing effective adaptation and mitigation strategies in a warming world. As we delve into the complexities of climate change and its impact on extreme weather events, it is crucial to establish a clear understanding of what constitutes an "extreme" event. In climate science, an event is deemed extreme when its value exceeds or falls below a specified threshold. These thresholds can vary, leading to different interpretations of extremes, even when the same terms are used.

13.13.1 Definition of Extremes

In climate science, an event is deemed extreme when its value exceeds or falls below a specified threshold. These thresholds can vary, leading to different interpretations of extremes, even when the same terms are used. For instance, two primary

Fig. 13.16 Eight weather megadisasters struck Earth in 2020–2022. Clockwise, from upper left: Hurricane Ian (Florida, 2022); drought (United States, 2022); drought (Europe, 2022); Hurricane Ida (Louisiana, 2021); flooding (Zhengzhou, China, 2021); flooding (Germany and Belgium, 2021); winter weather (Texas, 2021); and Hurricane Laura (Louisiana, 2020). (Photo credits are at the bottom of this post. Credit: https://yaleclimateconnections.org/)

metrics are often applied to identify hot or warm days. One employs relative thresholds, defining extremely hot days as those exceeding the 90th percentile of maximum daily temperatures for specific calendar days over a reference period, which can occur year-round with seasonal impacts. The other metric uses a fixed threshold, such as 35 °C, where exceeding this temperature can pose health risks, varying by location and local adaptation.

Both metrics serve to analyze trends in hot and warm events, yet they represent distinct phenomena driven by different processes and mechanisms. Changes in extreme events are evaluated from two perspectives: frequency changes, assessing how often a specific magnitude occurs, and magnitude changes, examining variations in the intensity of extreme events for a given return period. The probability of such events is linked to their rarity; rarer events typically experience more significant shifts in probability, while changes in magnitude do not always correlate as closely with rarity.

These perspectives are interconnected but carry different implications for stakeholders. Research shows that changes in the magnitude of temperature extremes often align linearly with rising global surface temperatures, while changes in frequency may exhibit nonlinear patterns. Context matters when assessing extreme events; if similar damage occurs when a fixed threshold is exceeded, the focus should be on frequency changes. However, if crossing the threshold becomes commonplace, it can lead to a saturation effect on probability changes. Adaptation to climate change can also shift relevant thresholds, though these considerations are rarely integrated into projections of changes in extremes. The framing of extremes—including definitions and the questions posed in the literature—will significantly influence the evaluations in this chapter.

13.13.2 The Escalating Frequency of Extreme Weather Events

Weather extremes, including temperature spikes, heavy precipitation, river floods, droughts, and intense storms such as tropical cyclones, are becoming more frequent and severe due to changes in the global climate system. These events, along with compound extremes—where multiple extreme events occur simultaneously or in close succession—pose significant challenges to ecosystems and societies. The analysis of these extremes reveals clear patterns that highlight the growing impact of global warming, especially over the last several decades.

Temperature Extremes: For example, temperature extremes have seen a marked increase, with more frequent heat waves that affect large regions and persist for longer durations. These temperature anomalies have far-reaching effects on agriculture, energy demand, and human health. Similarly, heavy precipitation events have intensified, leading to more frequent pluvial floods—where excessive rainfall causes flash flooding—and river floods. The increased intensity and frequency of these events can overwhelm infrastructure, disrupt livelihoods, and lead to significant economic losses. Since 1950, the frequency and intensity of hot extremes, such as heat waves, have risen globally, while cold extremes have decreased (virtually certain). These trends are evident in over 80% of regions, though some areas lack sufficient data to confirm. Human-induced greenhouse gas emissions are the main driver of these changes (virtually certain), with regional factors like land use, aerosols, and natural variability influencing local trends.

Hot extremes will continue to increase, and cold extremes will decrease with rising global temperatures. Even if global warming stabilizes at 1.5 °C, the intensity of extreme heat events is projected to double at 2 °C and quadruple at 3 °C. Mid-latitude and semi-arid regions are expected to experience the highest temperature increases during the hottest days, while the Arctic will see the most significant rise in cold extremes. Extreme heat events will become more frequent and intense as global warming progresses.

Heavy Precipitation and Floods. The frequency and intensity of heavy precipitation events have likely increased globally, particularly over regions with good observational data. Significant increases have been observed across North America, Europe, and Asia. Human-induced climate change, particularly due to greenhouse gas emissions, is likely the primary cause of these global trends. Evidence of human influence has emerged in several regions, especially in North America, Europe, and Asia. As global warming continues, heavy precipitation is expected to become more frequent and intense. At 4 °C of warming, very rare heavy precipitation events will occur more often and with greater intensity, both globally and regionally. The intensity increase will be proportional to atmospheric moisture capacity, rising about 7% per 1 °C of warming. This trend will be nonlinear, with rarer events doubling or tripling in frequency. Regional variations will depend on factors, such as local warming and storm dynamics.

River Floods: Significant trends in peak streamflow have been observed in some regions over recent decades, particularly in cold regions where snowmelt contributes to river floods, with peak streamflow occurring earlier (high confidence). Global hydrological models project that more land areas will experience an increase in river floods than a decrease (medium confidence). However, regional changes in river floods are more uncertain due to complex hydrological processes, land cover changes, and human water management.

Droughts: In many regions across the globe, droughts have also worsened, especially in areas that are already prone to arid conditions. The intensification of droughts is driven by shifting precipitation patterns and higher evaporation rates due to rising temperatures. In some regions, these prolonged dry spells lead to reduced water availability, increased risk of wildfires, and severe impacts on agriculture and food security. Droughts can be classified into different types, each with distinct impacts and responses to rising greenhouse gas concentrations. Agricultural and ecological droughts result from insufficient soil moisture, often worsened by increased atmospheric evaporative demand, while hydrological droughts arise from a lack of runoff and surface water.

Human-induced climate change has contributed to increased agricultural and ecological droughts in some regions due to rising evapotranspiration driven by higher temperatures, reduced humidity, and increased radiation. Precipitation trends play a lesser role globally but have influenced meteorological droughts in certain regions. Rising agricultural and ecological droughts have been observed on all continents, except for one region where drought conditions have decreased. As global warming intensifies, more regions will experience worsening agricultural and ecological droughts, even if warming is stabilized at 2 °C. At 4 °C of warming, nearly half of the world's inhabited regions are expected to see increased drought conditions. These droughts are primarily driven by increased evapotranspiration. Additionally, many regions are projected to face more severe hydrological droughts as global warming progresses. Although increased atmospheric CO_2 may enhance plant water-use efficiency, there is low confidence that it will significantly mitigate the severity of droughts. Rising global temperatures, particularly above 4 °C, will reduce the efficiency of land carbon sinks due to soil moisture limitations and prolonged drought conditions.

Extreme Storms, Including Tropical Cyclones: Storm activity, particularly tropical cyclones, has shown patterns of increased intensity, with stronger winds and heavier rainfall. While the frequency of such storms may not always increase, the severity of the strongest storms has escalated, contributing to catastrophic damage, particularly in coastal regions. These storms, in conjunction with rising sea levels, compound the risk of severe flooding and erosion, threatening both natural habitats and human settlements.

In a warming world, the intensity and rainfall rates associated with tropical cyclones (TCs), extratropical cyclones, and atmospheric rivers are increasing globally. Human-induced climate change likely contributes to extreme TC rainfall, with peak rain rates rising by about 7% per 1 °C of warming due to increased water vapor and wind intensity. Over the past four decades, the proportion of Category 3–5 TCs has likely grown, with shifts in their peak intensity location and a slowdown in translation speed over the United States. While the total number of TCs may decrease or remain stable as global warming continues, the intensity of TCs, particularly wind speeds, is expected to rise. For extratropical cyclones, future changes in wind speeds are expected to be modest, but storm tracks could shift poleward, causing regional impacts. Severe convective storms may see more intense rainfall, and in the United States, the frequency and duration of spring storms are projected to increase, though evidence from other regions is limited.

Compound events, such as the simultaneous occurrence of heat waves and droughts, or heavy rainfall following extended dry periods, present additional challenges. These multivariate extremes can amplify the impacts of individual events, creating a cascade of environmental and societal stresses. For example, a prolonged drought followed by intense rainfall can lead to severe soil erosion and heightened flood risk.

13.13.3 Effects of Large-Scale Circulation on Changes in Extremes

Atmospheric large-scale circulation patterns significantly influence regional climate and the characteristics of extreme weather events, including their magnitude, frequency, and duration. Changes in these patterns, driven by external forces, introduce uncertainty in the response to extreme events. For example, the El Niño–Southern Oscillation (ENSO) affects precipitation patterns, causing droughts in some areas and heavy rainfall in others. The Hadley circulation impacts where tropical and extratropical cyclones occur, influencing extreme precipitation and winds. Additionally, changes in land–ocean heat contrasts affect monsoon circulation, leading to heavy precipitation in East Asia.

While some robust changes in circulation patterns have been observed, such as the widening of the Hadley circulation and poleward shifts in cyclone tracks since the 1980s, there is still considerable uncertainty regarding future changes. Although the amplitude of ENSO variability may remain stable, extreme ENSO events are projected to increase in frequency, impacting droughts and tropical cyclones.

In summary, large-scale atmospheric circulation patterns are crucial drivers of local and regional extremes, but uncertainties regarding future changes hinder precise projections of these extremes.

13.13.4 Effects of Regional-Scale Processes and Forcings in Extremes

Changes in extreme weather events at local and regional levels are significantly influenced by regional processes, feedback mechanisms, and external forcings, such as land-use changes and aerosol concentrations. These factors, alongside large-scale circulation patterns, create complex interactions that can asymmetrically affect temperature distributions, particularly elevating the intensity of hot extremes.

Land use impacts regional extremes through various means, including agricultural practices and land cover alterations, such as deforestation. However, climate models often inadequately represent these processes, particularly the effects of forest cover on temperature cycles and irrigation practices. Changes in land use are particularly critical in low-emissions scenarios that involve extensive modifications to land for biofuels, reforestation, or food production.

Aerosol forcing, driven by regional emissions, also significantly affects temperature and precipitation extremes. For instance, increased aerosol concentrations from the 1950s to the 1980s contributed to regional cooling, followed by a warming phase associated with reduced aerosol levels since the 1990s. This reduction has been linked to an increase in the frequency of hot extremes in several regions.

Regional feedback mechanisms, particularly soil moisture feedback, are vital in exacerbating temperature extremes, leading to greater warming than the global average. Soil moisture and land surface conditions influence circulation patterns and contribute to the variability observed in climate model projections for temperature extremes. The interaction between soil moisture and precipitation also plays a critical role in extreme weather development, although models sometimes fail to accurately represent these interactions. Additionally, extreme events can mutually amplify each other; for example, heat waves can intensify droughts through increased evapotranspiration, leading to further drying and hotter temperatures.

In summary, regional processes, feedback from land use and aerosols, and interactions between soil moisture, temperature, and snow/ice albedo significantly modulate changes in extreme weather events. These mechanisms generally contribute to a higher rate of warming for extreme temperatures compared to mean temperatures, although confidence in the accurate representation of these processes in climate models is only medium.

The observed trends in these weather extremes are strongly linked to human-induced climate change, primarily driven by greenhouse gas emissions. Since the mid-twentieth century, the frequency and intensity of these extreme events have escalated, with clear evidence pointing to anthropogenic causes. Temperature extremes have shown a stronger attribute to human activity, with some of the hottest recorded events unlikely to have occurred without the influence of human-induced warming. As global temperatures continue to rise, regional changes in the intensity and frequency of weather extremes are expected to scale accordingly. Even small increases in global temperatures, such as half a degree Celsius, can significantly affect the frequency and severity of extreme weather events. This has already been observed in many regions, where even slight warming has led to noticeable increases in extreme temperatures, heavy precipitation, and storm intensity.

Future projections indicate that the occurrence of unprecedented extreme events will rise with ongoing global warming. As temperatures reach higher thresholds, the likelihood of rare and severe weather events increases. These projected changes underscore the importance of mitigating greenhouse gas emissions and enhancing resilience to the impacts of a warming climate.

13.14 Risks and Vulnerability from Climate Change

Climate-related risks to both natural and human systems are influenced by several factors, including the magnitude of global warming, geographic location, development levels, and the effectiveness of adaptation and mitigation strategies. The Global Framework for Climate Services (GFCS), established by the World Meteorological Organization (WMO) in 2011, addresses these risks by focusing on priority sectors such as agriculture, water management, health, and disaster risk reduction, while also covering sectors like energy, transport, and tourism.

The GFCS aims to integrate science-based climate information and forecasting into planning and policy at global, regional, and national levels.

A key priority is enhancing infrastructure to deliver comprehensive and advanced climate services. Understanding climate and weather impacts is crucial for assessing risks and building resilience. Effective forecasting of climate variability and its links to extreme weather events provides valuable information for spatial planning and risk management.

Weather and climate services are essential for societal benefit, offering:

1. Weather Forecasts and Alerts: Timely updates to protect health and safety
2. Flood and Drought Prediction and Monitoring: To mitigate impact and prepare responses
3. Natural Disaster Preparedness and Response: Enhancing resilience to disasters
4. Public Health Monitoring and Disease Control: Preventing and managing health risks
5. Fire Risk Assessment and Management: Reducing the impact of wildfires
6. Decision Support for Economic Sectors: Including water resources, agriculture, and transport

These services support basic human needs and foster socioeconomic progress by integrating weather and climate data for risk assessment. As climate-related risks and vulnerabilities increase, it is crucial to incorporate them into spatial planning systems. This involves adapting conventional planning practices to modern approaches that reflect the dynamic nature of climate change.

Developing and implementing effective spatial plans for climate impacts will help manage risks and enhance resilience. This process should be supported by institutional and legal frameworks that facilitate the integration of climate information into planning and decision-making.

Adaptation and mitigation are essential strategies for tackling climate change, focused on preventing negative effects and reducing vulnerability to its impacts. As shown in Fig. 13.17, many species adapt to survive in specific climatic conditions. Effective decision-making plays a crucial role in this process; analytical approaches that assess and manage expected climate risks can help limit the repercussions of climate change. However, without additional efforts in mitigation and adaptation, projections indicate that global warming by the end of the twenty-first century could lead to severe, widespread, and irreversible impacts.

Long-term sustainable development enhances future options and improves readiness to mitigate risks associated with climate change. Increasing the number of mitigation strategies is vital for limiting global warming to below 2 °C. Achieving effective climate action will require significant reductions in greenhouse gas emissions over the coming decades, targeting near-zero CO_2 emissions. Importantly, no single strategy can address climate change in isolation; a combination of interdisciplinary approaches, policies, cooperation, and diverse measures is necessary.

In terms of sector-specific mitigation strategies, the energy sector needs major transitions, including a reduction in fossil fuel use, widespread electrification, improved energy efficiency, and the adoption of alternative fuels such as hydrogen. Urban areas present unique opportunities for emission reductions through compact, walkable designs, the electrification of transportation, low-emission energy sources, and carbon removal technologies.

For industries, reducing emissions can be achieved by enhancing material efficiency, increasing recycling efforts, and minimizing waste. Agriculture, forestry, and land use can also contribute significantly to emissions reductions and carbon dioxide removal. Sustainable practices in these sectors support biodiversity and aid in adapting to climate change while ensuring the availability of food, water, and wood supplies.

Fig. 13.17 Adaptation to climate change. (**a**) Polar bears adapt by foraging on land and swimming longer, but shrinking sea ice is expected to increase food scarcity and reduce populations. (**b**) On the western Antarctic Peninsula, warming turns snow to rain and mud. Older penguins' waterproof feathers help them stay warm and swim efficiently, while younger ones face greater risks

Key outcomes for climate action include doubling the rate of emissions reduction by 2030 to make meaningful progress, addressing financial shortfalls to close investment gaps, and aligning climate actions with broader sustainable development objectives.

The Working Group III report emphasizes that the evidence for climate change is robust and urgent action is needed. Achieving the 1.5 °C limit requires immediate and comprehensive reductions in emissions across all sectors. With the right policies, infrastructure, and technologies, a 40–70% reduction in greenhouse gas emissions by 2050 is feasible, supporting the transition to a more sustainable and resilient future. To limit global warming to about 1.5 °C, global greenhouse gas emissions must peak before 2025 and decline by 43% by 2030. Methane emissions also need to be reduced by about one-third. Even with these efforts, it is almost inevitable that the 1.5 °C temperature threshold will be exceeded temporarily, but it is expected to decrease by the end of the century if these targets are met. Global temperature rise will stabilize when carbon dioxide emissions reach net zero, which for 1.5 °C means achieving net zero carbon dioxide emissions globally by early 2050.

For 2 °C, net zero CO_2 would need to be achieved by the beginning of 2070. Accelerated climate action is critical for sustainable development. Solutions such as increasing urban green spaces, including parks and open areas, and implementing urban agriculture can help absorb and store carbon while reducing the risk of flooding and mitigating heat island effects.

The pathways to address climate change require significant reductions in emissions over the next few decades, aiming for nearly zero CO_2 emissions. While many adaptation and mitigation options exist, no single approach can fully address the issue. Effective implementation necessitates an interdisciplinary approach, encompassing policies, cooperation, and actions across all levels. Mitigation strategies are available in every major sector, often in a more cost-effective manner when using an integrated approach that combines efforts to reduce energy use and greenhouse gas emissions. Climate change poses a substantial threat to sustainable development, yet it also presents opportunities to link mitigation, adaptation, and other societal goals through comprehensive responses.

Although climate change is a reality, its impact on precipitation, evaporation, surface runoff, and flood risk remains uncertain. Other effects of global warming include the melting of ice, species extinction, an increase in the range of diseases, and the retreat of glaciers. Glaciers, large masses of ice that move over land and form in areas where snow accumulates rather than melts or sublimates, are particularly sensitive to climate change and serve as indicators of global warming (Fig. 13.18).

Water availability plays a significant role in these temperature influences. In modern climate research, the European Space Agency (ESA) has conducted two missions over 3 years to better understand these impacts. The first mission aimed to measure the soil's water content around the globe every 3 days to a depth of 2 m. This research ties into the processes of photosynthesis and plant growth, which are crucial for understanding how carbon dioxide is released and absorbed, particularly by plants and oceans.

Fig. 13.18 Patagonia Glacier—The Patagonia Glacier, located in the southern regions of South America, is one of the world's most iconic and extensive ice fields. This glacier, like many others globally, is highly sensitive to changes in climate, making it a significant indicator of global warming. The glacier's retreat, which has accelerated in recent decades, highlights the profound impact of rising temperatures on ice masses. The figure illustrates the glacier's current state, emphasizing the broader implications for sea-level rise and freshwater availability as these glaciers continue to diminish

The second mission focused on measuring changes in the salt content of surface seas, which is essential for improving our understanding of the fundamental drivers of global ocean circulation patterns. Ocean salinity changes depend on the balance of freshwater through evaporation and precipitation, and in polar regions, on the melting and freezing of ice. To conduct this research, a satellite (SMOS) was launched from the Plesetsk Cosmodrome in northern Russia on November 2. As the first space-based instrument to measure Earth's water and ocean salinity, SMOS aims to fill critical gaps in our knowledge of the planetary life cycle and the impact of climate change on the movement of water across the Earth, air, and seas.

13.15 Global Climate Change and Health

Global climate change has the potential to affect human health in several ways. As temperatures rise and precipitation patterns shift, the habitats of disease vectors like mosquitoes and ticks expand, altering the distribution and seasonality of diseases such as malaria, dengue fever, and Lyme disease. This poses new health challenges in regions previously unaffected by these diseases.

Climate change disrupts ecosystems, leading to loss of biodiversity and changes in the availability of natural resources, such as clean water and food. These disruptions can exacerbate malnutrition, increase exposure to pollutants, and reduce the capacity of ecosystems to buffer against natural disasters.

The rising frequency and intensity of extreme weather events—including droughts, floods, heat waves, cold waves, forest fires, extreme storms, and floods—pose direct and indirect threats to human health. These events can lead to injury, death, displacement, and outbreaks of waterborne and vector-borne diseases. Additionally, air pollution, often exacerbated by these events, contributes to respiratory and cardiovascular diseases, further compromising public health.

Global climate change has wide-ranging impacts on the environment, and these changes can have significant implications for human health. The effects of climate change on health are complex and multifaceted, affecting different aspects of well-being. It is obvious that the climate of our planet is changing, and these changes are influencing the health and well-being of people (Fig. 13.19). Recent reports based on statistical indicators and scientific knowledge show that the climate is warming at a global level. As the planet warms, ocean and sea levels rise, floods and droughts become more frequent and intense, and heat waves, tropical cyclones, storms, and cold waves grow more severe. Weather and climate are inextricably linked to some of the most basic determinants of human health, such as clean air and water, adequate food and shelter, and the occurrence and spread of disease. Heat and cold waves, tropical cyclones, floods, and droughts threaten many lives and increase disease transmission every year.

Factors indirectly related to weather and climate—such as food insecurity and non-communicable diseases like cardiovascular and respiratory diseases resulting from exposure to poor air quality—also cause death and disease in many people.

Additionally, the increase in infectious diseases caused by water and insects, due to favorable conditions especially brought about by climate variability, results in significant societal and economic costs. Understanding the relationship between

Fig. 13.19 Climate change and health effects: This figure illustrates the various ways in which climate change impacts human health, including the changing geographic range and seasonality of infectious diseases, the disruption of natural ecosystems, and the increased frequency of extreme weather events, all of which contribute to adverse health outcomes

climate and health is fundamental for taking preventive action against climate-related health risks. Vulnerable populations—children, adults, and those with existing health problems—are exposed to heightened health risks from the impacts of climate change (Fig. 13.20). A major challenge for healthcare facilities is accessing, recognizing, understanding, interpreting, and applying available climate information. Furthermore, climate services institutions often do not fully address the needs and role that climate services can play in supporting public health.

Climate change has both direct and indirect effects on human health. Direct effects occur because of changes in the intensity and frequency of extreme weather events. Indirect effects are felt through changes in the frequency of diseases caused by insects or changes in water and air quality. Although global warming may in some cases offer local benefits, such as fewer winter deaths in temperate climates and increased food production in certain areas, the overall health effects of climate change are negative.

Climate change impacts the social determinants of health, including clean air, safe drinking water, sufficient food, and secure shelter. The World Health Organization (WHO) estimates that between 2030 and 2050, climate change is expected to cause approximately 250,000 deaths per year, due to factors such as malnutrition, the spread of diseases like malaria and diarrhea, and heat stress. Areas with weak health infrastructure—primarily in developing countries—will be the least able to cope and prepare to meet these challenges without assistance.

The impact of climate change on human health is unequivocal and depends on many factors, including the effectiveness of public health and safety systems, risk behavior, age, gender, and economic status of the affected populations. Reducing greenhouse gas emissions through better transportation, food choices, and energy use can improve health, particularly by reducing air pollution. The impacts vary by region, population sensitivity, degree and duration of exposure to climate change, and the ability of society to adapt to these changes.

Fig. 13.20 The youngest population—children—is a very sensitive category to climate change

13.15.1 Weather/Climate Extremes and Health

As a result of climate change, extreme weather events such as heat waves, storms, tornadoes, floods, droughts, and other severe weather phenomena are becoming more frequent and intense. These events have a significant negative impact on human life and health, leading to injuries, illnesses, emotional trauma, and even death. Weather extremes refer to unusual weather events that are exceptionally severe compared to the historical norms for a given area. The frequency and intensity of some of these extreme events are already increasing and are expected to continue to rise in the future.

This growing threat underscores the need for greater infrastructure preparedness to handle public health emergencies. Implementing early warning systems, efficient evacuation protocols, and effective sanitation measures can significantly reduce the impact of these extreme events on human health. Such measures assume that medical and emergency response systems will function efficiently, and that timely and effective adaptation strategies will be developed to address these challenges.

Heat waves are caused by large, stagnant masses of hot air that persist over a region for an extended period. They often develop in areas near bodies of water, such as oceans and bays, which can influence the formation of these extreme weather events. Certain regions are more prone to frequent and severe heat-related events. A heat wave is typically characterized by prolonged periods of excessively hot weather, often accompanied by high humidity, and can last for several days. These conditions create significant health risks, especially for those who are exposed to heat without adequate protection (Fig. 13.21).

Heat waves can vary in duration and intensity. They may consist of a longer period with higher-than-normal air temperatures or a shorter period with exceptionally high temperatures, as noted by the Intergovernmental Panel on Climate Change (IPCC) in 2018. In the mid-latitude regions of the Northern Hemisphere, heat waves generally occur in summer, often under stable atmospheric conditions and existing high-pressure systems.

The health risks associated with heat waves are substantial, making it crucial to take timely protective measures, particularly for vulnerable populations such as the elderly, the chronically ill, children, and those who work outdoors. During physical activity, the body generates heat, which must be dissipated to maintain a safe internal temperature. Normally, this heat is lost through contact with cooler air and perspiration, which cools the body as it evaporates. However, weather conditions play a significant role in this process.

For example, in windy conditions, sweat evaporates more quickly, allowing the body to cool down faster. Conversely, high humidity slows down evaporation, leading to a rise in body temperature. Extreme temperatures can directly contribute to an increase in deaths from cardiovascular and respiratory diseases, highlighting the growing public health problem posed by excessive heat.

Climate variability has profound social and health implications, as it directly affects ecosystems and, combined with increased human pressure on the environment, seriously impacts biodiversity and the transmission of infectious diseases. Disease transmission can be categorized into direct and indirect pathways. Direct

Fig. 13.21 Typical hot summer weather and occurrence of heat waves. Heat Islands are created in urban areas, where temperatures are several degrees higher than in surrounding regions. This extreme heat can pose significant dangers to health, particularly in densely populated cities

human transmission typically involves physical contact, droplet transmission, and contagion. In contrast, indirect transmission to humans mainly occurs through vectors or other pathogenic microorganisms.

Vector-borne infectious diseases are a significant category of transmissible diseases. These diseases, caused by bacteria, viruses, or parasites, require vectors—typically blood-feeding insects such as mosquitoes, ticks, lice, fleas, and midges—to transmit them between hosts. Before reaching a larger number of human or animal hosts, these pathogens spend time in vectors, where they undergo an incubation period necessary for the vector to become infectious. The infected vector then transmits the pathogen to the host, usually through a bite or sting.

Vectors, as organisms lacking mechanisms for thermoregulation (body heat), are highly dependent on external environmental factors such as temperature and humidity. These conditions are critical for their reproduction and activity, making vector-borne diseases seasonal. In countries with temperate continental climates, vectors are most active from spring to autumn.

Many vector-borne infectious diseases are endemic, meaning they are native to specific areas where conditions, including the presence of suitable infectious reservoirs and vectors, favor the maintenance of the pathogen. Numerous studies have provided evidence linking the occurrence of vectors to climate anomalies, particularly those driven by the El Niño phenomenon. Changes in temperature and precipitation patterns also lead to shifts in microclimatic features, causing some areas to become more prone to droughts, while others experience floods. These conditions serve as a basic prerequisite for the spread of infectious diseases through polluted water, air, and food.

In summary, climate variability and the disruption of the natural environment have significant consequences for health. In the future, these changes are expected to cause disease and suffering for millions of people worldwide (Fig. 13.22).

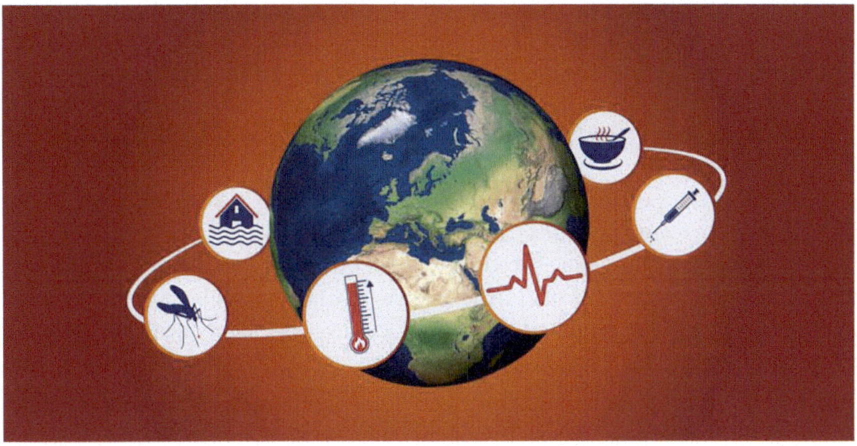

Fig. 13.22 Climate variability and health. A visual representation illustrating the dynamic interplay of environmental factors such as temperature fluctuations, precipitation changes, and extreme weather events and their potential impacts on human health and well-being. The figure highlights how climate variability can influence the spread of vector-borne diseases, disrupt ecosystems, and affect overall health outcomes

Long-term climate change poses a significant threat by exacerbating existing issues and undermining future health systems, infrastructure, social security, and the availability of essential resources such as food, water, and other ecosystem services crucial for human health. To address these challenges effectively, a detailed understanding of the nature and biological cycles of disease vectors is essential. This knowledge is crucial for implementing timely interventions to manage vector-borne diseases and their spread.

The negative impact of infectious diseases on health is closely linked to various stressors, including inadequate sanitation, limited access to clean drinking water and food, and the overall quality of public health services. To mitigate or reduce mortality associated with infectious diseases transmitted through vectors, air, water, or food, further research is needed to clarify specific aspects related to these diseases.

Future epidemiological studies focusing on these specific issues will provide a more accurate assessment of their full health impact. This research will be instrumental in enhancing public health measures and developing effective mitigation strategies to address and prevent the health effects related to these diseases, thereby improving overall resilience and response capabilities in the face of climate change.

13.15.2 The Upside of Climate Change: Potential Benefits and Challenges

Climate change may lead to some potential benefits, including enhanced growth of certain plant species due to rising temperatures and increased carbon dioxide levels. This could improve agricultural yields in regions with favorable water and soil conditions. Warmer winters may also reduce cold-related mortality, particularly among vulnerable populations such as the elderly, resulting in fewer health issues associated with extreme cold. Additionally, longer growing seasons can boost agricultural productivity, allowing for the cultivation of crops that previously could not thrive in those areas. Milder winters may also encourage outdoor activities and overall well-being.

However, these potential benefits come with important caveats. Increased temperatures during warmer months can exacerbate heat-related health risks, with more frequent and intense heat waves leading to higher mortality rates among vulnerable populations. The advantages of climate change may not be evenly distributed; while some regions could see benefits, others may face severe droughts, sea-level rise, and ecosystem disruptions. Moreover, climate changes can threaten biodiversity and disrupt ecosystems, potentially harming food chains and vital ecosystem services. In summary, while climate change may present some positive effects, these are often overshadowed by significant challenges and risks. The overall impact on health and well-being is complex, highlighting the need to consider both potential benefits and drawbacks when planning for future climate scenarios and implementing adaptation strategies.

13.15.3 Future Perspectives on Climate Change: Unveiling the Complexities Ahead

The future of climate change presents substantial global issues that are critical for the survival of our planet and the sustainability of life and ecosystems. As we navigate an era marked by rising temperatures and shifting climate patterns, the implications for biodiversity, human health, and global economies become increasingly urgent.

One of the most pressing challenges is the degradation of ecosystems, which play a vital role in maintaining the balance of our planet. Climate change threatens habitats, leading to loss of biodiversity and the extinction of species. This decline disrupts ecosystem services that are essential for human survival, such as clean air and water, food production, and natural disease regulation. Protecting and restoring these ecosystems is crucial for maintaining the health of our planet.

Additionally, the impacts of climate change disproportionately affect vulnerable communities around the world. Social inequalities often dictate the ability of individuals and communities to adapt to changing conditions. Marginalized populations frequently bear the brunt of climate-related disasters, facing increased risks to their health, livelihoods, and safety. Addressing these inequalities is essential for fostering resilience and ensuring equitable access to resources and opportunities. The increasing frequency and severity of extreme weather events pose further threats to global stability. Natural disasters, such as hurricanes, droughts, and floods, disrupt food and water supplies, displace populations, and strain emergency response systems. Developing robust infrastructure that can withstand these events is vital for safeguarding communities and mitigating long-term impacts. Moreover, the intersection of climate change and public health cannot be overlooked. Rising temperatures contribute to the spread of infectious diseases and exacerbate respiratory and cardiovascular conditions due to air pollution. Climate change also affects food security, as shifting agricultural patterns can lead to crop failures and increased malnutrition. Comprehensive strategies that prioritize public health in the context of climate action are essential for promoting well-being and resilience.

Finally, addressing climate change requires global cooperation and governance that transcend national boundaries. Climate action cannot be achieved in isolation; it demands collective efforts, innovative solutions, and a commitment to sustainability from all sectors of society. Engaging diverse stakeholders—governments, businesses, communities, and individuals—is critical to developing and implementing effective climate policies.

In summary, the future of climate change is characterized by significant challenges that threaten the survival of Earth and the sustainability of its ecosystems. A holistic approach that integrates environmental, social, and economic considerations is necessary to navigate these challenges. By prioritizing the health of our planet and its inhabitants, we can work toward a resilient future that ensures the survival of life in all its forms.

References

Arrhenius, S. (1896). On the influence of carbonic acid in the air upon the temperature of the ground. *Philosophical Magazine and Journal of Science, 41*(5), 237–276.

Bennett, J. R., & O'Neill, B. C. (2021). *Climate change: Science and solutions*. Oxford University Press.

Benton, M. J. (2019). The mass extinction that changed the world: What happened during the Permian-Triassic extinction? *Geology Today, 35*(2), 55–60.

Berger, A., Loutre, M. F., & Mélice, J. L. (2010). Equatorial insolation: From precession harmonics to eccentricity frequencies. *Climate of the Past, 6*(4), 539–575.

Boulanger, J., & Gassmann, X. (2023). *Climate change and global warming: A multidisciplinary perspective*. Springer.

Bretherton, I., Ridgeway, D., & Cassidy, J. (1990). Assessing internal working models of the attachment relationship: An attachment story completion task for 3-year-olds. In M. T. Greenberg, D. Cicchetti, & E. M. Cummings (Eds.), Attachment in the preschool years: Theory, research, and intervention (pp. 273–308). The University of Chicago Press.

Cai, W., Wang, G., Santoso, A., McPhaden, M. J., Wu, L., Jin, F. F., & Takahashi, K. (2020). Climate impacts of the El Niño–Southern Oscillation on global land precipitation. *Nature Climate Change, 10*(1), 43–49.

Cohen, J., et al. (2022). The impact of climate variability on weather extremes. *Journal of Climate, 35*(12), 4532–4553.

Crutzen, P. J., & Stoermer, E. F. (2000). The Anthropocene. *Global Change Newsletter, 41*, 17–18.

Gates, R. D., Baghdasarian. G. and Muscatine, L. (1992) Temperature streşs causes host-cell detachment in symbiotic cnidarians-implications for coral bleaching. Biol Bull, 182(3): 324–332. http://dx.doi.org/10.2307/1542252.

Geiger, R., & Pohl, R. (1998). Klima der Erde: Eine Einführung in die Klimatologie.

Graham, N. E., et al. (2023). Machine learning approaches to seasonal climate prediction. *Climate Dynamics, 61*(3), 987–1001.

Hartmann, D. L. (2016). *Global physical climatology*. Elsevier.

Higgins, R. W., & Shi, W. (2000). Dominant factors responsible for interannual variability of the summer, monsoon in the southwestern United States. *J. Climate*, 13, 759–776.

Hoffman, P. F., Schrag, D. P., & Kump, L. R. (2017). Snowball earth: The story of the frozen planet. In *The Earth system: Physical and chemical dimensions of global environmental change* (pp. 239–261). Springer.

Houghton, J. T., Ding, Y., Griggs, D. J., Noguer, M., van der Linden, P. J., & Dai, X. (2001). *Climate change 2001: The scientific basis*. Cambridge University Press.

IPCC. (2021). *Sixth assessment report*. Intergovernmental Panel on Climate Change.

IPCC. (2023a). *Climate change 2023: Synthesis report*. IPCC report.

IPCC. (2023b). In H. Lee & J. Romero (Core Writing Team) (Eds.), *Climate change 2023: Synthesis report. Contribution of working groups I, II, and III to the sixth assessment report of the intergovernmental panel on climate change* (pp. 35–115). IPCC, Geneva. https://doi.org/10.59327/IPCC/AR6-9789291691647.

IPCC. (2023c). Summary for policymakers. In H. Lee & J. Romero (Core Writing Team) (Eds.), *Climate change 2023: Synthesis report. Contribution of working groups I, II, and III to the sixth assessment report of the intergovernmental panel on climate change* (pp. 1–34). IPCC, Geneva. https://doi.org/10.59327/IPCC/AR6-9789291691647.001.

Köppen, W. (1936). *Das geographische System der Klimate*. Gebrüder Borntraeger.

Lüthi, D., et al. (2008). High-resolution carbon dioxide record over the past 800,000 years. *Nature, 453*(7193), 379–382.

Lyons, T. W., Reinhard, C. T., & Planavsky, N. J. (2014). The rise of oxygen in Earth's early ocean and atmosphere. *Nature, 506*(7488), 307–315.

Marcott, S. A., et al. (2013). A reconstruction of regional and global temperature for the past 11,000 years. *Science, 339*(6124), 1198–1201.

Met Office. (2023). *Climate change: The facts*. Met Office Climate

NASA. (2023). *Climate change: How do we know?* NASA Climate

National Oceanic and Atmospheric Administration (NOAA). (2022). *The climate system: An overview*. NOAA website

National Research Council (NRC) (2002). Down to Earth: Geographic Information for Sustainable Development in Africa. Washington, DC: The National Academies Press.

NOAA. (2023). *Climate change impacts*. NOAA Climate

Peel, M. C., Finlayson, B. L., & McMahon, T. A. (2007). Updated world map of the Köppen-Geiger climate classification. *Hydrology and Earth System Sciences, 11*(5), 1633–1644.

Rosen, R. A. (2023). *Climate change and society: Vulnerability and resilience in a changing world*. Harvard University Press.

Saha, S., et al. (2023). Oceanic anomalies and climate variability: The role of Atlantic and Indian oceans. *Geophysical Research Letters, 50*(4), e2023GL102331.

Steffen, W., Crutzen, P. J., & McNeill, J. (2007). The Anthropocene: Are humans now overwhelming the great forces of nature? *AMBIO, 36*(8), 614–621.

Stern, N. (2021). *Why are we waiting? Logic, urgency, and promise of tackling climate change*. Harvard University Press.

Trenberth, K. E. (2010). Changes in precipitation with climate change. Climate Research, 47, 123–138. https://doi.org/10.3354/cr00953.

Warner, T. T. (2011) Numerical Weather and Climate Prediction, Cambridge University Press, Cambridge, UK. ISBN: 978-0-521-51389-0. Hardback, 526 PP.

Zachos, J. C., Pagani, M., Sloan, L. C., Thomas, E., & Billups, K. (2001). Trends, rhythms, and aberrations in global climate 65 Ma to present. *Science, 292*(5517), 686–693.

Zalasiewicz, J., et al. (2010). The new world of the Anthropocene. *Geological Society, London, Special Publications, 337*(1), 35–43.

Chapter 14
The New Era of Environmental Monitoring: Insights into Weather, Water, and Climate

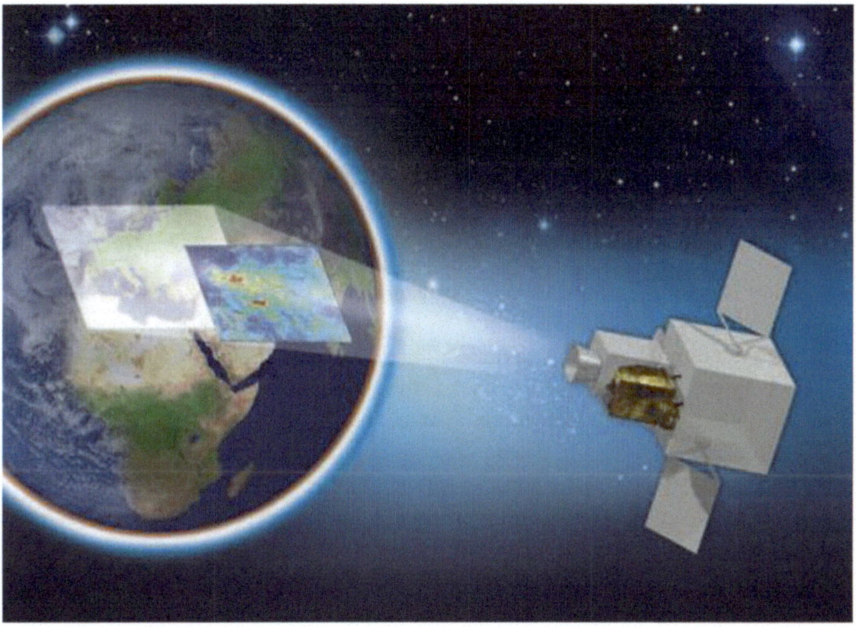

14.1 Global Earth and Environmental Monitoring

Atmospheric observations are essential for a wide range of critical activities, including real-time weather analysis, forecasting, and issuing warnings for natural disasters related to weather, water, and climate (WMO, 2015). Geophysical and

© The Author(s), under exclusive license to Springer Nature Switzerland AG 2025
V. Spiridonov et al., *Atmospheric Perspectives*,
https://doi.org/10.1007/978-3-031-86757-6_14

meteorological monitoring is conducted not only to support these functions but also to facilitate local, weather-dependent operations, such as those at airports, in construction, transportation, agriculture, and healthcare (Wallace & Hobbs, 2006a, b). Additionally, meteorological and climate data play a pivotal role in advancing climate and climate change research (Trenberth et al., 2013).

Since the advent of meteorological satellites in 1957/1958, which led to the establishment of the World Weather Watch (WWW) in 1963, meteorology has made remarkable strides in the accuracy and variety of weather forecasting services (Kalnay, 2003). However, the ongoing rise in global temperatures is expected to result in higher sea levels, more intense extreme weather events, and shifts in precipitation patterns and regimes (IPCC, 2021). These evolving societal challenges, driven by the growing impacts of climate change, necessitate further development of the Earth observation network. This includes upgrading global space- and surface-based observing systems and adopting a new, integrated approach that leverages the latest scientific and technological advancements (Hollmann et al., 2013).

Global Earth and environmental monitoring of weather, water, and climate involve systematic observation, analysis, and interpretation of various environmental parameters to understand and predict atmospheric and hydrologic processes. This monitoring is crucial for assessing the state of the planet, detecting changes, and making informed decisions about resource management, disaster preparedness, and climate change mitigation (Asrar & Dozier, 2014). The following sections provide an overview of the key aspects of global monitoring in these domains.

1. *Weather monitoring* relies on satellite observations, which provide a comprehensive overview of Earth's atmosphere by capturing data on cloud cover, precipitation, temperature, and atmospheric composition. Ground-based observations, such as those from surface weather stations, radar systems, and weather balloons, collect essential data on temperature, humidity, wind speed and direction, and atmospheric pressure. These observations are crucial for local and global weather analysis. Supercomputers employ numerical weather prediction (NWP) models to simulate Earth's atmosphere, enabling meteorologists to produce short- to medium-range weather forecasts with increasing accuracy.
2. *Water monitoring* also benefits from satellite remote sensing, which tracks ocean currents, sea surface temperatures, and salinity, in addition to monitoring snow cover, river discharge, and changes in water bodies. Ground-based observations, including stream gauges, buoys, and water quality sensors, measure river flow, water levels, and key water quality parameters like pH, dissolved oxygen, and nutrient concentrations. Computer models simulate the movement and distribution of water within Earth's hydrological cycle, aiding in water resource management and flood prediction.
3. *Climate monitoring* involves the collection of long-term climate records from sources such as ice cores, tree rings, and sediment layers, providing insights into past climate variations. The Global Climate Observing System (GCOS), an international climate monitoring system, integrates satellite and in situ observa-

tions to deliver a comprehensive view of key climate variables. Climate models, which simulate Earth's climate system using mathematical–physical equations, are instrumental in helping global forecasting centers understand trends, project future changes, and assess the impacts of human activities on the climate.

14.2 Global Earth and Environment Observing System

The Global Telecommunication Observing System (GTS) forms an extensive and interconnected network designed to monitor and collect data on various environmental aspects of the Earth (Fig. 14.1). This system's primary objective is to provide precise and timely information that supports activities like weather forecasting, climate monitoring, disaster management, and scientific research. It encompasses a wide range of instruments and sensors deployed across the globe, such as weather stations, satellites, ocean buoys, radar systems, and more. These devices collect vital data on environmental parameters including temperature, humidity, atmospheric pressure, wind patterns, and ocean currents.

The information gathered by GTS is essential for improving weather forecasting models, enabling meteorologists to predict short-term weather patterns and long-term climate trends more accurately (WMO, 2017). This network also plays a key role in monitoring Earth's changing climate by continuously gathering data on

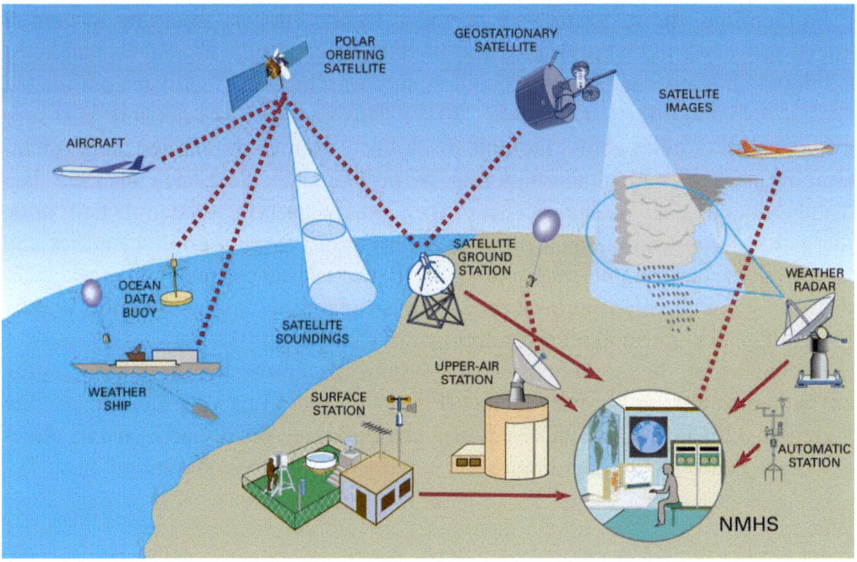

Fig. 14.1 Global Telecommunication System. The Global Telecommunication System (GTS) enables fast and reliable exchange of meteorological and environmental data worldwide. Operated by the WMO, it supports weather forecasting, disaster response, and climate monitoring by linking national meteorological services. (Source: WMO)

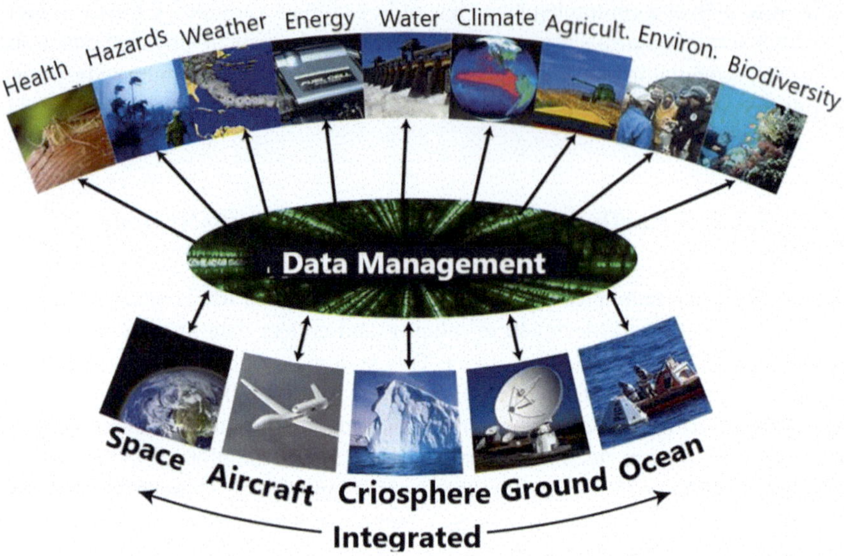

Fig. 14.2 Global system for monitoring the planet and the environment

critical variables such as temperature, greenhouse gas concentrations, and sea-level rise (IPCC, 2021). These measurements enhance our understanding of climate change and its consequences for ecosystems and human activities.

Furthermore, the real-time data provided by the Global Observing System is crucial for managing natural disasters, including hurricanes, tornadoes, tsunamis, and floods. It allows authorities to issue timely warnings and facilitate coordinated disaster response efforts (Lazo et al., 2011). This system's vast network also supports scientific studies across multiple disciplines, such as climate research, oceanography, and environmental monitoring, by providing accurate and consistent data (Trenberth et al., 2013). The success of the Global Observing System is built upon international cooperation, as it involves contributions from multiple countries and organizations, ensuring global coverage and data consistency. Technology advancements have continually improved the system's capabilities, allowing for the collection of more accurate and high-resolution data (Kalnay, 2003).

In conclusion, the Global Observing System is a cornerstone in understanding and managing the Earth's environment (Fig. 14.2). It delivers vital data for weather prediction, climate analysis, disaster response, and scientific research, and its effectiveness depends on ongoing innovation and international collaboration.

14.3 Meteorological Observation Systems

Meteorological measurements are broadly categorized into two methods: direct measurements and remote sensing systems, the latter including space-based observations. The Global Observing System (GOS) is structured to fulfill the varied requirements for meteorological data collection, integrating both surface-based and space-based subsystems.

The surface-based subsystem comprises several station types tailored to specific purposes, such as surface weather stations, upper-air stations, and climatological stations. These stations provide valuable data on parameters like temperature, humidity, wind speed, and atmospheric pressure. In contrast, the space-based subsystem utilizes multiple spacecraft equipped with advanced onboard instruments designed for atmospheric sounding, along with the necessary ground infrastructure for command, control, and data reception. These space-based observations are critical for global coverage, especially in remote areas where surface-based stations are sparse (Kalnay, 2003).

The concept of representativeness in meteorological observations refers to how well a given observation reflects the actual value of a variable over a specific area or for a particular purpose. For instance, synoptic observations typically represent a region of up to 100 km surrounding a station, although for more localized applications, the area of interest can be as small as 10 km or less (Stull, 1988).

Meteorological phenomena are classified by their horizontal scales, which range from micro- to planetary scales, according to Orlanski's (1975) classification:

- Microscale: Less than 100 m, relevant for detailed studies in agricultural meteorology, such as evaporation and crop–climate interactions.
- Local scale: From 100 m to 3 km, applicable to air pollution dispersion, tornadoes, and localized wind patterns.
- Meso-scale: Extending from 3 to 100 km, covering thunderstorms, sea breezes, and mountain breezes, as well as other localized weather systems.
- Large scale: Ranging from 100 to 3000 km, encompassing larger weather systems such as fronts, cyclones, and cloud clusters.
- Planetary scale: Larger than 3000 km, involving global phenomena such as long upper tropospheric waves and planetary circulation patterns (Orlanski, 1975).

Understanding these scales is crucial for developing and improving weather models, as different phenomena require varying levels of spatial and temporal resolution in data collection (Wallace & Hobbs, 2006a, b).

14.3.1 Basic Meteorological Elements

The current state and future behavior of the atmosphere are determined by a set of basic meteorological elements. These include air temperature, atmospheric pressure, humidity, wind speed and direction, cloudiness, precipitation, and visibility (or atmospheric transparency). Additional elements include soil temperature, surface water temperature, solar radiation, terrestrial radiation, and long-wave atmospheric radiation. Weather phenomena such as thunderstorms and snowstorms are also considered part of these basic meteorological elements.

The variations in these meteorological elements are driven by atmospheric processes and are fundamental in defining both weather and climate. Observations of these elements are conducted at radiosonde and meteorological stations, as well as meteorological observatories, using specialized instruments. At surface observing stations, the following elements are typically monitored:

- Air temperature
- Relative humidity
- Atmospheric pressure
- Wind speed and direction
- Precipitation
- Snow cover
- Solar radiation
- Soil temperature
- Evaporation

From these primary measurements, secondary elements such as dew point, freezing point, and others can be derived. There are instruments available for measuring all these elements, except for cloud type, which remains observationally classified.

14.3.2 *Standardization of Measurements*

The accuracy and consistency of meteorological data are crucial for a wide range of applications, including synoptic, climatological, and aeronautical purposes. With advancements in technology, many essential meteorological elements—such as temperature, humidity, and pressure—can now be measured automatically. However, to ensure that these measurements are uniform and comparable worldwide, they must adhere to internationally recognized standardized procedures and methods.

Meteorological observations conducted at stations around the globe are governed by the World Meteorological Organization (WMO), which has established rigorous guidelines for instrument calibration, measurement methods, and data reporting. Each country, through its National Meteorological Service, maintains a network of stations that follow these standards. The data collected are shared internationally through the Global Telecommunication System (GTS), ensuring that they meet the same high standards of accuracy and reliability (WMO, 2018).

The term "standard" in this context refers to the specific instruments, methods, and operational procedures employed to minimize measurement uncertainties and improve data quality. To maintain consistent accuracy, national and regional standards for meteorological instruments are implemented, including routine calibrations and systematic maintenance schedules. These protocols guarantee that the meteorological data obtained from different regions of the globe are comparable, consistent, and free from significant deviations due to instrument error or procedural inconsistencies (Harrison & Sturges, 2009).

The importance of standardization in meteorological measurements extends beyond just improving local forecasting efforts. It is critical for the international exchange of data that underpins global climate monitoring, disaster management, and aeronautical safety, making it a cornerstone of modern meteorological science.

14.3.3 Meteorological Instruments

Meteorological instruments are vital for collecting accurate data on various atmospheric parameters, which is critical for weather forecasting, climate research, and environmental monitoring. Figure 14.3 provides an overview of conventional instruments used to measure key meteorological elements, such as air temperature, relative humidity, snow cover depth, atmospheric pressure, wind speed and direction, precipitation, and evaporation. These instruments, including thermometers, barometers, anemometers, and rain gauges, have long been the foundation of meteorological measurements.

However, rapid advancements in technology have expanded the range of available tools, introducing modern, highly sophisticated devices that offer increased precision and broader measurement capabilities. These include digital sensors, acoustic and optical instruments, as well as infrared and laser-based devices. For instance, LIDAR (light detection and ranging) technology is now frequently used for studying atmospheric particles and wind patterns, while radar systems are instrumental in tracking precipitation and storm development. These cutting-edge tools enable meteorologists to observe atmospheric dynamics more effectively, allowing for real-time monitoring and enhanced accuracy in weather models (Brock & Richardson, 2001).

Modern instruments, by providing high-resolution and real-time data, are indispensable for addressing the complexities of weather systems and climate processes. Their use has significantly advanced our ability to predict severe weather events,

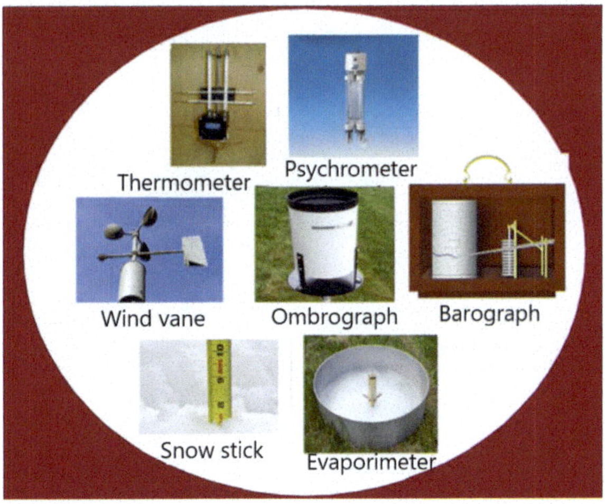

Fig. 14.3 Conventional meteorological instruments

assess long-term climate trends, and understand environmental changes (Stull, 2017). According to the WMO, effective meteorological measurement systems are crucial for improving the quality of weather forecasts and enhancing public safety (WMO, 2020). By integrating data from both traditional and modern instruments, meteorological science can continue to refine its forecasting models and contribute to better preparedness for natural hazards.

14.4 Special Techniques for Profiling the Boundary Layer and the Troposphere

Profiling the boundary layer and the troposphere requires specialized observational techniques to capture the dynamic and thermodynamic properties of the atmosphere. These methods are essential for understanding processes such as turbulence, convection, and the exchange of energy, moisture, and momentum between the Earth's surface, atmospheric boundary layer, and the free atmosphere (Fig. 14.4). Special profiling techniques have been developed to obtain high temporal and spatial resolution data needed for analysis, forecasting, and research on smaller meteorological scales and for various specialized applications. These techniques are broadly classified into:

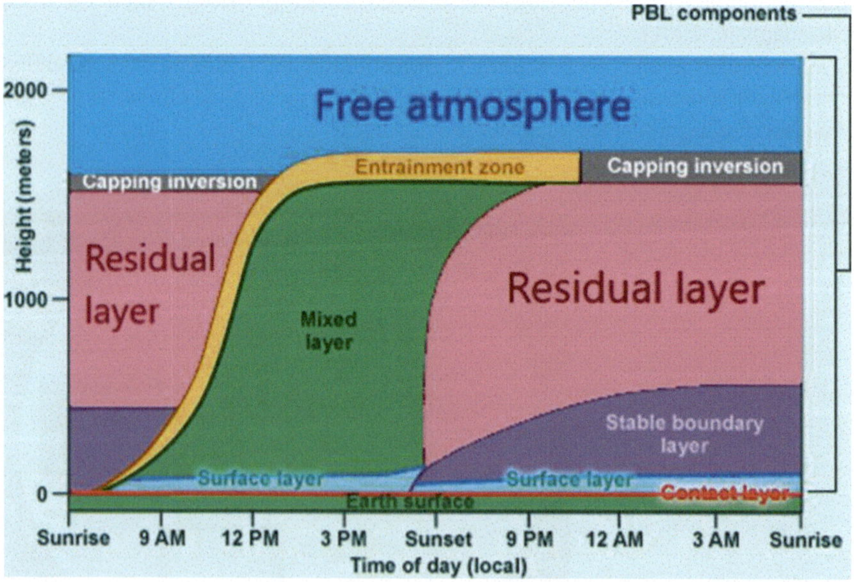

Fig. 14.4 Special profiling techniques, including remote sensing and in situ methods, provide essential data for atmospheric studies, with some systems capable of measuring the entire troposphere and others optimized for the lower troposphere, particularly the planetary boundary layer

(a) **Remote Sensing Techniques**:

- Doppler LIDARs: Utilize laser beams to measure wind speed, direction, and aerosol distribution in the boundary layer.
- Weather Radars: Provide high-resolution data on precipitation, cloud dynamics, and vertical wind velocities.
- SODARs (Sound Detection and Ranging): Analyze turbulence and wind patterns near the surface through sound wave scattering.
- Microwave Radiometers: Measure temperature and humidity profiles by detecting natural microwave radiation emitted by atmospheric gases.
- Satellite-Based Atmospheric Sounders: Extend profiling capabilities to the upper troposphere and beyond, offering global temperature, humidity, and trace gas data.

(b) **Direct Measurements (In Situ):**

- Radiosondes: Balloon-borne instruments that ascend through the atmosphere, providing vertical profiles of temperature, humidity, and wind parameters.

Some of these techniques, like radiosondes and satellite-based sounders, can profile the entire troposphere, while others, such as SODARs and Doppler LIDARs, are more focused on the lower troposphere, particularly within the planetary boundary layer (Fig. 14.3). By combining these techniques, researchers can achieve a comprehensive understanding of boundary layer and tropospheric dynamics, which is crucial for improved weather prediction, climate studies, and specialized applications like urban meteorology and renewable energy. One widely used technique is radiosonde balloon measurements, which provide vertical profiles of temperature, humidity, wind speed, and wind direction. Radiosondes ascend through the troposphere, transmitting data back to ground stations, and are essential for weather forecasting and atmospheric research. Remote sensing instruments like Doppler LIDARs and radars offer high-resolution measurements of wind and turbulence. Doppler LIDARs use laser light to measure wind profiles and aerosol distribution in the boundary layer, while weather radars provide detailed information on precipitation, cloud dynamics, and vertical velocity fields. Microwave radiometers are another critical tool, capable of detecting temperature and humidity profiles by measuring the natural emission of microwave radiation from atmospheric gases. Similarly, acoustic remote sensing systems, such as SODARs, monitor turbulence and wind structure in the lower atmosphere by analyzing the scattering of sound waves. For profiling the upper troposphere and beyond, satellite-based instruments, such as atmospheric sounders, capture global temperature, humidity, and trace gas distributions. These observations complement ground-based and airborne measurements, providing a comprehensive understanding of atmospheric dynamics. Combined with numerical modeling, these techniques allow researchers to explore boundary layer processes, monitor atmospheric stability, and improve weather predictions and climate studies. By employing a variety of profiling methods, scientists can gain a deeper understanding of the atmospheric structure and its interactions with the Earth's surface.

14.4.1 Remote Measurement-Sensing Systems

Modern remote sensing technology has its roots in the early use of cameras for topographic mapping in the 1840s. Initially, still photographs were taken from cameras fixed to balloons. The use of aerial photography expanded significantly during World War I when cameras were mounted on aircraft to provide revolutionary views of the land, impacting military strategies and operations. The advent of the space age marked a significant advancement in remote sensing technology, with satellites beginning to play a crucial role in capturing images of Earth's surface and observing other spacecraft. The term "remote sensing" was first coined in the 1950s by Evelyn Pruitt of the U.S. Office of Naval Research. Today, remote sensing is an invaluable tool for collecting meteorological measurements and monitoring the Earth's atmosphere from a distance. This technique utilizes a range of sensors and instruments deployed on satellites, aircraft, ground-based automatic weather stations, and other platforms to gather data on various atmospheric parameters. Remote sensing meteorological measurements enable scientists to collect critical weather-related data, which is essential for understanding and forecasting weather patterns, climate trends, and environmental changes (Gonzalez, 2018). Remote sensing involves gathering information about the Earth and its environment from afar, using sensors mounted on satellites, aircraft, drones, or other platforms (Fig. 14.5). This technology provides a comprehensive view of atmospheric conditions, contributing significantly to weather forecasting and environmental monitoring. For instance, satellite-based sensors can measure parameters such as sea surface temperature, atmospheric humidity, and cloud cover, allowing for the analysis of large-scale weather systems and climate phenomena (Chesnokov, 2019).

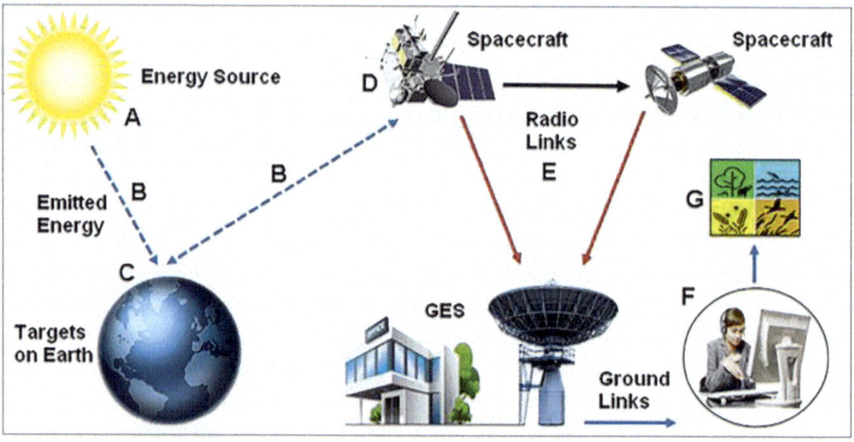

Fig. 14.5 Remote sensing process: this diagram illustrates the key stages of remote sensing, including data collection via sensors on satellites, aircraft, or ground-based platforms, transmission to processing centers, data analysis, and interpretation for applications in weather forecasting, climate monitoring, and environmental assessment

Additionally, remote sensing systems can integrate data from different sources, facilitating a more holistic understanding of the Earth's atmospheric processes. As technology advances, the accuracy and resolution of remote sensing data continue to improve, enhancing meteorological applications and climate research (WMO, 2020).

These sensors record data in different parts of the electromagnetic spectrum, allowing scientists and researchers to study the Earth's surface, atmosphere, and oceans. Remote sensing measurements play a critical role in studying the Earth's surface, atmosphere, and oceans. These sensors capture data across different parts of the electromagnetic spectrum, enabling scientists to gather valuable information for a variety of applications, including environmental monitoring, resource management, and disaster response. Optical remote sensing utilizes sensors that record visible and near-infrared light, allowing for land cover and use classification, monitoring of vegetation health, and supporting urban planning initiatives. Thermal infrared remote sensing, which measures the thermal radiation emitted by the Earth's surface, is commonly used for monitoring surface temperatures, detecting forest fires, and studying geothermal activity. Microwave remote sensing operates in the microwave portion of the electromagnetic spectrum and is particularly useful for applications such as measuring soil moisture, monitoring sea surface temperature, and detecting ice cover. LIDAR, or light detection and ranging, employs laser pulses to measure distances and create high-resolution 3D maps of the Earth's surface, commonly used for topographic mapping, forest canopy analysis, and urban modeling.

Radar systems, which use radio waves to detect and measure the characteristics of objects on the Earth's surface, are employed in weather forecasting, flood monitoring, and tracking moving objects like ships and aircraft. Satellite imagery, which provides high-resolution images from space, is widely used in disaster monitoring, urban planning, and agricultural management. Aerial photography, captured by aircraft equipped with cameras, allows for detailed images of specific areas and supports applications in cartography, land-use planning, and environmental assessment. Remote sensing of the atmosphere, conducted through instruments on satellites and ground stations, collects data on atmospheric parameters such as temperature, humidity, and composition, crucial for weather forecasting and climate research. In addition to atmospheric monitoring, ocean remote sensing uses specialized satellite sensors to track parameters like sea surface temperature, sea level, and ocean color. This data is valuable for climate studies, fisheries management, and coastal zone monitoring. Overall, remote sensing measurements provide critical insights into environmental changes, resource management, and natural disaster monitoring, supporting informed decision-making across a range of scientific, environmental, and societal applications.

14.4.2 Atmospheric Parameters

Remote sensing instruments are essential tools for measuring various atmospheric parameters, including temperature, humidity, pressure, wind speed, and wind direction at different altitudes. These measurements are crucial for developing accurate weather models and tracking atmospheric changes. One of the key applications of remote sensing is cloud and precipitation monitoring, which enables detailed observation of cloud cover, cloud types, and precipitation patterns. This information is vital for forecasting rainfall, snowfall, and severe weather events. In addition, remote sensing instruments measure radiative fluxes, such as solar radiation, outgoing longwave radiation, and energy fluxes at the Earth's surface. Understanding these radiative fluxes is fundamental to studying climate dynamics and the energy balance of the Earth. Satellite-based remote sensing also provides valuable data on sea surface temperatures (SSTs), which are crucial for monitoring ocean currents, tracking tropical storm development, and studying climate variability. Remote sensing technology also plays an important role in assessing vegetation health, land cover, and land-use changes. These measurements are valuable for agriculture, forestry, and environmental research, supporting sustainable land management practices. In the field of atmospheric chemistry, satellites detect pollutants and atmospheric gases, such as carbon dioxide, methane, and aerosols, providing data that helps to understand greenhouse gas concentrations and air quality. Remote sensing is also instrumental in monitoring and predicting extreme weather events such as hurricanes, tornadoes, and wildfires, supporting early warning systems and effective disaster management. Moreover, continuous data collection through remote sensing contributes to long-term climate monitoring, helping scientists study climate trends and variations over time. Overall, remote sensing meteorological measurements are indispensable for enhancing our understanding of Earth's atmosphere, weather dynamics, and climate systems, thus supporting improved weather forecasting, disaster preparedness, and climate change mitigation efforts.

14.5 Measurement of Solar Radiation

Solar radiation measurements are crucial for applications such as weather forecasting, solar energy system design, climate studies, and agriculture. Various methods and instruments are used to measure solar radiation, each with its strengths and limitations. Key methods and instruments include:

- Pyranometers measure global solar radiation, capturing both direct and diffuse components (Fig. 14.6a). They consist of a sensor that detects incoming solar radiation and converts it into an electrical signal. Pyranometers are sensitive to a broad range of solar wavelengths, making them suitable for various applications.

Fig. 14.6 (**a**) Pyranometer—an instrument for measuring global solar radiation. (**b**) Pyrheliometer—an instrument for measuring direct solar radiation

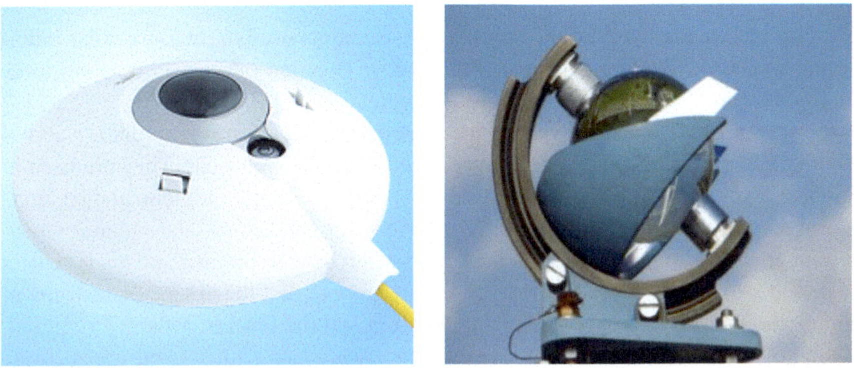

Fig. 14.7 (**a**) Pyrgeometer—an instrument for measuring longwave radiation emitted by the Earth's surface and atmosphere. (**b**) Heliograph—an instrument for measuring solar radiance by recording the duration and intensity of sunlight

- Pyrheliometers measure direct solar radiation (Fig. 14.6b). These instruments are oriented directly to the sun and use a tracking mechanism to continuously follow its movement. They are sensitive to wavelengths ranging from 280 to 3000 nm and consist of metal surfaces positioned perpendicular to solar radiation.
- Pyrgeometers: Pyrgeometers measure longwave radiation, including thermal radiation emitted by the Earth's surface and atmosphere (Fig. 14.7a). They are crucial for calculating net radiation and studying energy balance.
- Solar Duration Sensors: These instruments measure the duration of sunlight over a specific period, which is valuable for climate and meteorological studies. Common types include Campbell–Stokes sun recorders and electronic sun sensors. A heliograph, a type of solar duration sensor, records the duration and intensity of sunlight using a glass ball that concentrates sunlight onto a card, which then shows the number of sunshine hours in a day (Fig. 14.7b).

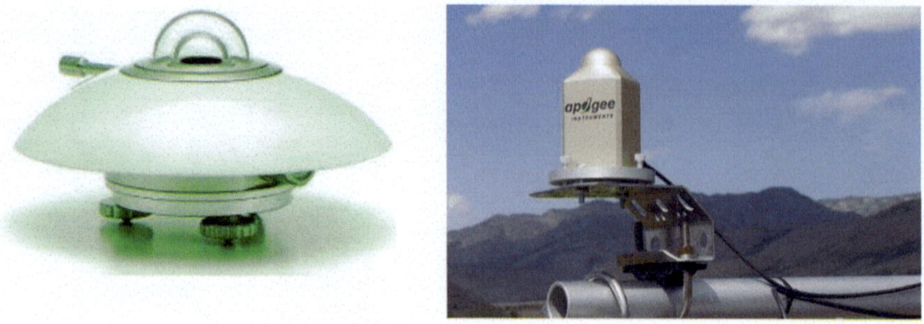

Fig. 14.8 (**a**) Solarimeter—a pyranometer used to measure combined direct and diffuse solar radiation. (**b**) Spectroradiometer—instrument for measuring solar radiation across different wavelengths

Solar meters are hand-held or portable instruments used to measure solar radiation at a specific location and time (Fig. 14.8a). They are often used for field measurements and solar power system evaluations.

Spectroradiometers are advanced instruments that measure solar radiation across a wide range of wavelengths, providing spectral data (Fig. 14.8a). They are used in research, especially to study the spectral composition of solar radiation and its interaction with materials. Data loggers are used to record and store solar radiation measurements over time. Specialized software is often used to process and analyze data collected by radiation sensors. It is important to note that the choice of instrument and method depends on the specific application and the level of accuracy required. Calibration and maintenance of these instruments are also critical to ensure accurate and reliable measurements of solar radiation.

14.6 Measurement of UV Radiation

Measuring solar UV radiation is crucial due to its detrimental effects on both the environment and human health, particularly with the depletion of the ozone layer increasing UV exposure at the Earth's surface. The UV spectrum is conventionally divided into three categories:

UV-A (315–400 nm): This wavelength range is just outside the visible spectrum and is less biologically active. Its intensity at the Earth's surface is not significantly affected by atmospheric ozone content.

UV-B (280–315 nm): This radiation is biologically active and its intensity is influenced by the atmospheric ozone column. It is known for its adverse effects on human skin, such as reddening and increased risk of skin damage.

UV-C (100–280 nm): This radiation is absorbed by the atmosphere and does not reach the Earth's surface naturally.

Fig. 14.9 Instrument for measuring UV radiation (radiometer)

The most advanced commercial instruments for measuring UV radiation are called "radiometers" (Fig. 14.9). These devices use holographic traps to scatter incoming energy and rely on dual monochromatic or blocking filters to isolate UV radiation from other wavelengths. Monochromatic output measurement typically employs a photoamplifier to ensure precise readings.

Database and Software: Data loggers capture and store solar radiation measurements over time, while specialized software processes and analyzes this data. The choice of instruments and methods depends on the specific application and required accuracy. Proper calibration and maintenance are essential to ensure accurate and reliable measurements of solar radiation.

14.7 Visibility Measurement

Meteorological visibility is a crucial parameter in weather measurements, typically estimated by an observer. It is influenced by various subjective and physical factors. The term "visibility" refers to the distance one can see through the atmosphere, which can be measured objectively as the meteorological optical range. Visibility can be assessed using telephotometric instruments known as transmissometers (Fig. 14.10). These instruments measure the attenuation of light over a specific distance, comparing the visibility of an object with that of the surrounding sky. While transmissometers are effective in measuring average attenuation coefficients, they are not commonly used for routine measurements. Instead, direct visual observations are often preferred. Transmissometers consist of two main components: a transmitter that emits modulated light from a constant power source and a receiver with a photodetector. Halogen lamps are frequently used as light sources, and modulation helps to minimize distortion from sunlight. The transmission factor is calculated based on the output from the photodetector, allowing for the determination of the attenuation ratio over the meteorological optical distance.

Fig. 14.10 Transmissometer for measurement of horizontal and vertical visibility. The figure shows a transmissometer, which measures visibility by detecting the transmission of light over a set distance, commonly used in aviation and meteorology

14.8 Upper-Air Measurements

Meteorological altimetry measurements, often referred to as upper-air measurements, involve the use of radiosondes to gather data about atmospheric conditions from ground level up to approximately 30 km. These measurements are essential for weather forecasting, climate research, and atmospheric studies. Radiosondes, equipped with sensors, measure key meteorological parameters such as pressure, temperature, and humidity at various altitudes (Fig. 14.11).

A radiosonde is typically attached to a balloon and includes a radio transmitter that sends data back to an observation station. These devices can also be launched from aircraft or rockets. Radiosondes provide detailed measurements over land but offer limited coverage over oceans. To complement this, aircraft measurements are used, especially in remote areas. Satellite measurements, although lower in vertical resolution, are useful for analyzing temperature and water vapor distribution over oceans and other regions with sparse upper-air soundings.

Accurate vertical profiling of temperature and water vapor is crucial for understanding atmospheric stability and improving weather forecasts. Radiosonde data are also vital for studying environmental pollution and climate change with altitude.

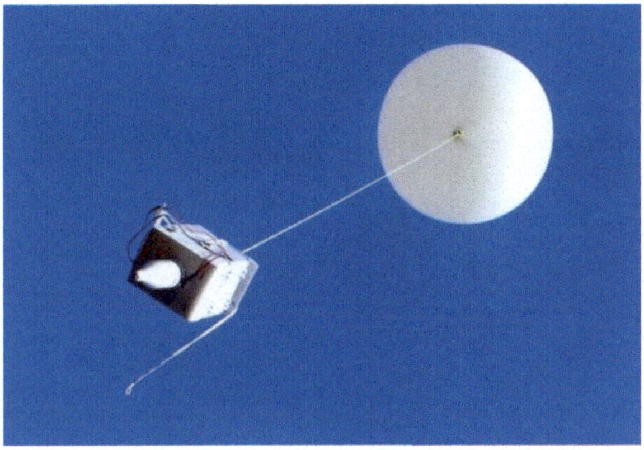

Fig. 14.11 Radiosonde: an instrument carried by a balloon to measure atmospheric pressure, temperature, and humidity at various altitudes. Data is transmitted to a ground station for weather forecasting and climate studies

14.9 Light Detection and Ranging (LIDAR)

Light detection and ranging (LIDAR) is a remote sensing technology that utilizes laser pulses to measure distances and gather detailed information about objects and their surroundings. LIDAR systems emit laser pulses to determine the range to Earth's surface, creating precise three-dimensional maps of terrain and features. The primary components of a LIDAR system include a laser emitter, a scanner, and a GPS receiver (Fig. 14.12). LIDAR data acquisition is commonly performed using aircraft and helicopters.

There are two main types of LIDAR: topographic and bathymetric. Topographic LIDAR employs near-infrared lasers for land mapping, while bathymetric LIDAR uses water-penetrating green light to measure seafloor and riverbed elevations. These systems provide accurate and flexible surveying capabilities for both natural and man-made environments (Mallet & Bretar, 2009). LIDAR technology is crucial for detecting wind patterns at various altitudes and is expected to play a significant role in future satellite-based wind measurements (Rocca et al., 2020). While LIDAR is often associated with mapping and autonomous vehicles, it also has valuable applications in meteorological measurements. For instance, LIDAR can be used to profile atmospheric conditions, including aerosol distribution and cloud heights, enhancing our understanding of weather dynamics and climate processes (Duncan et al., 2018).

14.9.1 Cloud and Aerosol Profiling

LIDAR systems are employed to analyze the vertical distribution of clouds and aerosols in the atmosphere (Fig. 14.13). By emitting laser pulses and measuring the return time of the scattered light, LIDAR provides detailed data on cloud height,

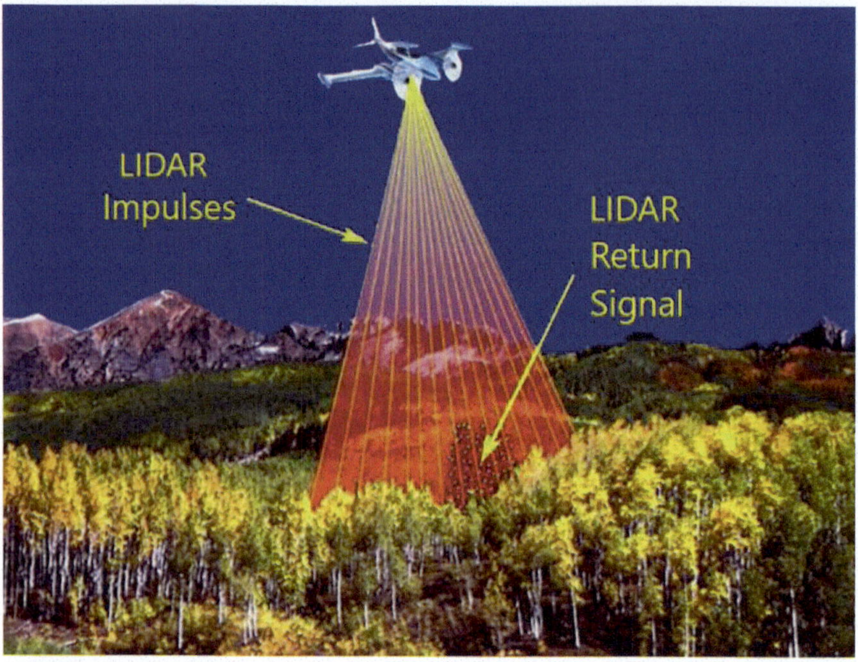

Fig. 14.12 LIDAR sensor: the LIDAR sensor operates by emitting laser pulses and measuring their travel time as they are reflected off objects, providing information about the direction and distance to the target

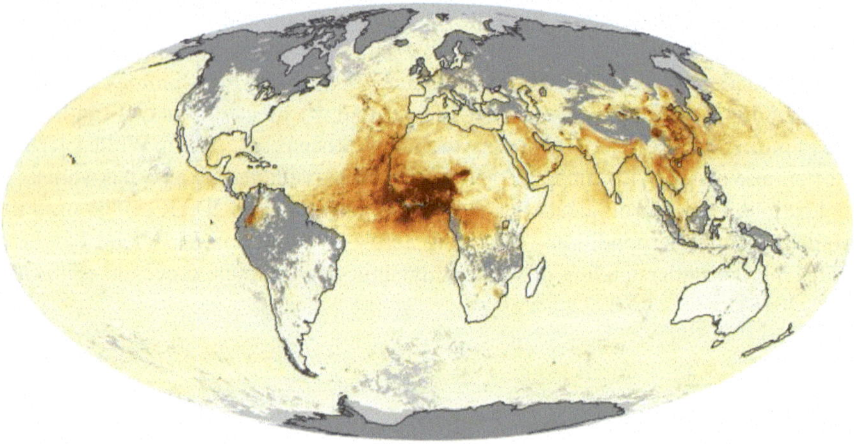

Fig. 14.13 Optical thickness of aerosols and fire detection. (Credit: NASA Earth Observatory)

thickness, and density, as well as aerosol concentration. This information is essential for accurate weather forecasting and climate research.

14.9.2 Wind Profiling

LIDAR technology offers a powerful tool for measuring wind profiles throughout the atmosphere (Fig. 14.14). By employing laser light, LIDAR systems can capture detailed information about wind characteristics at variowus altitudes. Specifically, Doppler LIDAR, which utilizes the Doppler shift of the scattered light to determine velocity, excels in measuring both wind speed and direction across different heights. This capability is crucial for enhancing our understanding of atmospheric dynamics, as it allows for the detailed observation of wind patterns and their variations. Accurate wind profiling is essential for predicting severe weather events, such as thunderstorms and tornadoes, by providing insights into wind shear and turbulence that can precede these phenomena. Moreover, wind profiling plays a significant role in optimizing wind energy production by assessing wind resource availability and variability at turbine hub heights, thereby informing the placement and operation of wind farms. The data obtained from LIDAR wind profiling also contributes to

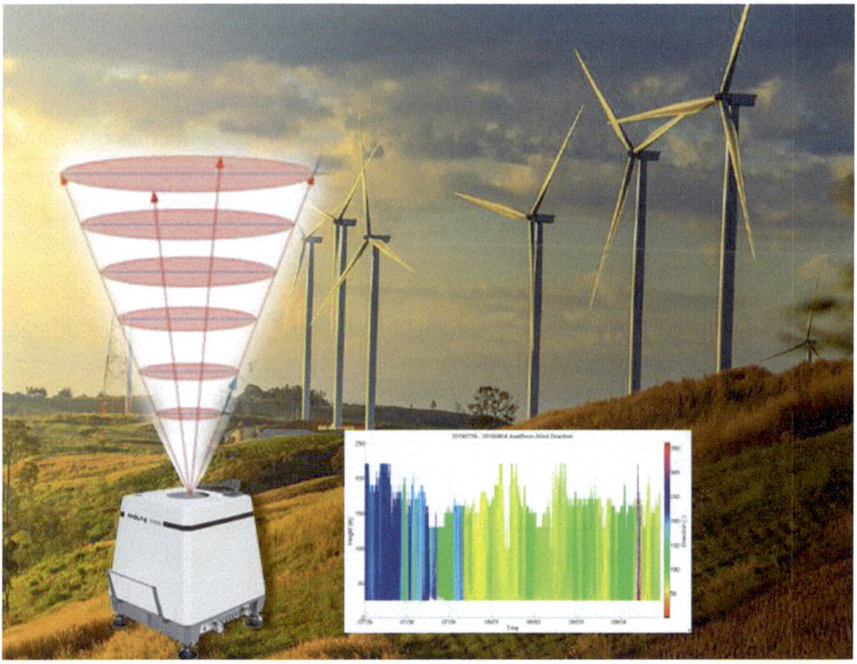

Fig. 14.14 Wind field measurements by scanning multilidar system. (Credit: https://uol.de/en/we-sys/research/lidar-wakes)

improving weather forecasting models and enhancing climate research, offering a comprehensive view of atmospheric behavior over time.

14.9.3 Precipitation Measurement

LIDAR can estimate rainfall intensity by analyzing the scattered light from raindrops. By measuring the intensity of reflected laser beams, it can provide valuable information on rainfall patterns and rainfall rates, which is useful for flood forecasting, hydrology, and water resources management. LIDAR measurements of precipitation usually involve the use of specialized LIDAR systems, such as Cloud LIDAR or Doppler LIDAR, which are capable of detecting and characterizing precipitation particles in the atmosphere. Here's how LIDAR can be used to measure precipitation:

(a) Detection of Clouds and Particles: LIDAR systems can detect the presence of clouds and precipitation particles in the atmosphere by emitting laser pulses and measuring the time it takes for the light to return to the sensor. The size, shape, and density of the particles can be inferred from the feedback signal, allowing differentiation between rain, snow, hail, and other forms of precipitation.
(b) Precipitation Intensity Estimation: LIDAR can provide information on the intensity of precipitation, known as rain rate. By analyzing the backscattered light from precipitation particles, researchers can estimate the rate at which rain falls at a particular location.
(c) Vertical Profiling: LIDAR can create vertical profiles of precipitation in the atmosphere. This means it can provide information about the height at which precipitation occurs, which is crucial for understanding the dynamics of weather systems.
(d) Precipitation Type Identification: LIDAR can help distinguish between different types of precipitation, such as rain, snow, sleet, or hail, based on the characteristics of the return signal. This information is valuable for weather forecasting and climate research.
(e) Climate Research and Climate Studies: An essential technology for climate monitoring is laser radar (light detection and ranging), or LIDAR, which scientists have used since the 1960s to study atmospheric particles and clouds. LIDAR's short pulses of laser light detect particles or gases in the atmosphere much like radar bounces radio waves off rain in clouds. The telescope collects and measures the reflected laser radiation, generating a map of the structure of the atmosphere. Researchers can then determine the location, distribution, and nature of atmospheric particles and clouds and, under special circumstances, molecular structure. Data obtained by LIDAR measurements of precipitation can be used for climate studies and atmospheric process research. Long-term LIDAR measurements can provide insight into precipitation patterns and trends over time, which is important for understanding the impacts of climate change.

While LIDAR can offer valuable data for precipitation measurements, its application for global precipitation monitoring is limited by several factors. Typically,

LIDAR measurements are confined to specific areas or research campaigns rather than being deployed on a global scale. This limitation arises because LIDAR systems often require direct line-of-sight to the atmospheric targets, making it challenging to cover extensive areas continuously. Consequently, for comprehensive global precipitation monitoring, satellite-based systems equipped with radar and radiometer instruments, such as those on the Global Precipitation Measurement (GPM) Mission, are more commonly utilized. These satellites provide broader and more continuous coverage, enabling the observation of precipitation patterns across the globe. Despite these limitations, LIDAR technology remains valuable for targeted applications and research scenarios. It offers detailed insights into the type, intensity, and vertical distribution of precipitation, which can be critical for understanding localized weather phenomena and improving weather forecasts. LIDAR is particularly useful for studying the atmospheric boundary layer, the lowest part of Earth's atmosphere. This layer, which extends from the surface up to a few kilometers, is crucial for weather and air quality predictions. LIDAR can provide detailed information on temperature inversions, turbulence, and mixing processes within the boundary layer, enhancing our understanding of local meteorological conditions and pollutant dispersion. LIDAR also plays a significant role in detecting and tracking volcanic ash clouds in the atmosphere (Fig. 14.15). By analyzing the light scattered by ash particles, LIDAR can track the movement and concentration of volcanic

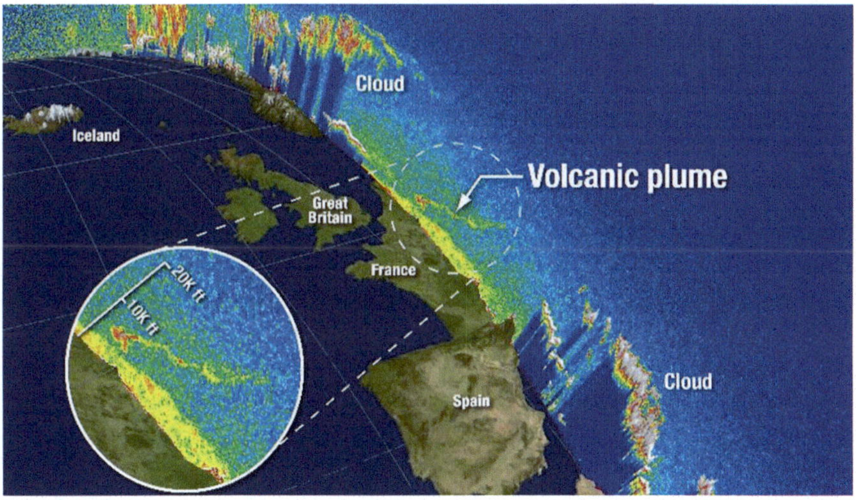

Fig. 14.15 Observation of the spread of the volcanic dust cloud over the European continent in 2010, originating from Iceland's Eyjafjallajökull volcano. This image was captured using the Cloud-Aerosol LIDAR and Infrared Pathfinder Satellite Observations (CALIPSO) satellite. The CALIPSO mission, a collaboration between NASA and CNES, utilizes LIDAR technology to provide detailed insights into the distribution and movement of aerosol particles and cloud cover in the atmosphere. The volcanic ash cloud caused significant disruption to air travel across Europe, highlighting the importance of satellite-based observation in monitoring and managing the impacts of volcanic eruptions. (Credit: NASA)

ash, which is essential for aviation safety. Volcanic ash clouds can severely impact aircraft engines and flight operations, so timely and accurate detection is critical for mitigating the associated hazards and ensuring safe air travel. Overall, while LIDAR is a powerful tool with specific applications, satellite-based systems are preferred for global-scale monitoring due to their broader coverage and continuous data collection capabilities.

Temperature and Humidity Profiling: Some LIDAR systems are equipped to measure temperature and humidity profiles in the atmosphere. These systems utilize advanced technology to detect variations in temperature and moisture content at different altitudes, providing essential data for meteorologists. Understanding temperature and humidity profiles is crucial for assessing atmospheric stability, which influences cloud formation, precipitation, and overall weather patterns. By offering insights into these parameters, LIDAR helps predict severe weather events, such as thunderstorms and hurricanes, and contributes to more accurate weather forecasts. Additionally, the data from LIDAR systems play a vital role in monitoring climate change and enhancing our comprehension of atmospheric processes. In summary, LIDAR technology is invaluable for obtaining detailed and precise information on various atmospheric conditions, including cloud characteristics, wind patterns, and precipitation, thereby supporting both short-term weather predictions and long-term climate studies.

14.10 Sound Detection and Ranging (SODAR)

Sound detection and ranging (SODAR) is a remote sensing tool used to measure atmospheric disturbances by analyzing the dispersion of sound waves (Fig. 14.16). It assesses wind speed and direction at various altitudes above the Earth's surface by emitting sound waves and analyzing the reflected signals. SODAR systems provide crucial data on the vertical profile of wind conditions and atmospheric turbulence.

One of the key benefits of SODAR is its ability to measure at higher altitudes compared to traditional weather towers, which are typically limited to around 150 m. In contrast, SODAR systems can extend beyond this limit, enabling accurate measurements at greater heights (Sasaki et al., 2004). Additionally, SODAR is relatively simple to deploy and install, making it an efficient option for both temporary and long-term monitoring applications. This ease of use facilitates rapid data collection and analysis. While SODAR is highly effective for profiling wind conditions, it is often used in combination with other techniques, such as radio sounding, which employs motion tracking or free balloons to gather data at even higher altitudes (Müller et al., 2016). Together, these technologies provide a comprehensive view of wind patterns and atmospheric conditions, significantly enhancing weather forecasting and atmospheric research. In summary, SODAR is a versatile and efficient tool for measuring wind speed and direction at elevated heights. Its advantages, including higher altitude coverage and ease of installation, make it a valuable component in understanding atmospheric dynamics when combined with other remote sensing methods.

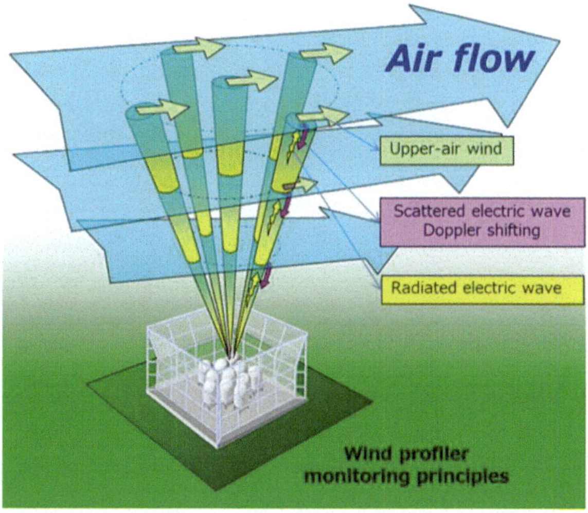

Fig. 14.16 SODAR, which stands for sound detection and ranging, is a meteorological instrument utilized as a wind profiler. It operates by measuring the scattering of sound waves caused by atmospheric turbulence. This allows SODAR to determine wind speed and direction at various heights just above the Earth's surface. The ability to provide vertical wind profiles makes SODAR a valuable tool for studying atmospheric conditions and enhancing weather forecasts. (Credit: https://www.jma.go.jp/)

14.11 Measurements at Automatic Weather Stations (AWS)

An Automatic Weather Station (AWS) is defined as a meteorological station where measurements of various meteorological elements are conducted automatically (Fig. 14.17). These stations are equipped with sensors that continuously collect data on atmospheric parameters, such as temperature, humidity, wind speed, and precipitation. The measurements are transmitted in real time to a central data acquisition unit for further processing and analysis.

AWS systems operate through external sensors that capture instrumental measurements and relay this information to the central processor. This data can be processed either locally at the current AWS or remotely at a central processing unit within a network. The integration of various devices and measuring units within an AWS enables a flexible approach to data collection and processing. Automatic meteorological stations are designed to fulfill a range of needs. They can provide simple assistance to human observers by offering real-time data and alerts or serve as fully automated systems that replace traditional manual observations. The ability to continuously monitor and record meteorological conditions makes AWS invaluable for weather forecasting, climate monitoring, and environmental research. In summary, AWS systems offer a comprehensive solution for automated weather observations, enhancing the accuracy and efficiency of meteorological data collection and analysis.

Fig. 14.17 Automatic Weather Station (AWS). This station features integrated sensors for continuous measurement of meteorological elements, with data transmitted automatically to a central data acquisition unit. The system is designed to provide accurate and real-time weather observations with minimal manual intervention. (Source: IRI Columbia)

14.12 Observing the Earth and the Environment from Space

Satellite monitoring of Earth and its environment is essential for protecting our planet in numerous ways. Space-based observations provide crucial data on environmental parameters such as temperature, humidity, sea level, and air quality. This information aids in monitoring climate change, assessing ecosystem health, and detecting environmental trends and anomalies. Satellites play a vital role in climate change analysis by providing data that helps scientists understand the impact of human activities on the environment and develop strategies to mitigate climate change. They also enhance our ability to predict and manage natural disasters, such as hurricanes, wildfires, and earthquakes, by detecting early signs and providing timely information for early warning systems and disaster management efforts. Weather forecasting benefits significantly from satellite data, which provides accurate and timely information essential for agriculture, transportation, and disaster preparedness. Additionally, satellite observations support natural resource management by monitoring forests, water bodies, and agricultural lands, promoting sustainable practices and preventing resource depletion. Ocean and coastal monitoring are improved through satellite data that track sea surface temperatures, ocean currents, and coastal erosion. This information is crucial for marine ecosystem conservation and coastal management. Satellites also monitor air quality and pollution, offering valuable insights into pollutant sources and distribution, which are key for reducing air pollution and protecting public health. The conservation of biodiversity is supported by satellite observations that track habitat changes, deforestation, and wildlife movements, contributing to effective conservation and ecosystem management. Space-based observations also aid land-use planning and urban development by providing data on urban growth, infrastructure, and land-use changes, which help in creating more sustainable and resilient cities. Satellite technology facilitates global connectivity, including Internet access, which is vital for information sharing, disaster response coordination, and environmental issue dissemination. Furthermore, space-based observations support scientific research across various fields, enhancing our understanding of Earth's systems and processes. International cooperation is crucial for sharing and analyzing satellite data and fostering collaborative efforts to address global environmental challenges. Overall, observing Earth from space provides essential data and insights that help us understand, mitigate, and adapt to environmental changes and challenges, highlighting the importance of satellite technology in safeguarding our planet.

14.13 Satellite Measurements and Observations

Weather, water, and climate monitoring are crucial for protecting the planet and ensuring a sustainable future for generations to come. Satellite measurements play a key role in these efforts by utilizing remote sensing techniques to monitor meteorological variables and support observations of the Earth's surface, troposphere, and stratosphere (Kallberg and Jansson, 2019). Satellite measurements refer to the use of satellite systems to gather data on various atmospheric and surface parameters. These systems are continually evolving, with new technologies expected to become operational shortly. Typical meteorological satellites orbit Earth at altitudes ranging from approximately 850 km to about 36,000 km, providing images and quantitative information about surface properties and the lowest 20 km of the atmosphere (Fig. 14.18).

Satellites equipped with sensors offer several advantages over ground-based observation systems. They provide comprehensive coverage and the ability to capture satellite weather images, which are invaluable for diagnosing and analyzing meteorological features. Cloud images, for instance, offer critical insights into weather patterns and atmospheric conditions.

Currently, meteorological satellites measure a range of variables with varying degrees of resolution and accuracy, including:

- Temperature profiles and cloud top temperatures, sea surface temperatures, and land temperatures
- Moisture profiles, providing information on atmospheric humidity and water vapor
- Cloud and ocean surface wind patterns, crucial for understanding weather systems and ocean currents
- Liquid and total water intensity of precipitation, which is essential for hydrological studies and weather forecasting
- Total radiation and albedo, which help in assessing the Earth's energy balance and surface reflectivity
- Cloud type and the height of cloud tops, aiding in weather analysis and forecasting.
- Total ozone levels, important for monitoring the ozone layer and ultraviolet radiation

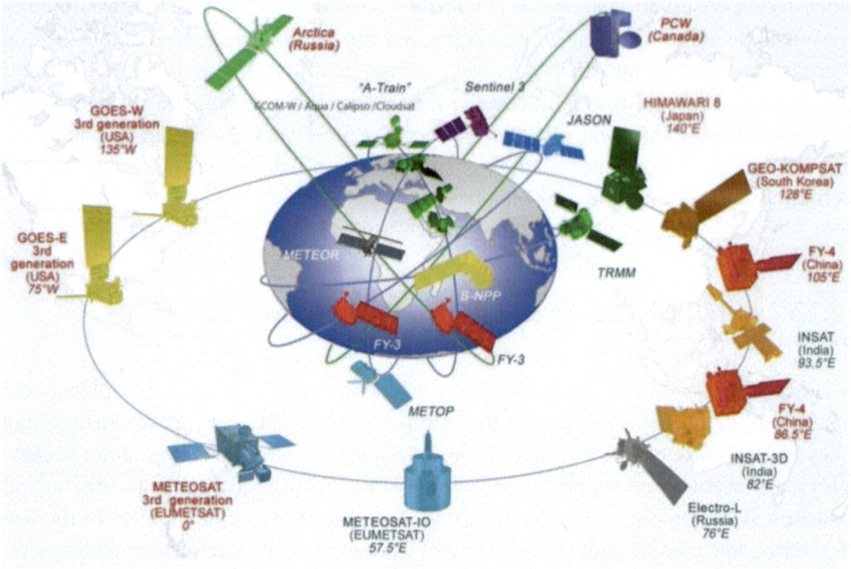

Fig. 14.18 Global satellite coverage. This figure shows the extent of observation provided by satellites in different orbits: low Earth orbit (LEO) satellites at approximately 850 km offer detailed views of the Earth's surface and lower atmosphere, while geostationary satellites at around 36,000 km provide continuous, broad coverage of the equatorial region. This coverage is crucial for accurate weather forecasting and climate monitoring

- Ice and snow coverage and height, relevant for climate studies and understanding seasonal changes

In addition to meteorological variables, satellites also measure non-meteorological variables that are operationally significant. These include vegetation coverage, volcanic dust, and other environmental factors. The ability to monitor these variables enhances our understanding of various natural phenomena and supports effective management and response strategies. Overall, satellite measurements provide essential data for weather forecasting, climate monitoring, and environmental management. They offer a comprehensive view of atmospheric and surface conditions, supporting both immediate weather analysis and long-term climate studies (WMO, 2019).

14.13.1 Current Satellites

The Meteosat Second Generation (MSG) satellites play a critical role in providing real-time images and weather forecast data for Earth. This advanced series of weather satellites builds on the successful design of the original Weather-Sat satellites, incorporating new technologies to enhance their capabilities. Among the MSG satellites, the SEVIRI (Spinning Enhanced Visible and Infrared Imager) radiometer, specifically MSG-2, is notable for its comprehensive observational capabilities. Equipped with 12 channels, SEVIRI scans the Earth every 15 min, delivering detailed imagery and meteorological data across various wavelengths. This frequent scanning allows for timely updates and accurate monitoring of weather conditions. The visible channel of SEVIRI is particularly distinguished by its high resolution, providing data with a resolution of 1 km. This fine detail is crucial for analyzing small-scale weather phenomena and atmospheric conditions. In contrast, the other channels offer data with a resolution of 3 km, which is still highly valuable for broader atmospheric observations. Overall, the Meteosat Second Generation satellites, with their advanced imaging technology and frequent data acquisition, are essential tools for obtaining precise and real-time information about the state of the atmosphere. They support a wide range of applications, including weather forecasting, climate monitoring, and environmental assessment, enhancing our ability to understand and respond to atmospheric changes.

In addition to obtaining data on cloud systems across different regions of the globe, numerous applications have been developed to measure and monitor various atmospheric phenomena and processes using satellites. These advancements have significantly improved our ability to detect and understand the development of cyclones and other small-scale atmospheric disturbances. Satellites are now instrumental in predicting and monitoring the formation and evolution of tropical cyclones, hurricanes, typhoons, and tornadoes. Furthermore, satellites are employed to predict and monitor wildfires, air pollution, low-layer clouds or fog, solar eclipses, and related atmospheric phenomena.

The European Organisation for the Exploitation of Meteorological Satellites (EUMETSAT) plays a key role in these observations. EUMETSAT operates several satellites that provide essential data for weather forecasting, climate monitoring, and environmental observation. The organization manages the geostationary satellites Meteosat 10 and 11 over Europe and Africa and Meteosat 9 over the Indian Ocean. It also oversees two Metop satellites in polar orbit, which are part of the Joint Polar System (JPS) shared with the US National Oceanic and Atmospheric Administration (NOAA). Additionally, EUMETSAT participates in the Jason cooperative sea level monitoring missions (Jason-3 and Jason-CS/Sentinel-6), which involve collaboration between Europe and the United States (Fig. 14.19). Moreover, EUMETSAT is responsible for the four Sentinel missions of the Copernicus space component within the EU, which are dedicated to monitoring the atmosphere, oceans, and climate. In collaboration with the European Space Agency (ESA), EUMETSAT currently operates the Sentinel-3 and Sentinel-6 maritime missions. The Sentinel-3 mission satellite provides valuable data on sea surface temperature, ocean color, sea-level rise, and land surface temperature. EUMETSAT also plans to launch the Sentinel-4 mission, focused on monitoring trace gas concentrations in the Earth's atmosphere, and the Sentinel-5 mission, which will concentrate on atmospheric composition and air quality.

The capability to monitor Earth from space has been significantly enhanced with the launch of the next generation of EUMETSAT weather satellites and a new fleet

Fig. 14.19 Meteorological satellites. This figure depicts various key satellites in meteorological and environmental observation: the Meteosat Second Generation MSG-2, Sentinel-3, Metop, and Jason-3. These satellites provide comprehensive data for weather forecasting, climate monitoring, and oceanographic studies, each offering unique capabilities for observing different aspects of Earth's atmosphere and surface

Fig. 14.20 MTG-1 METEOSAT Third Generation satellite. This figure shows the MTG-1, the first satellite of the Meteosat Third Generation series, launched on December 13, 2022. The MTG-1 represents a significant advancement in meteorological satellite technology, designed to provide enhanced observations and data for improved weather forecasting, climate monitoring, and environmental assessment

of Copernicus Sentinel satellites. The first satellite of the Meteosat Third Generation (MTG) series, the MTG-I imaging satellite, was successfully launched on December 13, 2022 (Fig. 14.20). This new generation of satellites represents a major advancement in our ability to observe and understand atmospheric and environmental changes.

The MTG system is the most complex and innovative meteorological geostationary system ever built. Its fleet will consist of two imaging satellites and one sounding satellite in orbit 36,000 km above Earth. MTG's range of innovative instruments will provide meteorologists with a continuous stream of land, ocean, and atmospheric data for Europe, Africa, and beyond. The Meteorological Satellite Third Generation (MTG) is a six-satellite system that includes four imaging satellites equipped with lightning imaging and detection capabilities, as well as two sounding satellites offering advanced infrared and ultraviolet observations. This system is designed to provide critical information on atmospheric chemistry and air quality, supporting both climate and meteorological applications.

14.14 Radar Observations and Measurements

Radar, an acronym for radio detection and ranging, operates on the principle of emitting radio waves in sequences and detecting the waves reflected by objects. By measuring the time between emission and detection, radar can calculate the distance to an object. The intensity of the radar signal is influenced by the concentration, size, and type of the particles, such as snow, hail, or rain. Radar data is typically displayed as reflectivity in decibels (Lee & Crum, 2017). This technology is crucial for weather monitoring, enabling meteorologists to detect precipitation intensity and movement, which enhances forecasting accuracy (World Meteorological Organization [WMO], 2019).

Doppler radar, which utilizes the Doppler effect, determines the speed and direction of precipitation by measuring changes in the frequency of the returned signal. The most advanced meteorological radar technology today is dual-polarization Doppler radar (Fig. 14.21a). This technology enhances precipitation estimates by improving the ability to distinguish between different types of precipitation, such as heavy rain, hail, snow, and sleet. It also aids in detecting non-meteorological echoes, including ground clutter, chaff, anomalous propagation, birds, and tornado debris. It helps identify the melting layer, commonly referred to as the bright band. NEXRAD (next-generation weather radar) illustrated in Fig. 14.21b is a network of advanced Doppler radar systems across the United States, operated by the National Weather Service. It detects precipitation, measures storm intensity, and tracks severe weather events like tornadoes and hurricanes, providing critical data for forecasts and warnings.

14.15 Special Ocean Measurements

The modern Chikyu platform, developed as part of advanced programs for specialized measurements in both the atmosphere and deep ocean floors, represents a significant leap in oceanographic research technology. This state-of-the-art vessel is designed to conduct meteorological and geophysical research in extreme oceanic environments, including drilling up to approximately 7 km beneath the seabed at water depths exceeding 4 km (Fig. 14.22). The Chikyu's sophisticated capabilities enable it to probe the Earth's crust and study oceanic phenomena with

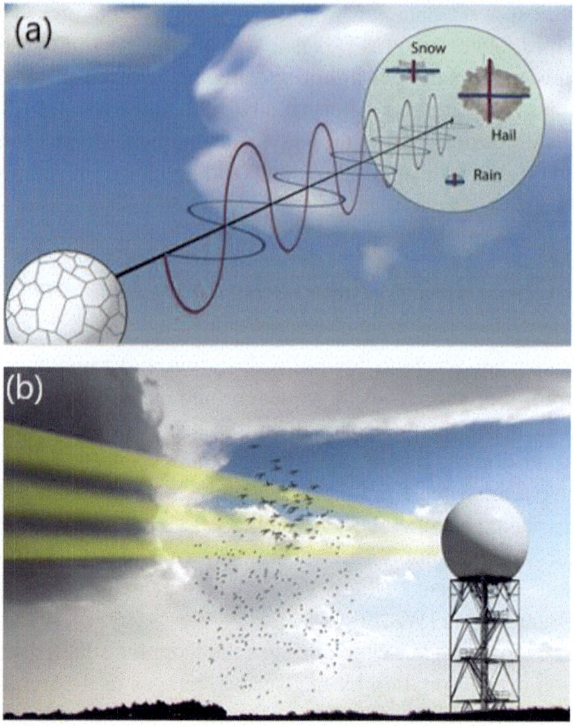

Fig. 14.21 Dual-polarization Doppler radar. (**a**) Schematic view of dual-polarization Doppler radar, which uses both horizontal and vertical polarization to provide detailed information about precipitation types and enhance accuracy in weather observations. (**b**) NEXRADs, next-generation radars, represent advanced Doppler radar systems with dual-polarization capabilities for improved precipitation measurement and weather analysis. (Credit: NOAA National Severe Storms Laboratory)

unprecedented detail. Its drilling operations allow scientists to gather core samples from deep below the ocean floor, providing critical insights into the geological processes that shape our planet. This data is instrumental in understanding seismic activity, volcanic processes, and the formation of hydrothermal vents. Moreover, the Chikyu platform plays a crucial role in investigating major oceanic phenomena such as El Niño and La Niña events. These quasi-periodic warming and cooling patterns in the tropical Pacific Ocean have profound effects on global weather and climate. By studying the ocean's subsurface structure and the interactions between tectonic plates and oceanic systems, the Chikyu helps researchers uncover the mechanisms driving these climatic events. The platform's advanced instrumentation and capabilities support research into how such phenomena influence weather patterns, ocean circulation, and even terrestrial climate systems.

Chikyu's contributions extend beyond immediate scientific discovery. Its findings enhance our understanding of oceanic processes that affect global climate systems, contribute to the assessment of natural hazards, and aid in the sustainable management of ocean resources. As a mobile and versatile research platform, the Chikyu represents a critical tool for advancing our knowledge of the deep ocean and its complex interactions with Earth's atmospheric and geological systems.

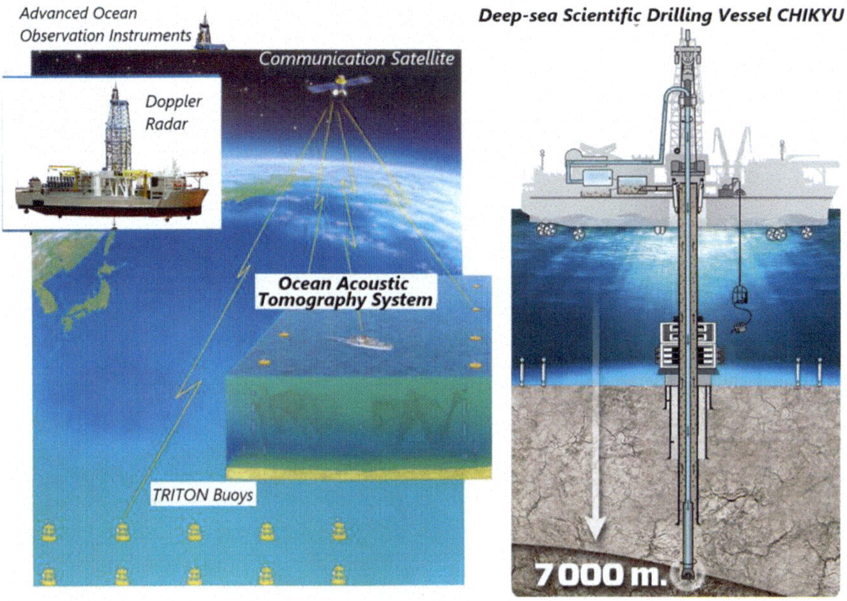

Fig. 14.22 Advanced instruments for ocean floor observation and measurement using the CHIKYU platform. The image showcases the sophisticated drilling equipment and scientific instruments employed by CHIKYU to conduct deep-sea research and geophysical measurements. These tools are designed for extensive underwater exploration, allowing for the collection of core samples and the study of oceanic and geological processes

14.16 Aircraft Measurements of the Atmosphere

Aircraft play a pivotal role in atmospheric research by providing a unique platform for collecting in situ measurements, remote sensing, and atmospheric sampling. These airborne systems are equipped with a range of advanced instruments that gather critical data on various atmospheric parameters.

In Situ Measurements: Aircraft equipped with sensors and sampling equipment can measure atmospheric variables such as temperature, humidity, pressure, and wind speed directly within the atmosphere. This real-time data is essential for understanding vertical profiles of the atmosphere, which are crucial for weather forecasting and climate studies.

Remote Sensing: Modern aircraft are also outfitted with remote sensing technologies, such as LIDAR and radar systems, that can scan and map atmospheric conditions over large areas. These instruments provide detailed information on cloud formation, aerosol distribution, and other atmospheric phenomena.

Atmospheric Sampling: Aircraft can collect air samples from different altitudes to analyze the composition of atmospheric gases, including trace gases and pollut-

Fig. 14.23 Aircraft laser measurements of the Greenland ice cap

ants. This sampling is vital for studying air quality, greenhouse gas concentrations, and chemical reactions occurring in the atmosphere.

The integration of these capabilities allows aircraft to contribute valuable data for various applications, including weather prediction, climate modeling, and environmental monitoring. By flying through different layers of the atmosphere and covering diverse geographic regions, aircraft enhance our understanding of atmospheric processes and their impacts on weather patterns and climate systems. The special aircraft measurements utilize the airborne laser technology to conduct high-resolution measurements of ice thickness and surface topography on Greenland's ice cap (Fig. 14.23). The laser system provides precise data crucial for understanding ice dynamics, assessing changes in ice volume, and studying the impacts of climate change on polar ice regions.

14.17 NASA Earth Monitoring Observatory

NASA's Earth Observatory is a crucial online resource provided by the National Aeronautics and Space Administration (NASA) dedicated to disseminating information, images, and data about Earth's environment. Many NASA satellites study the Earth and its climate. The NASA Aura satellite monitors Earth's atmosphere, focusing on air quality and pollution (Fig. 14.24). Equipped with advanced instruments, Aura measures key pollutants like ozone, carbon monoxide, and nitrogen dioxide, providing vital data to track global air pollution, study climate interactions, and improve environmental policies. Its primary mission is to share the scientific research and insights generated by NASA regarding the dynamic systems of our planet and the impact of human activities on the environment. By offering a comprehensive view of Earth's changing climate, weather patterns, ecosystems, and other Earth System phenomena, the Earth Observatory aims to educate the public, support educators, and inform decision-makers.

The Earth Observatory website hosts a vast array of resources, including articles, images, animations, and interactive tools that cover topics such as climate change,

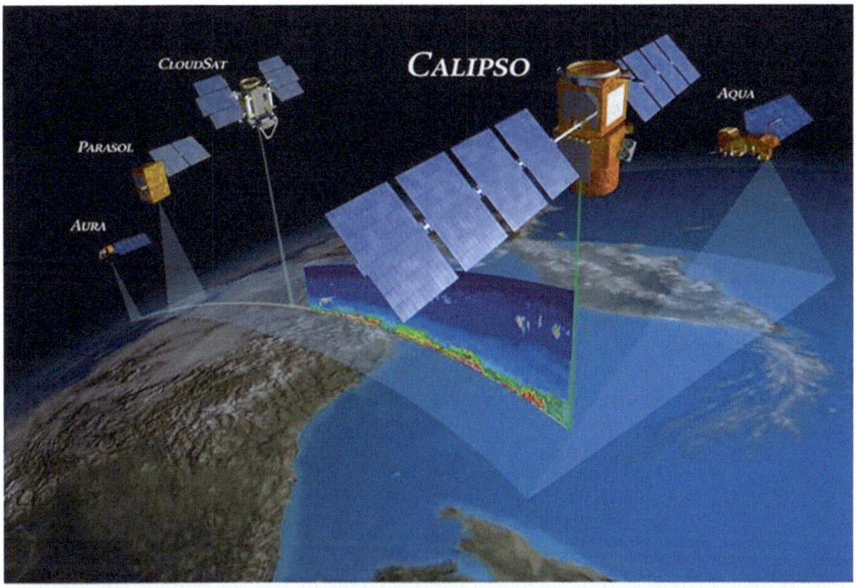

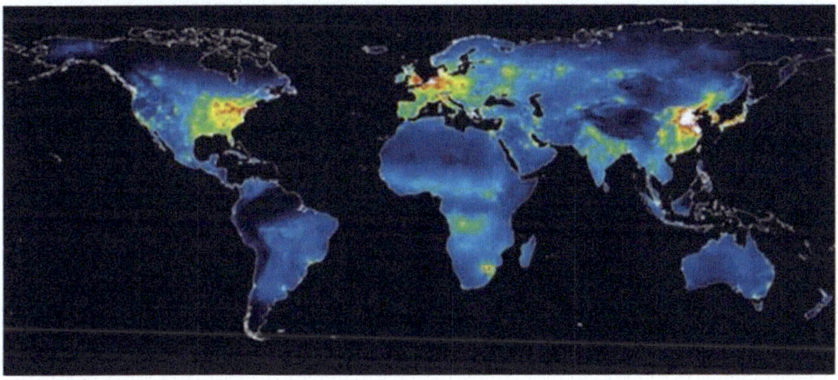

Fig. 14.24 The Aura satellite is designed to answer questions about monitoring the planet's air pollution from space and changes in our life-sustaining atmosphere. (Credit: NASA)

weather events, natural disasters, air pollution, land use, and atmospheric composition. One of the key instruments featured is the Ozone Monitoring Instrument (OMI), which measures nitrogen dioxide (NO_2) from space. NO_2, a significant air pollutant emitted by burning fossil fuels and biomass, is tracked through annual average maps to visualize its distribution and changes over time. The site employs a combination of engaging visuals and data visualization tools to help users grasp complex Earth science concepts. It provides detailed visualizations of satellite data and other NASA missions, offering insights into major events like hurricanes, wildfires, volcanic eruptions, and their impacts on the planet. By presenting these events in the context of broader Earth system processes, the observatory highlights the interconnectedness of global phenomena. Additionally, the Earth Observatory regularly updates its content with articles and research findings from NASA's scientific missions, including those involving satellites, aircraft, and ground-based observations. These updates contribute to a deeper understanding of Earth's environment and support informed policy-making. Overall, NASA's Earth Observatory is a valuable resource for exploring Earth's environmental science, offering educational materials and real-time insights into our planet's health and dynamics.

14.17.1 Forest Fire Detection

Something is always burning on Earth. Wildfires can be triggered by lightning, human activity, or intentionally controlled burns for managing farmland, pastures, and clearing vegetation. While fires can generate significant smoke pollution, release greenhouse gases, and degrade ecosystems, they also play a crucial role in many ecosystems. In certain environments, such as boreal forests and grasslands, periodic fires are essential for plant reproduction and ecosystem health. NASA and other space agencies worldwide leverage satellite technology to monitor and observe wildfires. NASA's Fire Information for Resource Management System (FIRMS) provides real-time information on active fires through satellite data. FIRMS uses data from the Moderate Resolution Imaging Spectroradiometer (MODIS) aboard NASA's Terra and Aqua satellites to detect and monitor fires globally (Fig. 14.25).

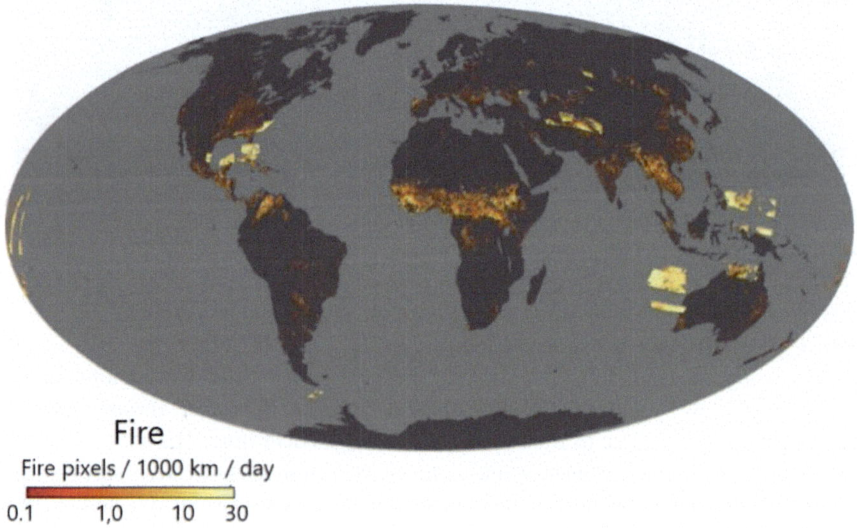

Fig. 14.25 A global view of active fires every month, obtained with observations from the Moderate Resolution Imaging Spectroradiometer (MODIS) on NASA's Terra satellite. (Source: NASA Earth Observatory)

The MODIS sensors are capable of detecting thermal anomalies associated with fire and providing detailed information about the location, size, and intensity of wildfires. These satellite-based tools are invaluable for managing fire response efforts, studying fire behavior, and assessing the impact of fires on air quality and ecosystems. FIRMS and MODIS data support emergency services, aid in environmental research, and help policymakers make informed decisions about fire management and land use. By providing comprehensive and timely information, NASA's fire detection systems contribute to our understanding of fire dynamics and their effects on the planet.

14.17.2 Measurements of Aerosols

NASA's Earth Observatory provides invaluable information and imagery on various aspects of Earth's environment, including aerosols. Aerosols are tiny solid or liquid particles suspended in the atmosphere that significantly influence climate, weather, and air quality. These particles can range in size from nanometers (ultrafine particles) to micrometers (larger particles), affecting their atmospheric behavior, their ability to scatter or absorb sunlight, and their potential health impacts when inhaled. The Moderate Resolution Imaging Spectroradiometer (MODIS) on NASA's Terra satellite plays a crucial role in measuring aerosol sizes and distributions. Figure 14.26 illustrates the global distribution of different aerosol sizes as detected by MODIS. The map highlights regions with varying aerosol characteristics: green areas indicate dominance of larger particles, red areas represent plumes of smaller particles, and yellow areas show regions with a mix of particle sizes. Gray areas denote regions where data was not collected.

A prominent pattern observed is that in the Southern Hemisphere, especially over oceans, larger aerosol particles predominate. This is largely due to natural sea salts, which are a significant source of large aerosols. In contrast, the Northern Hemisphere, with its greater landmass, exhibits a higher abundance of smaller aerosols. These smaller particles are often associated with human activities, such as fires and industrial pollution, as well as natural sources like wildfires and dust storms. Desert regions, such as the Sahara Desert in northern Africa and the Arabian Peninsula, are notable sources of large aerosols due to frequent dust storms. Conversely, areas with significant fire activity, such as the Amazon during the dry season or northern Canadian forests during summer, are characterized by smaller aerosols. Additionally, industrial pollution contributes to smaller aerosols in overdeveloped regions like the Eastern United States and Europe, particularly in summer months when pollution levels peak. NASA's Earth Observatory provides detailed data and visualizations that help researchers understand these patterns and their implications for climate and air quality.

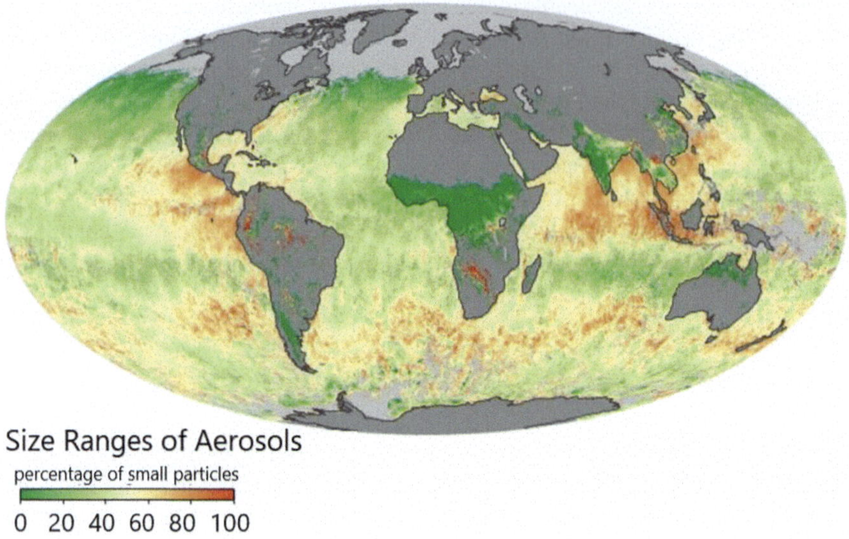

Fig. 14.26 Global representation of the percentage distribution of different aerosol sizes monthly, obtained with observations from the Moderate Resolution Imaging Spectroradiometer (MODIS) on NASA's Terra satellite. (Source: NASA Earth Observatory)

14.18 Methods for Detection of Lightning and Thunder

The detection of lightning and thunder relies on a combination of advanced technologies and observational techniques to monitor the electrical and acoustic phenomena associated with thunderstorms. Ground-based lightning detection networks use sensors to identify electromagnetic pulses produced by lightning strikes (Fig. 14.27). These systems detect very low-frequency (VLF) and low-frequency (LF) radio waves, enabling precise identification of both cloud-to-ground and intracloud lightning events. Such networks provide near real-time data and are crucial for weather forecasting, safety, and research. *Satellite-based detection systems* complement ground-based networks by monitoring lightning activity from space. Equipped with optical and infrared sensors, satellites like the Geostationary Lightning Mapper (GLM) track the intensity and frequency of lightning, offering a broader perspective of storm dynamics and activity over large, remote, or oceanic regions. *Acoustic detection systems* focus on thunder, the sound produced by the rapid expansion of air around a lightning channel. Microphones or infrasound detectors record the low-frequency sound waves generated by thunder, providing insights into the characteristics of the associated lightning strikes. *Optical and infrared sensors* detect visible light and heat from lightning strikes, often used in satellites or ground stations for monitoring.

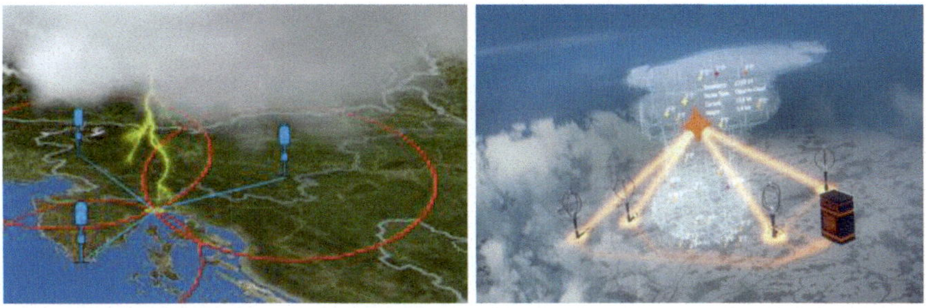

Fig. 14.27 Detection methods for lightning and thunder, including ground-based networks for electromagnetic pulses, satellite systems for optical and infrared monitoring, and acoustic systems for thunder analysis, providing comprehensive insights into storm activity and dynamics

Together, these methods create a comprehensive approach to monitoring and understanding lightning and thunder, thereby enhancing meteorological research and improving public safety measures during severe weather events.

14.19 Urban Meteorological Measurements

Urban meteorological measurements are essential for understanding the unique climatic and atmospheric conditions of cities, which differ significantly from rural or natural environments due to factors such as high population density, extensive built-up areas, and human activities. These measurements focus on capturing a wide range of atmospheric parameters, including temperature, humidity, wind speed and direction, solar radiation, precipitation, and air quality (Fig. 14.28). Cities are characterized by unique microclimatic phenomena, such as the urban heat island (UHI) effect, where urban areas exhibit higher temperatures than surrounding rural regions due to heat retention by buildings and paved surfaces. Additionally, urban environments can alter wind patterns, increase turbulence, and trap pollutants, leading to localized air quality issues. Accurate meteorological measurements are critical for studying these phenomena and their impacts on human health, energy consumption, and ecosystem services.

To capture the complexity of urban climates, measurements are taken at various spatial and temporal scales. Traditional weather stations are complemented by dense sensor networks installed at multiple levels, such as street canyons, building rooftops, and urban parks, to account for spatial heterogeneity (Netzband and Jürgens, 2010). High-resolution remote sensing technologies, including satellite imagery, Doppler radar, and LIDAR, further enhance data collection by providing three-dimensional views of atmospheric conditions.

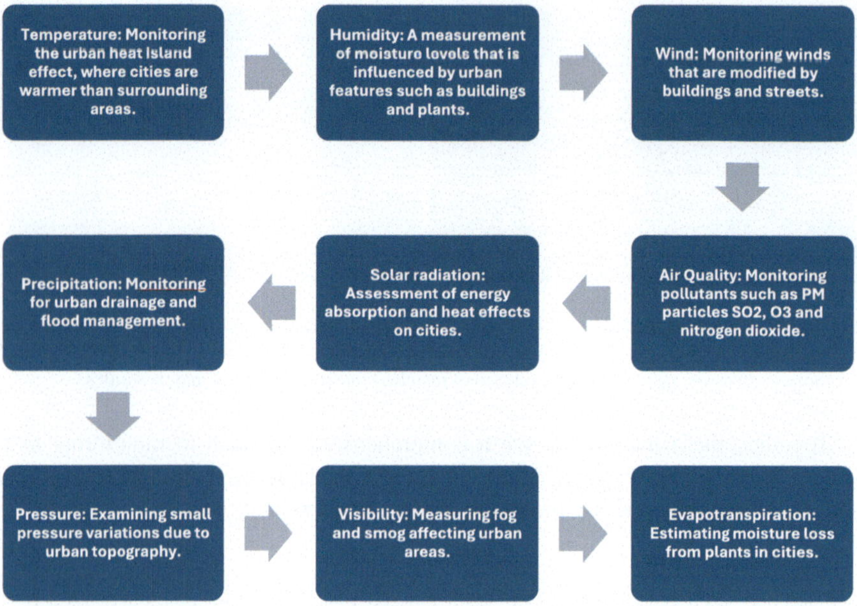

Fig. 14.28 Urban meteorological measurements focus on monitoring weather conditions through monitoring of various meteorological parameters in cities, where human activity and infrastructure affect the local climate

Urban meteorological observations are increasingly integrated with advanced modeling tools to predict weather and climate trends within cities. These measurements support a range of applications, including urban planning, public health initiatives, disaster management, and the mitigation of climate change impacts, especially with heat waves and urban heat islands (Fig. 14.29). They also play a critical role in designing sustainable and resilient urban environments by informing policies on green infrastructure, energy efficiency, and air quality management. In summary, urban meteorological measurements are a cornerstone of modern urban climate research, providing the detailed data necessary to address the unique challenges posed by urbanization and its effects on local and regional climates.

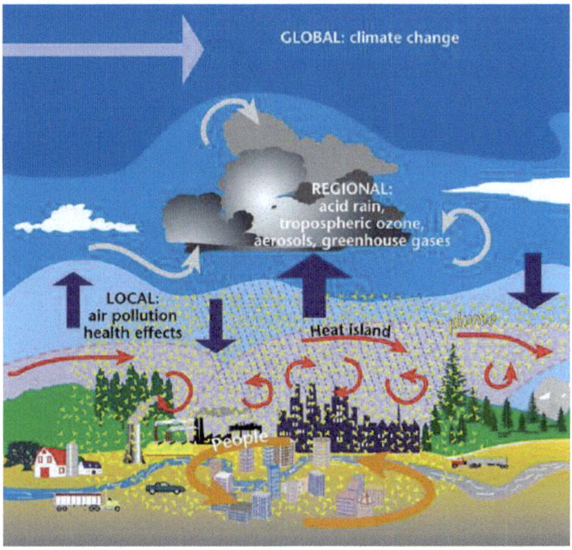

Fig. 14.29 Urban meteorological measurements capture atmospheric parameters such as temperature, wind, and air quality, enabling the study of phenomena like the urban heat island effect and supporting urban planning, climate resilience, and environmental management

14.20 Road Meteorological Measurements

Road meteorological measurements are critical for monitoring and managing weather-related conditions that impact road safety, traffic efficiency, and infrastructure maintenance (Fig. 14.30). These measurements provide real-time data on atmospheric and surface conditions along roadways, ensuring timely responses to adverse weather events and enhancing the safety of drivers and pedestrians. Key parameters monitored in road meteorology include air temperature, humidity, wind speed, and direction, precipitation type and intensity, visibility, and road surface conditions such as temperature, wetness, and icing potential (Fig. 14.31). These measurements are typically conducted using specialized sensors installed at Road Weather Information System (RWIS) stations, strategically located along highways, urban roads, and bridges. RWIS stations integrate atmospheric, pavement, and subsurface sensor data to provide comprehensive weather insights.

Advanced technologies, such as infrared sensors, laser-based visibility sensors, and non-contact road surface thermometers, allow for precise detection of critical conditions like black ice, snow accumulation, or reduced visibility due to fog. These technologies are complemented by mobile weather measurement systems mounted on maintenance or emergency vehicles, which collect dynamic, on-the-go data across larger areas. Real-time road meteorological data are transmitted to traffic management centers, where they are analyzed to inform decisions such as road closures, de-icing operations, or adjustments to speed limits. Weather-responsive traffic management systems integrate this data with traffic flow information to optimize operations during severe weather events. In addition to operational benefits, road meteorological measurements support research on the impacts of weather on traffic patterns, accident risks, and infrastructure durability. This data also informs

Fig. 14.30 Road meteorological measurements, including air and surface conditions, visibility, and precipitation monitoring, enable real-time weather management to enhance road safety and traffic efficiency

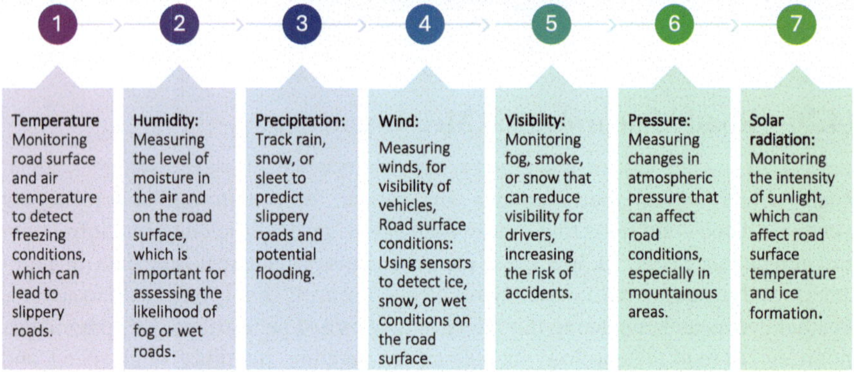

Fig. 14.31 Road meteorological measurements focus on monitoring weather conditions and key meteorological parameters that directly affect road safety, traffic flow, and maintenance

the design of climate-resilient road networks by identifying vulnerabilities and guiding the implementation of adaptive measures. By providing accurate, localized, and timely information, road meteorological measurements play a vital role in minimizing weather-related disruptions, ensuring safer travel, and improving the resilience of transportation systems under changing climatic conditions.

14.21 Mobile Meteorological Drones: A New Era of Precision in Meteorology

Across the globe, extreme weather events such as thunderstorms, high winds, fog, and hail have a profound impact on economies and public safety. Traditionally, weather prediction models have struggled with accuracy due to insufficient data collection from the atmospheric boundary layer, a crucial zone for understanding local weather phenomena. Meteorological drones offer a cutting-edge solution to this challenge by filling the data gaps in these models. Meteodrones, specially designed and patented for meteorological research, represent a breakthrough in the collection of atmospheric data. These drones are equipped to capture high-resolution measurements of key weather variables like temperature, humidity, air pressure, and wind speed. This capability allows for direct and detailed observation of the lower and middle layers of the atmosphere, significantly improving the accuracy of weather forecasts when integrated into prediction models (Müller et al., 2020; Fig. 14.32). Meteorological drones, also known as weather or atmospheric drones, are UAVs outfitted with advanced instruments to measure atmospheric conditions in real-time. They gather data on temperature, humidity, pressure, wind, and even radiation, providing detailed information that surpasses traditional methods like weather balloons or ground stations (Graham et al., 2021). Some drones are also equipped with sensors to track precipitation, cloud cover, and other meteorological factors, allowing them to fly through targeted areas of the atmosphere for more precise data collection (Fig. 14.33).

These drones have a wide array of applications, from refining weather forecasts to monitoring climate change, natural disasters, and air quality. The data they collect is instantly relayed to ground-based stations or satellite networks, enabling meteorologists to rapidly analyze and interpret the information. Meteorological drones come in diverse forms, ranging from fixed-wing models to quadcopters, with the choice of drone tailored to the mission's specific needs, such as flight range, endurance, and payload capacity. However, using these drones in weather research comes with challenges like adverse conditions, limited battery life, and regulatory limitations. Ongoing research aims to address these obstacles, enhancing the reliability of

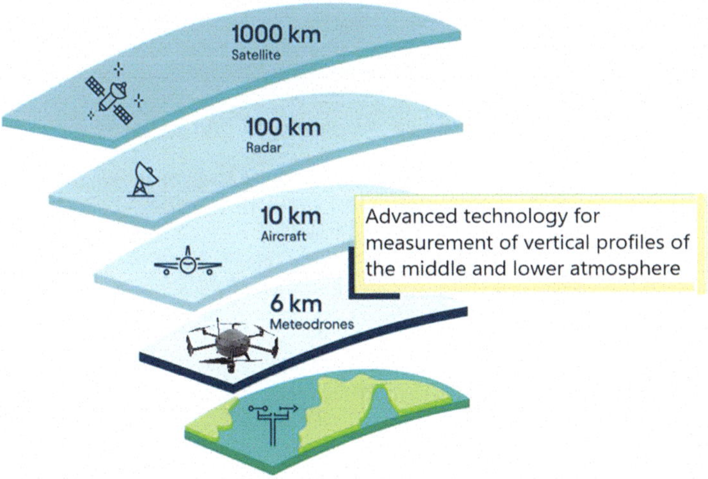

Fig. 14.32 Meteorological drones are regularly used for measuring the profile of meteorological elements in the lower and middle layers of the atmosphere

Fig. 14.33 A small drone flies through the clouds to measure atmospheric processes over Southern England. (Source: NCAS)

drone operations. As technology advances, meteorological drones are expected to become even more integral to atmospheric research. By delivering valuable data from the Earth's boundary layer, these drones are poised to dramatically improve weather models and deepen our understanding of climate systems.

References

Asrar, G. R., & Dozier, J. (2014). *Climate science for serving society: Research, modeling and prediction priorities*. Springer.

Brock, F. V., & Richardson, S. J. (2001). *Meteorological measurement systems*. Oxford University Press.

Chesnokov, G. (2019). *Remote sensing of the environment: An Earth resource perspective*. Academic Press.

Duncan, B. N., et al. (2018). Lidar measurements of aerosol and cloud properties: A review. *Remote Sensing, 10*(10), 1620. https://doi.org/10.3390/rs10101620

Gonzalez, R. C. (2018). *Digital image processing*. Pearson.

Graham, J. D., Smith, R. C., & Hinton, C. (2021). The role of drones in modern meteorology: Enhancements to weather data collection. *Meteorological Applications, 28*(3), e2052.

Harrison, R. M., & Sturges, W. T. (2009). *An introduction to global environmental issues*. Routledge.

Hollmann, R., et al. (2013). The ESA climate change initiative: Satellite data records for essential climate variables. *Bulletin of the American Meteorological Society, 94*(10), 1541–1552.

IPCC. (2021). *Climate change 2021: The physical science basis*. Cambridge University Press.

Kallberg, P., & Jansson, M. (2019). *Remote sensing and global climate change*. Wiley.

Kalnay, E. (2003). *Atmospheric modeling, data assimilation and predictability*. Cambridge University Press.

Lazo, J. K., et al. (2011). Economic value of improved weather forecasts. In *The Oxford handbook of climate change and society*.

Lee, D. K. H., & Crum, J. C. (2017). *Radar meteorology: A first course* (2nd ed.). Springer.

Mallet, C., & Bretar, F. (2009). Full waveform lidar data processing: A review. *Remote Sensing, 1*(4), 1050–1090. https://doi.org/10.3390/rs1041050

Müller, T., et al. (2016). A comparison of SODAR and LIDAR wind profiles in the lower troposphere. *Journal of Atmospheric and Oceanic Technology, 33*(2), 313–324. https://doi.org/10.1175/JTECH-D-15-0057.1

Müller, A., Bäumer, D., & Klein, P. (2020). Advancements in meteorological drone technology: Applications and implications. *Atmospheric Measurement Techniques, 13*(2), 789–805.

Netzband, M., Jürgens, C. (2010). Urban and Suburban Areas as a Research Topic for Remote Sensing. In: Rashed, T., Jürgens, C. (eds) Remote Sensing of Urban and Suburban Areas. Remote Sensing and Digital Image Processing, vol 10. Springer, Dordrecht. https://doi.org/10.1007/978-1-4020-4385-7

Orlanski, I. (1975). A rational subdivision of scales for atmospheric processes. *Bulletin of the American Meteorological Society, 56*(5), 527–530.

Rocca, F., et al. (2020). The role of lidar in satellite-based wind measurements. *Atmospheric Measurement Techniques, 13*(5), 2581–2592. https://doi.org/10.5194/amt-13-2581-2020

Sasaki, K., et al. (2004). Remote sensing of wind profiles with SODAR. *Journal of Applied Meteorology, 43*(3), 526–537. https://doi.org/10.1175/JAM2146.1

Stull, R. B. (1988). *An introduction to boundary layer meteorology*. Springer.

Stull, R. B. (2017). *Practical meteorology: An algebra-based survey of atmospheric science*. University of British Columbia.

Trenberth, K. E., et al. (2013). *Climate system modeling*. Cambridge University Press.

Wallace, J. M., & Hobbs, P. V. (2006a). *Atmospheric science: An introductory survey* (2nd ed.). Academic Press.

Wallace, J. M., & Hobbs, P. V. (2006b). *Atmospheric science: An introductory survey*. Elsevier.

WMO. (2015). *WMO guidelines on multi-hazard impact-based forecast and warning services*. World Meteorological Organization.

WMO. (2017). *The global observing system for climate: Implementation needs*. World Meteorological Organization.

WMO (World Meteorological Organization). (2019). *Manual on codes: International codes* (WMO-No. 306). WMO manuals.

World Meteorological Organization (WMO). (2018). *Guide to meteorological instruments and methods of observation* (WMO-no. 8).

World Meteorological Organization (WMO). (2020). *Guide to meteorological instruments and methods of observation* (WMO-no. 8).

Summary

Atmospheric Perspectives: Unveiling Earth's Environmental Challenges provides a profound exploration of the interconnected systems that shape our planet, weaving together science, technology, and humanity's collective responsibility. As the title suggests, this work delves into the atmospheric and environmental stressors defining the challenges of our age, with particular emphasis on climate change, resource depletion, and ecological imbalance. It invites readers to grasp the intricate workings of Earth's systems while confronting the pressing need for sustainable solutions. This book opens with a reflection on Earth's extraordinary complexity and beauty, setting a tone of reverence and urgency. Climate change, arguably the most significant challenge of the twenty-first century, is a recurring theme. Rising global temperatures, intensifying extreme weather events, and shifting ecosystems are analyzed not as isolated phenomena but as manifestations of deeper systemic imbalances. By examining these issues through the lens of interconnected Earth systems, this book underscores the urgency of addressing the causes and consequences of a warming planet. Central to the narrative is the concept of environmental stressors—factors that disrupt the equilibrium of natural systems and threaten biodiversity, human health, and economic stability. These stressors include deforestation, ocean acidification, and air pollution, each exacerbated by unsustainable human activities. The book's detailed exploration of these topics highlights the cascading effects of human-induced changes, such as how deforestation reduces carbon sequestration, accelerates climate change, and diminishes habitat for countless species.

The Anthropocene, a term denoting a new geological era characterized by profound human impact on Earth's systems, is introduced as a critical framework for understanding these challenges. This era underscores how anthropogenic activities—from industrial emissions to land-use changes—have accelerated global warming and destabilized climate systems. This book elaborates on how this unprecedented influence necessitates a paradigm shift in how humanity interacts with the planet. The hydrosphere and cryosphere's vulnerability to climate change receives particular attention, with chapters detailing the alarming trends of glacier retreat, rising sea

© The Editor(s) (if applicable) and The Author(s), under exclusive license to Springer Nature Switzerland AG 2025

V. Spiridonov et al., *Atmospheric Perspectives*,
https://doi.org/10.1007/978-3-031-86757-6

levels, and ocean warming. These phenomena are reshaping coastlines and altering marine ecosystems and global weather patterns. This book emphasizes that these changes are not distant concerns; they are unfolding now, with profound implications for food security, freshwater availability, and geopolitical stability. Similarly, the atmosphere's role as both a protective shield and a dynamic system influencing weather and climate is meticulously examined. The thinning of the ozone layer, the proliferation of greenhouse gases, and the increasing prevalence of particulate matter are presented as critical challenges. The narrative connects these atmospheric changes to human health, illustrating how air pollution disproportionately affects vulnerable populations and exacerbates existing inequalities. The book's treatment of natural hazards—from hurricanes and tornadoes to wildfires and droughts—highlights the dual role of these events as natural phenomena and as amplifiers of human vulnerability. It delves into the mechanisms driving these hazards, such as the interplay between atmospheric and oceanic circulation systems, and how climate change is intensifying their frequency and severity. Importantly, this book advocates proactive measures, such as early warning systems and resilient infrastructure, to mitigate their impact.

A key strength of the work is its forward-looking perspective, particularly in the chapters discussing environmental monitoring and technological innovation. Satellite-based systems, such as NASA's FIRMS and MODIS, are presented as essential tools for tracking changes in Earth's systems and informing policy decisions. "These technologies, combined with cutting-edge data analysis and predictive modeling, illuminate pathways toward mitigating environmental challenges and adapting to a changing climate." Yet, this book does not shy away from the limitations of technology, emphasizing the need for systemic change. It calls for a shift in societal values toward sustainability, urging readers to consider their role in fostering an equitable and resilient future. This message is woven throughout the narrative, culminating in a compelling argument for collective action at individual, community, and global scales.

The book *Atmospheric Perspectives* is more than an academic text; it is *a call to action*. By blending scientific rigor with accessible prose, it bridges the gap between expert knowledge and public understanding. It challenges readers to engage deeply in the complexities of our planet, appreciate its fragility, and commit to its preservation. This work resonates with a broad audience, from students and educators to policymakers and environmental advocates. Its multidimensional approach—encompassing the geosphere, hydrosphere, biosphere, and atmosphere—provides a holistic view of the Earth system, making it an invaluable resource for those seeking to understand and address the defining challenges of our time. In doing so, it inspires a sense of stewardship, urging humanity to embrace its responsibility as caretakers of the only home we have ever known.

Index

© The Editor(s) (if applicable) and The Author(s), under exclusive license to Springer Nature Switzerland AG 2025
V. Spiridonov et al., *Atmospheric Perspectives*,
https://doi.org/10.1007/978-3-031-86757-6

D

E

F

The manufacturer's authorised representative in the EU is Springer Nature Customer Service Centre GmbH, Europaplatz 3, 69115 Heidelberg, Germany. If you have any concerns regarding our products, please contact ProductSafety@springernature.com

Printed and bound by CPI Group (UK) Ltd, Croydon, CR0 4YY
01/10/2026
02225935-0016